6 02300

PERGAMON INTERNATIONAL LIBRARY
of Science, Technology, Engineering and Social Studies
The 1000-volume original paperback library in aid of education, industrial training and the enjoyment of leisure
Publisher: Robert Maxwell, M.C.

Relativity: The Theory and its Philosophy

FOUNDATIONS & PHILOSOPHY OF SCIENCE & TECHNOLOGY

General Editor: MARIO BUNGE, *McGill University, Montreal, Canada*

This series has three goals. The editor formulates them as follows:

(1) To encourage the systematic exploration of the foundations of science and technology

(2) To foster research into the epistemological, semantical, ontological and ethical dimensions of scientific and technological research

(3) To keep scientists, technologists, scientific and technological administrators and policy makers informed about progress in the foundations and philosophy of science and technology

Some Titles in the Series

AGASSI, J.
The Philosophy of Technology

BUCHTEL, H.
The Conceptual Nervous System

BUNGE, M.
The Mind-Body Problem

HATCHER, W.
The Logical Foundations of Mathematics

SIMPSON, G.
Why and How: Some Problems and Methods in Historical Biology

WILDER, R.
Mathematics as a Cultural System

Pergamon Journals of Related Interest

STUDIES IN HISTORY AND PHILOSOPHY OF SCIENCE*

Editor:

Prof. Gerd Buchdahl, Department of History and Philosophy of Science, University of Cambridge, England

This journal is designed to encourage complementary approaches to history of science and philosophy of science. Developments in history and philosophy of science have amply illustrated that philosophical discussion requires reference to its historical dimensions and relevant discussions of historical issues can obviously not proceed very far without consideration of critical problems in philosophy. *Studies* publishes detailed philosophical analyses of material in history of the philosophy of science, in methods of historiography and also in philosophy of science treated in developmental dimensions.

* Free specimen copies available on request.

Relativity: The Theory and its Philosophy

by

ROGER B. ANGEL
Concordia University, Montreal, Canada

PERGAMON PRESS
OXFORD · NEW YORK · TORONTO · SYDNEY · PARIS · FRANKFURT

UK	Pergamon Press Ltd., Headington Hill Hall, Oxford OX3 0BW, England
USA	Pergamon Press Inc., Maxwell House, Fairview Park, Elmsford, New York 10523, USA
CANADA	Pergamon of Canada, Suite 104, 150 Consumers Road, Willowdale, Ontario M2J 1P9, Canada
AUSTRALIA	Pergamon Press (Aust.) Pty. Ltd., P.O. Box 544, Potts Point, NSW 2011, Australia
FRANCE	Pergamon press SARL, 24 rue des Ecoles, 75240 Paris, Cedex 05, France
FEDERAL REPUBLIC OF GERMANY	Pergamon Press GmbH, 6242 Kronberg-Taunus, Hammerweg 6, Federal Republic of Germany

First edition 1980

British Library Cataloguing in Publication Data

Angel, Roger B
Relativity. – (Foundations & philosophy of science & technology). – (Pergamon international library).
1. Relativity (Physics)
I. Title II. Series
530.1′1′01 QC173.585 79–41551

ISBN 0–08–025197–8 (Hardcover)
ISBN 0–08–025196–X (Flexicover)

Printed and bound in Great Britain by
William Clowes (Beccles) Limited, Beccles and London

FOR
UNNI AND VICKI

Contents

Preface

MY principal aim in writing this book was to help to remedy the situation which has long been of concern to university teachers of the philosophy of physics—namely, that few students are sufficiently prepared in both philosophy and physics either to have access to the serious literature or to undertake serious work of their own in this field. Too often, the student of philosophy must avail himself of the popular literature. Consequently, he is unable to distinguish the genuine scientific content of a physical theory from the particular philosophical axe which its expositor happens to be grinding. I, too, have axes to grind. However, I have made an effort to keep the two as distinct as one may reasonably expect.

I attempt to provide a completely self-contained treatment of the philosophical foundations of the theory of relativity. By that, I naturally do not intend that the book contains everything that needs to be known, or even a significant fraction thereof. I merely mean that it should not be necessary for the average reader to master other works in order to attain a reasonable understanding of everything in this one. I have assumed that the reader dimly recalls the rudiments of elementary algebra, including the use of exponents, and those of elementary geometry. I have also taken for granted a passing familiarity with the notation of first-order logic. On this meagre basis, I attempt to provide a survey of the most essential mathematical techniques and concepts which seem to me to be indispensable to an understanding of the foundations of both the special and general theories of relativity. In short, the book includes a crash course in applied mathematics, ranging from elementary trigonometry to the classical tensor calculus. Needless to say, much more attention has been given to mathematical intuition than to rigorous presentation. This mathematical treatment has been divided between Chapter 1 and Chapter 7, so that the interest and patience of the reader would not be too severely tried. At the same time, I employ these mathematical tools in an exposition of the two theories, which I hope is an adequate basis for serious, independent philosophical reflection. Finally, I devote several chapters to the exposition and exploration of what I take to be among the more central and interesting philosophical problems and points of view which arise from these theories.

The level of treatment is between the merely popular and that of the high-grade treatise, whether scientific or philosophical. However, I hope that in bridging the gap between the two, I succeed in providing the industrious reader with direct access to the latter. That has been my goal, and the several peculiarities which characterize this book derive from that intention. I am firmly convinced that anyone who has serious aspirations in the field of philosophy, particularly those parts which touch on epistemology or metaphysics, should have a reasonably detailed knowledge of at least one scientific theory. The purpose of a textbook is to open intellectual doors. I shall be content if this book succeeds in so doing.

The preface of a first book provides its author with a unique opportunity to acknowledge his intellectual debts. Mine are too numerous to be mentioned in their entirety. In matters mathematical, I have found Wrede's *Introduction to Vector and Tensor Analysis* to strike an ideal middle ground between rigour and readability. Of the many books on relativity which I have studied, I have profited most from Bergmann's *Introduction to the Theory of Relativity*, Anderson's *Principles of Relativity Physics*, Adler, Bazin and Schiffer's *Introduction to General Relativity*, and Ohanian's *Gravitation and Spacetime*. But these are just a few of the books which are mentioned in the bibliographies which are appended to each chapter.

No writer in this field can fail to be aware of a debt to Professor Adolf Grünbaum, who is largely responsible for the current high level of interest in, and quality of, the philosophy of space and time. Even those who express disagreement with various of his conclusions, including myself, must allow that he has led the way to many of the more interesting highways and byways of this branch of philosophy. I owe a personal debt to many colleagues including Dr. Barry Frank of the department of physics and Dr. Harold Proppe of the department of mathematics, both of Concordia University, for their patient response to my many requests for advice and information.

My greatest intellectual debt is to my teacher, Professor Mario Bunge and to his copious writings and especially his *Foundations of Physics*. Although our styles are very different, those readers who are familiar with his work will recognize his influence on scores of pages throughout this book, including, I venture to say, those in which I disagree with him in matters of detail. It was he who first taught me how to understand science and how to philosophize about it. Indeed, it was he who made me fully aware that the former is a necessary condition of the latter. Whatever may be of merit in these pages is due to his profound influence on my philosophic outlook. The mistakes, of course, are of my own invention.

Finally, I express my gratitude to my wife Unni and my daughter Vicki, to both of whom this book is dedicated, for remaining cheerful through all of the author's ups and downs.

Montreal, June 1979

List of Logical Symbols

$\wedge$ = *conjunction* meaning “and”.
$\vee$ = *disjunction* meaning “either-or”.
$\rightarrow$ = *material implication* meaning “is a sufficient condition for”.
$\leftrightarrow$ = *biconditional* meaning “is a necessary and sufficient condition for”.
$\sim$ = *negation* meaning “not”.
(x) = *universal quantifier* meaning “for every x”.
$(\exists x)$ = *existential quantifier* meaning “there is an x such that”.
Px = *monadic predicate* meaning “x is P”.
Pxy = *dyadic predicate* meaning “x is P to y”.
$Pxyz$ = *triadic predicate* meaning, for example, “x is P to y and z”.

1

Mathematical Preliminaries

FOR the reader with little or no mathematical background, this introductory chapter may well prove to be the hardest in the book. In effect, he will be entering an entirely new conceptual world. Moreover, the subject-matter encompasses a great many fundamental mathematical concepts which are normally acquired, step by step, over a period of years. I strongly advise you to read this material quite slowly. The worked examples should be studied until you are able to reproduce them with only a minimum of reference to the text. In this way, you will find that you make rapid gains in both confidence and facility. Do not be discouraged in finding that you are unable to recall various details. They will gradually become fixed in your memory as you continue to refer back to them, as needed, in the course of reading the later chapters. At all times, you should be sufficiently motivated by the thought of acquiring one of the most valuable, and ultimately fascinating, keys to knowledge. Once you have obtained it, you will find that it opens incredibly many doors in both science and philosophy. For so high a return, the investment will prove relatively slight.

Sets, Relations, Functions

The most fundamental concept of classical mathematics is that of a *set*. The term *set* is roughly synonymous with those of class and collection. It refers to any arbitrary grouping of entities of any kind. Conventionally, sets are abstractly symbolized by upper-case Latin letters, e.g. A, B and C. When a set is finite and sufficiently small, it may also be displayed by enclosing the names of the objects which it comprises in braces. Thus, we may write: $A = \{1, 2, 3\}$, meaning that A is to symbolize the set consisting of the numbers 1, 2 and 3. The objects which are the members of a given set are called its *elements*. The relation of "belonging to" or "being a member of" a given set which holds between a set and any one of its elements is symbolized by the lower-case Greek epsilon (ε). Thus, the following expression is obviously true: $3\,\varepsilon\{1, 2, 3\}$. There is a close connection between being an element or member of a given set and having a property of a given kind. For example, to have the property yellow is to be an element of the set of all yellow objects. Thus, when the necessary and sufficient condition for membership in the set A is the possession of the property P, we may symbolize the set A by the expression $\{x \mid Px\}$, which simply means the set of all objects x having the property P. Finally, it should be noted that a set is not an *ordered* collection. That is to say, that the order in which the elements are taken has no effect on the set as such. Hence, $\{1, 2, 3\} = \{3, 1, 2\}$. By the same token, the repetition of one or more elements has no significance. Thus, $\{1, 2, 3\} = \{1, 2, 3, 2\}$.

The most important relation which may hold between two sets is the *subset* relation. Abstractly, A is a subset of B if and only if every element of A is an element of B. This relation is symbolized by $A \subseteq B$. Employing standard logical notation, we may write the defining expression for such a relation as

$$A \subseteq B \leftrightarrow (x)(x \varepsilon A \rightarrow x \varepsilon B)$$

Thus, we see that $\{1, 2\} \subseteq \{1, 2, 3\}$. Similarly, the set of ants is a subset of the set of insects, which is in turn a subset of the set of animals. The last example reveals an important property of the subset relation—namely, that it is transitive. In general, if $A \subseteq B$ and $B \subseteq C$, then $A \subseteq C$. The transitivity of the subset relation is sufficient to distinguish it from the membership relation, which lacks this property. For example, Plato is an element of the set of philosophers. The set of philosophers is an element of the set of learned professions. However, Plato is clearly not an element of the set of learned professions. That is simply to say that Plato is not a learned profession. Thus, $1 \varepsilon \{1, 2, 3\}$ but it is not the case that $1 \subseteq \{1, 2, 3\}$, although it is the case that $\{1\} \subseteq \{1, 2, 3\}$. The foregoing reveals the necessity for drawing a sharp distinction between 1 and $\{1\}$. The former is the number one while the latter is the set whose sole member is the number one. Clearly, such a set is not a number. Sets with but one member are called *unit sets* or *singletons*. Finally, it is to be noted that every set is a subset of itself, e.g. $\{1, 2, 3\} \subseteq \{1, 2, 3\}$. That is to say, that the subset relation is not asymmetric. However, we may define the asymmetric relation of *proper subset* symbolized by $\subset$. $A \subset B$ if and only if $A \subseteq B$ and $A \neq B$.

There are various operations which may be performed on sets. The reader will be familiar with the notion of operation in terms of the well-known operations of arithmetic such as addition and multiplication. Intuitively, these are operations on pairs of numbers such that the result of a given operation is another number. Similarly, operations on sets are operations performed on pairs of sets such that the result will be another set. The most elementary operation on sets simply consists in combining all of the elements of two sets to form a third set. Such an operation is called *union* and is symbolized by $\cup$. The defining expression for $\cup$ is $(x)(x \varepsilon A \cup B \leftrightarrow x \varepsilon A \vee x \varepsilon B)$. The *intersection* of A and B, symbolized $A \cap B$ consists in forming the set consisting of all elements which belong to both A and B. $(x)(x \varepsilon A \cap B \leftrightarrow x \varepsilon A \wedge x \varepsilon B)$. The *relative difference* of two sets, which is obtained by forming the set whose elements are all and only those elements which belong to the first set but not the second, is defined by $(x)(x \varepsilon A \sim B \leftrightarrow x \varepsilon A \wedge x \notin B)$
Examples:

$$\{1, 2, 3\} \cup \{3, 4, 5\} = \{1, 2, 3, 4, 5\}$$
$$\{1, 2, 3\} \cap \{3, 4, 5\} = \{3\}$$
$$\{1, 2, 3\} \sim \{3, 4, 5\} = \{1, 2\}$$

In order that these operations be defined for all arbitrary sets, it is necessary to introduce a distinguished set which is called the *empty* set. This is literally the set which has no members. Since a set is determined or defined simply in terms of its members, it follows that the empty set is unique. If two sets are empty, then they have precisely the same membership and are to be regarded as identical. While there may be philosophical objections to identifying the set of unicorns with the set of ten-mile-high buildings, such a treatment is adequate to mathematical contexts. The empty set is conventionally symbolized by $\varnothing$. Thus, $\{1, 2\} \cap \{3, 4\} = \{1, 2\} \cap \{\text{Plato}\} = \varnothing$. Sets whose intersection is $\varnothing$ are said to be *disjoint*.

So far we have discussed the very general notion of set. While most of our examples have referred to sets of numbers, this was simply for notational convenience. A set may consist of any sorts of objects whatsoever, including other sets. We now turn to the topic of *relations*, which will be treated as sets consisting of objects or elements of a more restricted kind. Specifically, we define a relation as a set of n-tuples (sometimes spelled "entuple"). An n-tuple is a sequence of n objects taken in a fixed order. Thus, we may speak of the first element, the second element and more generally of the ith element of the n-tuple, where $i = 1, 2, \ldots, n$. An n-tuple may be displayed in a manner similar to that in which one displays a set. However, the convention in this case is to enclose the elements (or coordinates) in pointed brackets. For example, $\langle 1, 2, 3 \rangle$ is the triple whose first element is one, whose second element is two, and so forth. Since an n-tuple is ordered, the order in which the elements are taken is, of course, significant. This is the principal difference between a set and an n-tuple. Thus, while $\{1, 2, 3\} = \{1, 3, 2\}$, $\langle 1, 2, 3 \rangle \neq \langle 1, 3, 2 \rangle$. We shall adopt the convention of using a lower-case indexed letter to symbolize an n-tuple. Thus, a_i is the n-tuple, where i ranges over all the positive integers from 1 to n. For example, if $a_i = \langle 1, 2, 3 \rangle$, then $a_1 = 1, a_2 = 2, a_3 = 3$. On the other hand, if $b_i = \langle 1, 3, 2 \rangle$, then $b_1 = 1, b_2 = 3, b_3 = 2$. In general $a_i = b_i$ only if $a_1 = b_1, a_2 = b_2, \ldots, a_n = b_n$. It is obvious that, unlike the case of a set, the repetition of elements is of significance. Thus, $\langle 1, 2, 3 \rangle \neq \langle 1, 2, 3, 2 \rangle$. It should perhaps be added that in a more elegant mathematical treatment of the subject, the notion of n-tuple is definable in terms of set and elementhood.

We are particularly interested in the notion of a *binary relation*, which will be treated as a set of couples. For example, suppose we have a set $A = \{1, 2, 3\}$. Then we may define on A a relation of 'less than' or "$<$" as follows:

$$< \; = \{\langle 1, 2 \rangle, \langle 1, 3 \rangle, \langle 2, 3 \rangle\}.$$

Conventionally, relations are abstractly symbolized by upper-case R with a numerical subscript. e.g. R_1, R_2, etc. The use of the upper-case is wholly appropriate since, on this interpretation, a relation is merely a special type of set. For example, 'husband of' is the set of all couples such that the first element or coordinate of each couple is the husband of the corresponding second element. It is interesting to note that just as the notion of set was seen to correspond to the logical notion of property or monadic predicate, so the notion of binary relation corresponds in precisely the same way to that of dyadic predicate. Thus, we may represent a binary relation—for example, 'brother of'—by the formal expression: $\{\langle x, y \rangle \mid Bxy\}$. That is the set of all couples, $\langle x, y \rangle$, such that x is the brother of y.

In so far as a binary relation is a *bona fide* set, all of the operations which are defined for sets in general are applicable to binary relations. Thus, we may speak of the intersection, union, etc., of two or more relations. However, in what follows we shall be more interested in those operations which are defined specifically for binary relations. Firstly, there is the operation of forming the *converse* of a relation. The converse of a relation R is denoted by $\breve{R}$. It may be defined as follows:

$$\langle x, y \rangle \, \varepsilon \, \breve{R} \leftrightarrow \langle y, x \rangle \, \varepsilon \, R.$$

In effect, then, the converse of R is formed by reversing the order of elements in each of the couples belongmg to R. Thus, the converse of "less than" ($<$) is "greater than" ($>$).

Referring to our earlier example, where "less than" was restricted to the set $A = \{1, 2, 3\}$, it is obvious that the converse of this relation would be $\{\langle 2, 1 \rangle, \langle 3, 1 \rangle, \langle 3, 2 \rangle\}$.

Another important operation on binary relations is the *relative product* of two binary relations. This is slightly more involved, so we shall begin with an intuitive account. Suppose we are given two relations R_1 and R_2. For example R_1 may be "sister of" and R_2 is "father of". If R_1 happens to contain the element $\langle$ Mary, John $\rangle$ and R_2 contains the element $\langle$ John, Alfred $\rangle$, we may infer that a certain relationship, R_3, must hold between Mary and Alfred. Specifically, Mary is the sister of Alfred's father, which is to say that Mary is the aunt of Alfred. Thus, we say that R_3 = "aunt of" is the relative product of R_1 = "sister of" and R_2 = "father of". The slash (/) is employed to denote this operation. Thus $R_1/R_2 = R_3$. We may define the relative product operation as follows:

$$\langle x, y \rangle \varepsilon R_1/R_2 \leftrightarrow (\exists z)(\langle x, z \rangle \varepsilon R_1 \wedge \langle z, y \rangle \varepsilon R_2)$$

To explain this, in words, we shall introduce some more terminology. We call the set of all first members of the couples in a relation R the *domain* of the relation, e.g. the domain of "son of" is the set of males. Similarly, the set of all second members of the couples of R is called the *codomain* of R. These sets are symbolized by $\mathscr{D}(R)$ and $\mathscr{C}(R)$ respectively. Now the relative product R_1/R_2 is formed by considering all those objects which are in the intersection of $\mathscr{C}(R_1)$ and $\mathscr{D}(R_2)$. Specifically, it is the set of couples $\langle x, y \rangle$ such that $\langle x, z \rangle \varepsilon R_1$ and $\langle z, y \rangle \varepsilon R_2$. A numerical example will serve to clarify this.

$$R_1 = \{\langle 1, 2 \rangle, \langle 3, 2 \rangle, \langle 4, 5 \rangle, \langle 7, 6 \rangle\}$$
$$R_2 = \{\langle 4, 7 \rangle, \langle 2, 7 \rangle, \langle 5, 6 \rangle\}$$
$$\mathscr{C}(R_1) = \{2, 5, 6\}, \ \mathscr{D}(R_2) = \{4, 2, 5\}$$
$$\mathscr{C}(R_1) \cap \mathscr{D}(R_2) = \{2, 5\}$$
$$R_1/R_2 = \{\langle 1, 7 \rangle, \langle 3, 7 \rangle, \langle 4, 6 \rangle\}$$

The reader should determine from this example that the relative product operation is not commutative, i.e. $R_1/R_2 \neq R_2/R_1$.

The most fundamental relation is the *Cartesian product* of two or more not necessarily distinct sets. This is the n-ary relation which is formed by taking all of the members of n sets in a fixed order. Again, a numerical example should suffice to clarify this notion. Let A = $\{1, 2, 3\}$ and $B = \{3, 4\}$. Then the Cartesian product $A \times B$ is

$$\{1, 2, 3\} \times \{3, 4\} = \{\langle 1, 3 \rangle, \langle 1, 4 \rangle, \langle 2, 3 \rangle, \langle 2, 4 \rangle, \langle 3, 3 \rangle, \langle 3, 4 \rangle\}$$

(It should be obvious that the operation of forming the Cartesian product of two or more distinct sets is not commutative.) Suppose that a particular domain of discourse refers to a specific basic set of entities such as the set of real numbers or the set of people. Then the Cartesian product of this set with itself constitutes the *universal* binary relation over the given domain. Any other binary relation which holds among the members of the domain is a *subrelation* of the universal relation.

We have now reached the point where we may explain the vital concept of a *function*. A function is a binary relation which has the property that every element of the domain is associated with a *unique* element, i.e. one and only one, of the codomain. For example, "child of" is a function, since every child is associated with a unique pair of parents. On the other hand, "parent of" is not a function, since it is not the case that every pair of parents is associated with a unique child. We may define a function, more abstractly, as a binary relation such that every distinct element, i.e. every couple, has a distinct first member. Since functions are a distinguished subset of binary relations we employ special symbols to represent them—namely, italicized letters, $f, g, h, \ldots$ with or without numerical subscripts. Moreover, the codomain of a function is usually called the *range* of the function. Due to its special character, a function is also called a many-one relation. The motivation for this terminology is readily discerned from the following diagrammatic representation of a functional relationship.

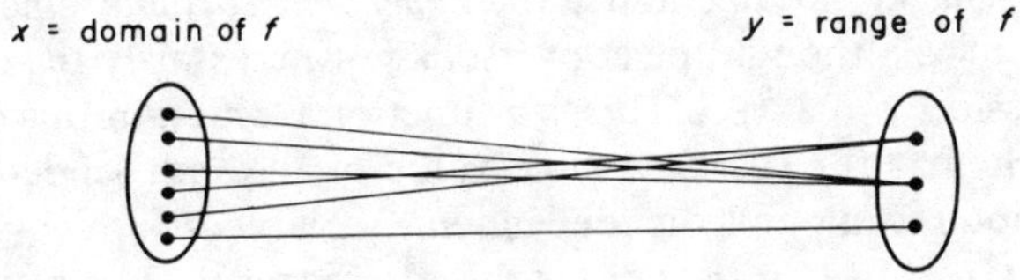

FIG. 1.1. The function $f\colon x \to y$

Clearly, every element of x is associated with one and only one element of y. Thus, the function f may be thought of as an operation which *maps* each element of the domain onto an element in the range. In many cases, the actual range of the function will be a subset of a given set, y. In this case, the function is said to be *into* y. However, if every element of y is in the range of the function, then the function is said to be *onto* y.

If the domain of a function is sufficiently small, then the function may be displayed in the manner of a relation. $f = \{\langle 1, 3\rangle, \langle 2, 7\rangle, \langle 6, 5\rangle\}$ is an example of a function. On the other hand, $\{\langle 1, 3\rangle, \langle 2, 7\rangle, \langle 6, 5\rangle, \langle 1, 8\rangle\}$ is not a function. This may be ascertained by applying our definition. In this case, the relation contains two distinct elements, $\langle 1, 3\rangle$ and $\langle 1, 8\rangle$ which have the same first member, in contravention of the definition. The reader should determine for himself that in general the union of two functions is not a function, whereas both the intersection and relative difference of two functions are also functions.

An important class of functions is the one of *invertible* functions. The inverse of a function f, denoted by f^{-1}, is a special case of the converse of a function. For example, referring to Fig. 1.1, you will perceive that although the relation between x and y is a function, the converse from y to x is clearly not. In other words, while f is many-one, $\breve{f}$ is one-many. However, in the case where each element of $\mathscr{D}(f)$ is associated with a distinct element of $\mathscr{R}(f)$, then the converse of f will also be a function. When a function has an inverse it is said to be a one-one function or, equivalently, an invertible function. Such a function will also be called a one-one correspondence.

Finally, we note that there is an operation called the *composition* of two functions, symbolized $f \circ g$, which is a specialization of the relative product operation. We may think of a function as an operation which 'carries' each element of the domain onto an element of the range. Now we may consider the case of two such operations carried out successively. The function g carries each element of x onto an element of z. Then the function f operates on the elements of z, carrying the latter onto elements of y. The resultant of this mode of combining g and f is a third function from x to y, which may be symbolized by $f \circ g = h$. For example, suppose that the domain of g is a set of human beings, and that g maps each element onto its corresponding chronological age. Then f could be the function which maps numbers onto the initial letter of their English name. Then h is a function from human beings into letters of the alphabet. It is easy to see that each human being will be associated with a unique letter, so that h is indeed a function.

Although the notion of function is extremely general in the sense that both domain and range may be sets of objects of any conceivable kind, our principal concern in this book will be with mathematical functions. Such functions may be represented by symbolic mathematical expressions which are abstractly of the form $f(x) = y$. In such contexts, x is said to be the *independent variable* and y the *dependent variable*, since every value of x determines a value of y, although the converse is not necessarily the case. Consider, for example, the expression $y = 2x + 3$. This is a function from x, in this case the set of real numbers, to y, such that for every possible value of the independent variable x there corresponds a unique value of the dependent variable y which is determined by multiplying the value of x by two and adding three to the result. The nature of this function may be pictured by constructing a table for a tiny subset of the domain with the corresponding values in the range of the function.

x	−5,	−4,	−3,	−2,	−1,	0,	1,	2,	3,	4,	5
$f(x) = y$	−7,	−5,	−3,	−1,	1,	3,	5,	7,	9,	11,	13

We may then draw the *graph* of f in a manner that most of you will remember from your schooldays. Such a graph is depicted in Fig. 1.2. The horizontal or X-axis represents the values of x, the vertical or Y-axis represents the values of y, and the sloping line represents the function f. The graph provides an excellent visualization of the behaviour of f. In particular, it should be noted that the graph of f is a straight line. Any such function is, therefore, called a *linear* function. The reader should determine for himself that any function of x, of the form $ax + b$, where a and b are arbitrary constants, will be linear. In the special case where $b = 0$, the function is said to be a homogeneous, linear function. To anticipate a later development which is crucial in the study of the theory of relativity, we may view the situation depicted in Fig. 1.2 from a somewhat different point of view.

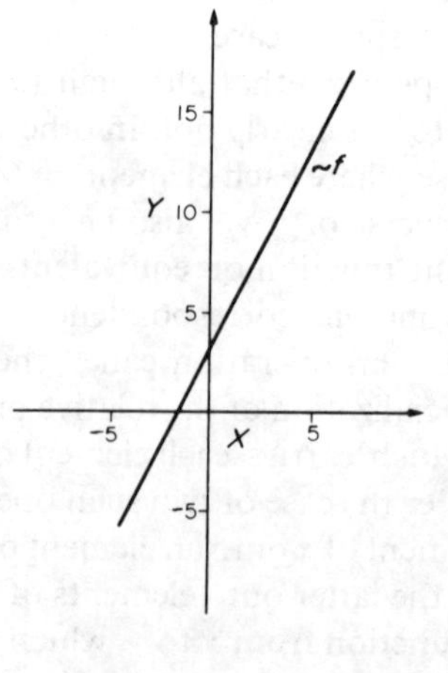

FIG. 1.2. The graph of $y = 2x + 3$

Instead of regarding f as an association of values of the variables x and y, we may think of f as an operation which *transforms* each value of x into the corresponding value of its 'image' under the transformation f. In this context, one generally denotes the transform of x not by y but by x'. Thus we may speak of the transformation: $x \to x' = 2x + 3$. This is an instance of a *linear transformation*. An excellent example of a linear transformation is that which transforms the centigrade scale into the fahrenheit scale, which is $C \to F = (\frac{9}{5})C + 32$.

N-tuples, Vectors, Matrices, and other Items

So far we have traversed territory which is fundamental to virtually every branch of mathematics. We shall now begin to narrow our treatment to those aspects of mathematics which are particularly applicable to physics in general and to relativity physics in particular.

In the concluding part of the previous section, we assumed that the reader was acquainted with the general idea of a Cartesian coordinate system. We employed this device in Fig. 1.2. A rectangular Cartesian coordinate system is a set of two or three straight lines which are mutually perpendicular and which intersect at a common point called the *origin*. These lines are called the axes of the system. The reader will probably recall that these axes are usually labelled X, Y, Z, corresponding to the three dimensions of space. However, for reasons that will eventually be clear, we shall label them X^1, X^2, X^3. All of the points on a given axis are assumed to be in one-one correspondence with the set of real numbers. Now such a system may be employed to represent the set of points in a three-dimensional space. Specifically, there will be a one-one correspondence between the points in the space and the triples $\langle x^1, x^2, x^3 \rangle$. Thus, every point of 3-space is associated with an n-tuple, specifically a triple. Accordingly, to investigate the nature of such a space, we should develop the treatment of the n-tuple which was introduced in the preceding section. Although we are principally concerned with the 3-space of experience, the generalization to spaces of n dimensions is so obvious that we shall adopt the more general approach. However, as an aid to intuition, the reader may mentally substitute triple for n-tuple in what follows.

S_n (intuitively n-dimensional space) is a set of n-tuples satisfying the following rules.

1. $\langle a_1, a_2, \ldots, a_n \rangle + \langle b_1, b_2, \ldots, b_n \rangle = \langle a_1 + b_1, a_2 + b_2 \ldots, a_n + b_n \rangle$. (More compactly, $a_i + b_i$, where $i = 1, 2, \ldots, n$.)
2. $k\langle a_1, a_2, \ldots, a_n \rangle = \langle ka_1, ka_2, \ldots, ka_n \rangle$, where k is an arbitrary real number.
3. $a_i = b_i$ if and only if $a_1 = b_1$, $a_2 = b_2, \ldots, a_n = b_n$.
4. $\langle 0, 0, \ldots, 0 \rangle$ is the zero element or n-tuple of S_n, and will be denoted by 0_i.
5. $a_i \varepsilon S_n$ and $b_i \varepsilon S_n$ implies $a_i + b_i \varepsilon S_n$.
6. $a_i + b_i = b_i + a_i$.
7. $(a_i + b_i) + c_i = a_i + (b_i + c_i)$.
8. $a_i + 0_i = a_i$.
9. For each $a_i \varepsilon S_n$ there is a unique element $-a_i \varepsilon S_n$ such that $a_i + (-a_i) = 0_i$.
10. $a_i - b_i = a_i + (-b_i)$.

The foregoing treatment of S_n will no doubt seem excessively abstract to many readers. A few remarks are perhaps in order. In the first place, we do not require the reader to memorize these laws but merely to get the feel of them. Secondly, the reader should examine them to see that all of the properties of n-tuples follow in a natural way from the fact that the components of an n-tuple are real numbers. Finally, we assure our reader that this abstract skeleton will shortly be supplied with some intuitive flesh.

We now turn to the vital geometric concept of the *vector*. It is customary in introductory treatments of vectors to depict them as arrows having a definite length and direction. Such arrows are then employed to depict or model various physical quantities. As we shall see, such a treatment is somewhat misleading. However, we shall sacrifice rigour for intuitive clarity and proceed in the traditional manner. Moreover, in the same spirit, we shall specialize the discussion to three-dimensional space. Let us imagine that our local space has been coordinatized by a Cartesian coordinate system as described earlier. We now imagine a line drawn from the origin (i.e. the point $\langle o, o, o \rangle = o_i$) to an arbitrary point in space corresponding to the point X^i, e.g. $\langle 1, 5, 3 \rangle$. This line may be regarded as an arrow whose tail is at the origin and whose tip is at the point $\langle 1, 5, 3 \rangle$. Clearly, it has a definite magnitude or length and a definite direction. An entity of this kind will be called a vector. The numbers 1, 5, 3 are the *components* of the vector. Geometrically, these components are the projections of the vector on the X^1, X^2, X^3-axes, respectively. Accordingly, we may think of a vector as a model or interpretation of the more abstract notion of an n-tuple or, in this case, triple, such that each component of the n-tuple corresponds to the geometric notion of the projection of the vector on the appropriate axis.

At this point, it will be helpful to provide a visual illustration of a vector. While it is possible to draw a three-dimensional diagram, it is clearer to deal with a two-dimensional one and there is no loss of generality.

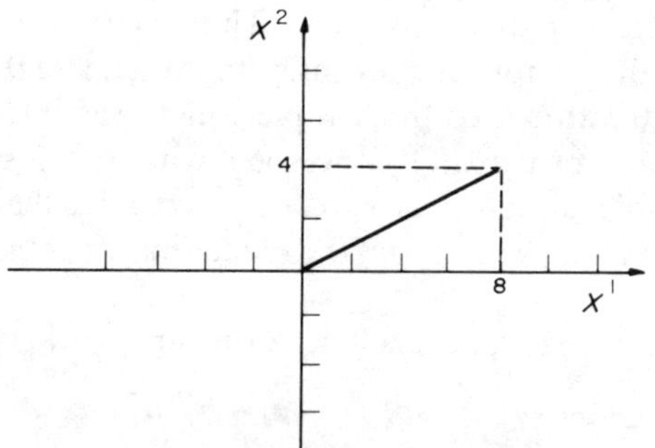

FIG. 1.3. Graphical representation of a vector and its components

Figure 1.3 depicts the vector whose components are $x^1 = 8$, $x^2 = 4$. The broken lines indicate how the vector is projected onto the two axes. It is important to note that the segment of X^1 from the origin to $x^1 = 8$ is itself a vector $\langle 8, 0 \rangle$. Similarly, the segment of X^2 from the origin to $x^2 = 4$ is the vector $\langle 0, 4 \rangle$. Since a vector is a type of n-tuple, it must obey the laws of n-tuples. In particular, we note that law 1 implies that the vector $\langle 8, 4 \rangle$ is the sum of the vectors $\langle 8, 0 \rangle$ and $\langle 0, 4 \rangle$.

We may generalize on the last point as follows. We define a *unit vector* along each coordinate axis. These may be denoted by $\hat{x}^1$, $\hat{x}^2$, $\hat{x}^3$, where $\hat{x}^1 = \langle 1, 0, 0 \rangle$, $\hat{x}^2 = \langle 0, 1, 0 \rangle$, $\hat{x}^3 = \langle 0, 0, 1 \rangle$. We call these unit vectors the *basis* or *base* vectors of the system. Clearly, any vector which lies along an axis may be represented as a multiple of the corresponding unit vector. Thus, $\langle 8, 0 \rangle = 8\langle 1, 0 \rangle$ and $\langle 0, 4 \rangle = 4\langle 0, 1 \rangle$. Moreover, since our original vector was found to be the sum of the latter vectors, it follows that any vector may be represented as a sum or linear combination of multiples of unit vectors. If we denote the three multipliers or coefficients by c_i and the unit vectors by $\hat{x}^i$, then an arbitrary vector A^i may be represented in the form $A^i = c_i\hat{x}^i$. The right-hand side of this expression is known in mathematical jargon as a linear form. The expression $A^i = c_i\hat{x}^i$ makes use of an extremely handy notational device known as the Einstein summation convention. Specifically, the convention is that whenever an index occurs twice in a single expression, it is to be summed over the repeated index. That is to say,

$$A^i = c_i\hat{x}^i = c_1\hat{x}^1 + c_2\hat{x}^2 + c_3\hat{x}^3.$$

We have characterized a vector as an object with a definite magnitude and direction. But, so far, we have not indicated how the magnitude of a vector is determined. In particular, the magnitude of an n-tuple is not defined. Referring once more to Fig. 1.3, we comprehend intuitively that the magnitude of the vector must be the distance from the origin to the point whose coordinates are (8, 4). The reader should recall from his high-school geometry that there is a well-known theorem—namely, the theorem of Pythagoras—which provides precisely the information we need. The vector in Fig. 1.3 is the hypotenuse of a right triangle whose other sides are of length 8 and 4. Thus, by virtue of the celebrated theorem, we compute the magnitude

$$A^i = (8^2 + 4^2)^{\frac{1}{2}} = \sqrt{80} = 8.944 \text{ approx.}$$

We have already remarked that distance was not defined in the n-tuple space S_n. Thus, now that distance has been defined our space is of a more specialized type. We call such a space a *metric* space and the expression $s^2 = (x^1)^2 + (x^2)^2 + (x^3)^2$ is called the *metric* or *metric form* of the space. Specifically, the above metric form, which may be generalized from 3 to n dimensions, is the well-known Euclidean metric.

It should be fairly obvious that the numerical components of an arbitrary vector will depend on the choice of coordinate system. Because our account of vectors has so far been defective in important respects, we have been obliged to choose a fixed coordinate system in terms of which every vector could be expressed. Nevertheless, the fact still remains that there are infinitely many Cartesian coordinate systems at our disposal. In fact, even if we consider only those Cartesian coordinate systems whose origins coincide with the tail of a specific vector, there are still infinitely many such systems at our disposal. For example, a second coordinate system may be obtained from the coordinate system of Fig. 1.3 by rotating the axes about the origin through an angle of θ degrees. Since there are infinitely many possible θ's there are infinitely many such coordinate systems. But all would serve equally well for expressing the vector of interest. However, to repeat, the actual numerical values of the components of the vector will vary with the choice of system. In general, in the new system, each new component will be a function of the three original components.

If we think of the vector as a *geometric object* which is held fixed at a point in some abstract space of its own, then the change of coordinate system corresponds to a transformation performed on the components of the vector. Such a transformation will involve for a vector in 3-space nine quantities. In short, any such transformation will be mathematically fairly complicated. Fortunately, however, the nine quantities are not all distinct. Still more fortunate than the foregoing is the fact that the transformation law, while complicated, is, nevertheless, linear in the components and there happens to be available a highly developed mathematical device for handling linear transformation laws. If we call our vector A^i, we may represent it in the second or *barred* coordinate system by $\bar{A}^i$. Hence, we are concerned with the nature of the transformation: $A^i \to \bar{A}^i$. The general form of such a transformation is the following.

$$\begin{aligned} A^1 \to \bar{A}^1 &= c_{11}A^1 + c_{12}A^2 + c_{13}A^3 \\ A^2 \to \bar{A}^2 &= c_{21}A^1 + c_{22}A^2 + c_{23}A^3 \\ A^3 \to \bar{A}^3 &= c_{31}A^1 + c_{32}A^2 + c_{33}A^3. \end{aligned}$$

This is the general form of the transformation of a vector by a rotation of the coordinate system without a displacement of the origin. Since the A^i are given, the components in the barred system may be computed from the nine coefficients. However, we are not immediately interested in the nature of these nine quantities but in something of a more fundamental nature.

We shall consider the nine quantities as a distinct mathematical object which may be exhibited in the following array.

$$\begin{pmatrix} c_{11} & c_{12} & c_{13} \\ c_{21} & c_{22} & c_{23} \\ c_{31} & c_{32} & c_{33} \end{pmatrix}.$$

Any such array of numbers is called a *matrix*. The example is that of a 3×3 *square* matrix. In modern mathematics, matrices are almost as ubiquitous as the very notion of number itself. A matrix is in certain respects a generalization of the n-tuple. We thought of an n-tuple as a single row (or column) of numbers. A matrix is a set of numbers which are ordered by both row and column. Just as we represented an n-tuple by its components a_i, so we may represent a matrix by its components a_{ij}, where the first index runs over the number of rows and the second over the number of columns. Thus the ijth component of the matrix is the component in the ith row and the jth column.

The theory of matrices is both extensive and fascinating. Moreover, it constitutes an ideal point of entry into many areas of both pure and applied mathematics. Our treatment will barely scratch the surface. However, since we shall find matrix operations and their generalizations cropping up throughout this book, it will be useful to devote more attention to these matters than to most of the other mathematical topics with which we shall deal. A firm grasp of the fundamentals of matrix theory is probably the single most important key to the understanding of the formal aspects of relativity theory. Moreover, it is a necessary condition for access to the technical literature, which any serious student of the philosophy of relativity theory will both need and desire.

A matrix with m rows and n columns is said to be of order $m \times n$. If $m = n$, i.e. if the matrix is square, then it is said to be a square matrix of order m. For example, $(1\ 7\ 5\ 1)$ is a matrix of order 1×4; (3) is a matrix of order 1. The transformation matrix c_{ij} which we encountered earlier is a square matrix of order 3. Matrices may be added, subtracted and multiplied by a constant in ways that follow rather trivially from the corresponding operations in ordinary arithmetic. However, addition and subtraction of matrices are only defined for matrices of the same order. Briefly, the sum of two matrices is the matrix whose elements are the sums of the corresponding elements of the given matrices. Example:

$$\begin{pmatrix} 5 & 7 & -3 \\ 8 & 14 & 2 \end{pmatrix} + \begin{pmatrix} -5 & 6 & 12 \\ 2 & -3 & 4 \end{pmatrix} = \begin{pmatrix} 0 & 13 & 9 \\ 10 & 11 & 6 \end{pmatrix}.$$

The difference of two matrices is defined in the same way.
Example:

$$\begin{pmatrix} 2 & 9 \\ -8 & 4 \\ 5 & -3 \end{pmatrix} - \begin{pmatrix} 9 & -6 \\ 5 & -2 \\ 4 & -3 \end{pmatrix} = \begin{pmatrix} -7 & 15 \\ -13 & 6 \\ 1 & 0 \end{pmatrix}.$$

The multiplication or *product* of two matrices is rather more complicated. However, if you learn the formal rule and apply it in a conscious manner, you should have no trouble. Multiplication of matrices is defined whenever the first matrix has the same number of columns as the second matrix has rows. In what follows, we shall extend the Einstein summation convention to repeated indices both of which occur as subscripts. The formal rule is simple enough to remember: $a_{ij}b_{jk} = c_{ik}$. However, its application may not be immediately obvious to the reader who is unaccustomed to the index notation. We remind the reader that the first index refers to the rows of the matrix and the second to the columns. Hence, a_{32} is the element of the matrix a_{ij} which occurs in the third row and the second column. Let us take the case where a_{ij} is of order 2×3 and b_{jk} is 3×4. The rule immediately tells us that the product matrix c_{ik} will be 2×4, since $i = 1, 2$; $k = 1, 2, 3, 4$. Thus, c_{ik} will be of the form

$$c_{ik} = \begin{pmatrix} c_{11} & c_{12} & c_{13} & c_{14} \\ c_{21} & c_{22} & c_{23} & c_{24} \end{pmatrix}.$$

Our task, now, is to determine from the rule how each of the elements is constructed from the elements of a_{ij} and b_{jk}. Let us see how we determine c_{23}. The indices tell us that $i = 2$ and $k = 3$. So the rule for this case is $c_{23} = a_{2j}b_{j3}$. But what about j? Since j is a repeated index it has a different significance from that of i and k. It does not take a particular value but indicates that a summation is to be performed over all possible values. Such repeated indices are called *dummy* indices. Their role is akin to that of bound variables in logic. Thus, finally we have

$$c_{23} = a_{21}b_{13} + a_{22}b_{23} + a_{23}b_{33}.$$

Note that the indices in each of the expressions on the right-hand side are so distributed that the two outer indices correspond to those of the product element and the two inner indices are always identical.

We shall better capture the flavour of matrix multiplication, however, by attempting a few concrete examples.

$$(5 \quad 4 \quad 6) \begin{pmatrix} 3 & 7 & -3 & 1 \\ 1 & -9 & 2 & 0 \\ 0 & 4 & 5 & 2 \end{pmatrix} = a_{ij}b_{jk}.$$

We note that a_{ij} is 1×3 and b_{jk} is 3×4, so that the product will be 1×4. The four elements are determined as follows:

$$\begin{aligned}
c_{11} &= a_{1j}b_{j1} = 5.3 \quad +4.1 \quad +6.0 = 19\\
c_{12} &= a_{1j}b_{j2} = 5.7 \quad +4.-9+6.4 = 23\\
c_{13} &= a_{1j}b_{j3} = 5.-3+4.2 \quad +6.5 = 23\\
c_{14} &= a_{1j}b_{j4} = 5.1 \quad +4.0 \quad +6.2 = 17.
\end{aligned}$$

Hence,

$$c_{ik} = (19 \quad 23 \quad 23 \quad 17).$$

Now we shall consider the case of the product of a 3×3 matrix with a 3×1 matrix. This example is of particular significance since it is the case which arises when a 3-vector is subjected to a coordinate transformation.

$$\begin{pmatrix} 1 & -4 & 1 \\ 5 & 1 & 2 \\ 2 & 3 & 7 \end{pmatrix}\begin{pmatrix} 2 \\ 1 \\ 3 \end{pmatrix}.$$

The product will obviously be 3×1, which makes rather good sense since the result of transforming a vector should be a vector of the same kind.

$$\begin{aligned}
c_{11} &= a_{1j}b_{j1} = 1.2 + -4.1 + 1.3 = 1\\
c_{21} &= a_{2j}b_{j1} = 5.2 + \quad 1.1 + 2.3 = 17\\
c_{31} &= a_{3j}b_{j1} = 2.2 + \quad 3.1 + 7.3 = 28.
\end{aligned}$$

Thus,

$$c_{ik} = \begin{pmatrix} 1 \\ 17 \\ 28 \end{pmatrix}.$$

Finally, we consider another case which has a particular physical application. It is the multiplication of two 3×3 matrices. This is the matrix representation of the result of performing two coordinate transformations in succession. The product will obviously be 3×3. Again, this is something we should expect. It signifies that the result of performing two coordinate transformations in succession is equivalent to the carrying out of a single coordinate transformation. Just think of a coordinate transformation as a one-one mapping and you will understand why.

$$\begin{pmatrix} 1 & 3 & -2 \\ -5 & 4 & 1 \\ 3 & 0 & 3 \end{pmatrix}\begin{pmatrix} 1 & 2 & 0 \\ 3 & 2 & 4 \\ 1 & -3 & 5 \end{pmatrix}$$

$$\begin{aligned}
c_{11} &= a_{1j}b_{j1} = \quad 1.1 + 3.3 + \quad -2.1 = \quad 8\\
c_{12} &= a_{1j}b_{j2} = \quad 1.2 + 3.2 + -2.-3 = \quad 14\\
c_{13} &= a_{1j}b_{j3} = \quad 1.0 + 3.4 + \quad -2.5 = \quad 2\\
c_{21} &= a_{2j}b_{j1} = -5.1 + 4.3 + \quad 1.1 = \quad 8\\
c_{22} &= a_{2j}b_{j2} = -5.2 + 4.2 + \quad 1.-3 = -5\\
c_{23} &= a_{2j}b_{j3} = -5.0 + 4.4 + \quad 1.5 = \quad 21\\
c_{31} &= a_{3j}b_{j1} = \quad 3.1 + 0.3 + \quad 3.1 = \quad 6\\
c_{32} &= a_{3j}b_{j2} = \quad 3.2 + 0.2 + \quad 3.-3 = -3\\
c_{33} &= a_{3j}b_{j3} = \quad 3.0 + 0.4 + \quad 3.5 = \quad 15.
\end{aligned}$$

Hence,

$$a_{ij}b_{jk} = c_{ik} = \begin{pmatrix} 8 & 14 & 2 \\ 8 & -5 & 21 \\ 6 & -3 & 15 \end{pmatrix}.$$

We now proceed to consider various types of matrices which will be of particular significance in physical applications. The matrix obtained by interchanging the rows and columns of a given matrix is called the *transpose* of the given matrix. The transpose of a_{ij} is symbolized by a'_{ij}. Thus, $a'_{ij} = a_{ji}$. A square matrix which is identical to its transpose is called a *symmetric* matrix. Such a matrix is symmetric in its indices, which is to say that any pair of indices may be interchanged freely. Thus, $a_{13} = a_{31}$, $a_{23} = a_{32}$ and so forth. In general, a square matrix is symmetric if and only if $a_{ij} = a_{ji}$. A matrix which is such that $a_{ij} = -a_{ji}$ is said to be *skew-symmetric*. The property of skew-symmetry obviously implies that $a_{11} = -a_{11}$, $a_{22} = -a_{22}$, etc. Clearly, this can only be realized if the diagonal elements, where $i = j$, are all zero. A particular case of a symmetric matrix is one in which all of the off-diagonal elements are zero. Such a matrix is called a *diagonal* matrix. If all of the diagonal elements of a diagonal matrix are equal, the matrix is called a *scalar* matrix. The motivation for this terminology is that multiplication by a *scalar* matrix has the same effect as scalar multiplication. A most important type of scalar matrix is that in which all diagonal elements are equal to 1. This is the *identity* or *unit* matrix. We employ the special symbol δ_{ij}, known as the Kronecker delta, for any such matrix. It may be defined as follows.

$$\delta_{ij} = 1 \text{ for } i = j$$
$$= 0 \text{ for } i \neq j.$$

The reader should attempt to verify the following important identity:

$$\delta_{ij}a_{jk} = a_{ik}.$$

Of fundamental importance is the *inverse* of a square matrix. The inverse of a_{ij} is denoted by a_{ij}^{-1}. It is defined by the following identity:

$$a_{ij}a_{ij}^{-1} = a_{ij}^{-1}a_{ij} = \delta_{ij}.$$

The general method for determining the inverse of a given matrix is rather complicated and involves certain mathematical notions with which we are not yet familiar, in particular that of the *determinant* of a square matrix. However, it happens that there is a particular class of square matrices, which looms very large in most applications short of those which arise in general relativity, for which the determination of the inverse is very simple indeed. These are called *orthogonal matrices*. The inverse of an orthogonal matrix is simply its transpose. Thus, a_{ij} is orthogonal if an only if $a_{ij}a'_{ij} = \delta_{ij}$.

We shall now take a quick and cursory look at the subject of determinants. With every square matrix a_{ij} we may associate a certain quantity which is called the *determinant* of a_{ij}, symbolized $|a_{ij}|$. Corresponding to a matrix of order n, there will be a determinant of order n. We shall begin our discussion with a simple treatment of the determinant of a 2×2 matrix.

Let

$$a_{ij} = \begin{pmatrix} 3 & 2 \\ 4 & 5 \end{pmatrix}.$$

Then

$$|a_{ij}| = \begin{vmatrix} 3 & 2 \\ 4 & 5 \end{vmatrix}.$$

The value of $|a_{ij}|$ is given by $a_{11}a_{22} - a_{12}a_{21}$. In other terms, it is the product of the left-right diagonal *less* the product of the right-left diagonal. In this instance, it is $3.5 - 2.4 = 7$. If the determinant is of order three this method of computing the value of a determinant will not work. However, a 3×3 determinant may be expanded in the following fashion.

$$\begin{vmatrix} D_{11} & D_{12} & D_{13} \\ D_{21} & D_{22} & D_{23} \\ D_{31} & D_{32} & D_{33} \end{vmatrix} = D_{11}\begin{vmatrix} D_{22} & D_{23} \\ D_{32} & D_{33} \end{vmatrix} - D_{12}\begin{vmatrix} D_{21} & D_{23} \\ D_{31} & D_{33} \end{vmatrix} + D_{13}\begin{vmatrix} D_{21} & D_{22} \\ D_{31} & D_{32} \end{vmatrix}.$$

We may describe the nature of the expansion in the following general way. We take the first element D_{11} and multiply it by the determinant which is obtained from $|D_{ij}|$ by removing the first row and the first column. Then we subtract the product of the element D_{12} and the determinant obtained from $|D_{ij}|$ by removal of the first row and the second column. Finally, we add the product of D_{13} and the determinant resulting from the removal of the first row and the third column. We evaluate the 2×2 determinants as before and finally obtain the expression

$$D_{11}(D_{22}D_{33} - D_{23}D_{32}) - D_{12}(D_{21}D_{33} - D_{23}D_{31}) + D_{13}(D_{21}D_{32} - D_{22}D_{31}).$$

Performing the indicated multiplications we obtain

$$D_{11}D_{22}D_{33} - D_{11}D_{23}D_{32} - D_{12}D_{21}D_{33} + D_{12}D_{23}D_{31} + D_{13}D_{21}D_{32} - D_{13}D_{22}D_{31}.$$

It may occur to the reader that the evaluation of a determinant is somewhat tedious. As a general rule, the more advanced we become the more we tend to avoid such tedious computations. In their place, we invent ingenious notational devices which express in a very compact form the result of performing the given operations. It is frequently found that the new notation acquires a life of its own in the sense that it obeys strict laws. Making use of such laws, we are able to carry out a computation without actually performing the indicated operations. In this respect, the more advanced mathematics is often simpler. We shall introduce such a notational device at this point. We do not really need it right away, although it is often encountered in the literature of physics. For the moment, it is simply a notation for the expansion of a determinant of order 3. It is called the *permutation symbol* and is defined in the following manner:

$$\varepsilon_{ijk} = \begin{cases} 0 \text{ if any two indices are equal.} \\ 1 \text{ for even permutations of the indices 1, 2, 3.} \\ -1 \text{ for odd permutations of the indices 1, 2, 3.} \end{cases}$$

An odd permutation of 1, 2, 3 is an interchange of two adjacent indices. An even permutation is the result of two odd permutations. The odd permutations are 213, 132, 321. The even permutations are 123, 312, 231. We may now represent the operation of evaluating the determinant $|D_{ij}|$ in the following way.

$$|D_{ij}| = \varepsilon_{ijk}D_{1i}D_{2j}D_{3k}$$

The value of this formula is that it is easy to remember and enables one to retrieve any or all terms in the expansion of $|D_{ij}|$ in a purely mechanical and foolproof (almost) way.

Moreover, and this is of significance, the expression is fully consistent with the summation convention. The more ambitious reader should verify this by performing the three indicated summations over i, j and k. (*Hint*: Remember that many of the terms will vanish in virtue of the definition of ε_{ijk}.)

It is appropriate at this point to mention two matters which are of fundamental importance. Firstly, a matrix whose determinant is equal to zero is said to be *singular*. The characteristic feature of a singular matrix is that it has no inverse. In terms of coordinate transformations this means that if the matrix of the coefficients of the transformation were singular, it would not be possible to reverse the effect of the transformation. In other words, the transformation carries the points of a space into the points of a subspace or space of fewer dimensions. For example, it is possible to map every point in a Euclidean plane onto corresponding points on a Euclidean line. But there would be no reverse mapping which would recover the original points in the plane.

The second point concerns the orthogonal matrices. From the fact that the determinant of a matrix is equal to the determinant of its transpose and the obvious fact that the determinant of a unit matrix is 1, it follows that the determinant of an orthogonal matrix must equal 1. This can be seen from the following considerations.

$$a_{ij}a'_{ij} = \delta_{ij}.$$

Whence,

$$|a_{ij}|\,|a_{ij}| = |\delta_{ij}| \text{ and } |a_{ij}| = 1.$$

This completes our somewhat involved discussion of the algebra of matrices.

A Trigonometric Digression

Up to this point we have managed to cover considerable mathematical ground with only one significant reminder of the reader's school mathematics. This was in part to assure those who learned only one thing about mathematics—namely, that they can't do it, that it is possible to start afresh and to progress a considerable way into the subject without repeating the painful secondary-school drill. However, we have reached the point where it is necessary to clothe our abstract ideas with something concrete. This will require the relearning of something that you may already have learned to hate—namely, elementary trigonometry. However, the remarkable thing about this topic is that whereas the schoolboy may have to endure the subject for an entire year, most of what is needed of it in higher mathematics can be learned in half a day. Of that, what we shall need can be acquired in a few minutes.

Figure 1.4 is a Cartesian coordinate system in which we have described a circle of radius r. A right-angled triangle with an arbitrary acute angle of θ has been constructed in each quadrant of the circle. The basis of trigonometry is to construct functions of θ into the real numbers and to study the interrelationships of such functions. We shall restrict our treatment to the three fundamental functions $\tan\theta$, $\sin\theta$, $\cos\theta$. Sin θ is defined as the ratio y/r. Cos $\theta = x/r$ and $\tan\theta = y/x$. It is easy to show on the basis of elementary geometrical considerations that these functions are independent of the size of the triangle. That is to say, that the functions have a determinate value for all values of θ.

It should be obvious that as θ increases in the interval from 0° to 90° that y (the height)

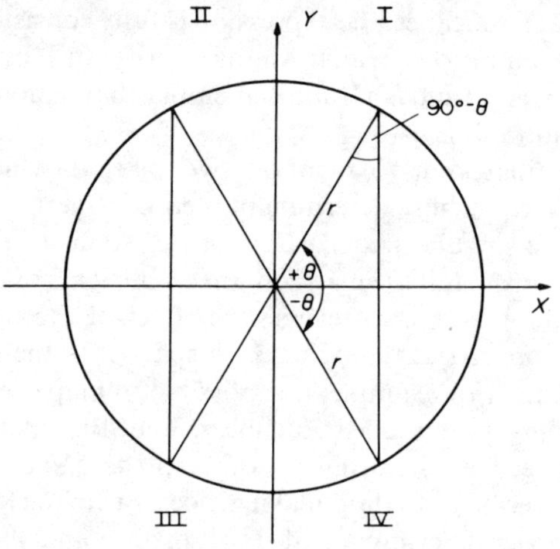

FIG. 1.4. The four quadrants

also increases so that $y/r = \sin\theta$ increases. At 90°, $y = r$ so that $\sin 90° = 1$. On the other hand, as θ increases x decreases so that $x/r = \cos\theta$ decreases and $\cos 90° = 0$. By similar reasoning, it may be seen that $\sin 0° = 0$, while $\cos 0° = 1$. It is important to understand the relationship between sine and cosine. For example, the Pythagorean theorem: $r^2 = x^2 + y^2$ may be exploited as follows.

$$\cos^2\theta = y^2/r^2,\ \sin^2\theta = x^2/r^2$$

$$\sin^2\theta + \cos^2\theta = x^2/r^2 + y^2/r^2 = \frac{x^2+y^2}{r^2} = 1.$$

Again, if we consider the other acute angle in Fig. 1.4—namely, $90° - \theta$, the roles of x and y are interchanged. Cos $(90-\theta) = y/r = \sin\theta$. Similarly, $\sin(90-\theta) = \cos\theta$. It should also be noted by reference to the construction in the IVth quadrant that since x is positive while y is negative that $\sin(-\theta) = -\sin\theta$, whereas $\cos(-\theta) = \cos\theta$. We discriminate between the behaviour of the two functions in the last regard by saying that sine is an odd function whereas cosine is an even function. It should also be remarked that, by definition, $\tan\theta = \sin\theta/\cos\theta$.

Finally, we remind the reader that there is an obvious connection between the trigonometric functions and our earlier discussion of the resolution of a vector (arrow) in terms of its projections along the X- and Y-axes. Specifically, we regard r as the vector or arrow. Since $\cos\theta = x/r$, the projection of r in the direction of X is $r\cos\theta$, while its projection in the direction of Y is $r\sin\theta = r\cos(90-\theta)$.

Vector Algebra

We now have at our disposal three important tools: the algebra of n-tuples, the algebra of matrices and trigonometry. These suffice for a treatment of vectors. However, the reader should be warned that the objects whose mathematical behaviour we are now going to study are not really vectors in the strict mathematical sense. They are what we have been picturesquely referring to as arrows. However, this is the approach which is almost universal in elementary introductions to vector algebra. While it is conceptually misleading, it is much more intuitive than the rigorous approach. However, once we have completed this part of the treatment, it will be an easy step to the appropriate generalization from the arrow to the vector.

Since a vector is, as we have seen, a type of *n*-tuple, all of the laws of *n*-tuples apply to vectors. If we represent an arbitrary vector in the form A^i, where each component represents the projection of the vector in the direction of an axis or equivalently in the direction of a unit base vector, then

$$|A^i| = ((A^1)^2 + (A^2)^2 + (A^3)^2)^{1/2}.$$

In short, the magnitude of a vector is given by the three-dimensional generalization of the Pythagorean theorem. A more compact version of this would be

$$|A^i|^2 = \delta_{ij} A^i A^j,\ i, j = 1, 2, 3, \text{ where } \delta_{ij} \text{ is the Kronecker delta.}$$

Another aspect in which the algebra of *n*-tuples must be supplemented is by a definition for the operation of multiplication of two vectors. In fact, we distinguish two very different types of vector multiplication. The first and simpler form is known variously as the *dot*, *scalar* or *inner* product. Using vector notation, which strictly we are not entitled to, the inner product is defined by

$$\mathbf{A} \cdot \mathbf{B} = |\mathbf{A}|\,|\mathbf{B}| \cos\theta, \text{ where } \theta \text{ is the angle between } \mathbf{A} \text{ and } \mathbf{B}.$$

The geometrical significance of the inner product is revealed by dividing both sides of the defining equation by $|\mathbf{A}|\,|\mathbf{B}|$.

$$\frac{\mathbf{A} \cdot \mathbf{B}}{|\mathbf{A}|\,|\mathbf{B}|} = \cos\theta.$$

That is to say, that the inner product of unit vectors in the direction of **A** and **B** is equal to the cosine between them. Since base vectors are perpendicular, $\hat{x}^1 \cdot \hat{x}^2 = \hat{x}^2 \cdot \hat{x}^3 = \hat{x}^3 \cdot \hat{x}^1 = 0$. Moreover, $\hat{x}^1 \cdot \hat{x}^1 = 1$, etc.

The inner product of *A* and *B* may be represented in matrix form by treating **A** as a 1×3 matrix and **B** as a 3×1 matrix. Then $\mathbf{A} \cdot \mathbf{B}$ is given by

$$(A^1\ A^2\ A^3)\begin{pmatrix} B^1 \\ B^2 \\ B^3 \end{pmatrix} = (A^1 B^1 + A^2 B^2 + A^3 B^3).$$

Note that the resultant is a 1×1 matrix which is just a special case of what we have called a *scalar* matrix. The most convenient notation for representing the inner product in terms of the components of vectors is $\delta_{ij} A^i B^j$. The reader should expand this expression to make sure that it agrees with the matrix representation.

The best way to get the feel of a vector operation is to associate it with a simple physical situation. You may recall from your school physics the formula:

$$\text{work} = \text{force} \times \text{distance}$$

Strictly speaking this formula is false if it is interpreted as a method of calculating the physical quantity called *work*. Intuitively, work is not a measure of effort expended but of useful or unwasted effort—effort in the right direction. For example, in moving an object from point 1 to point 2 we may exert a force on the object. However, a certain amount of the force will be wasted if it is not exerted in the direction from point 1 to point 2. Thus, two people may perform the same work at very different cost in terms of actual effort or expenditure of force. The work involved is simply the product of the component of the force in the direction in which the object is being moved and the distance moved. A diagram should clarify the situation.

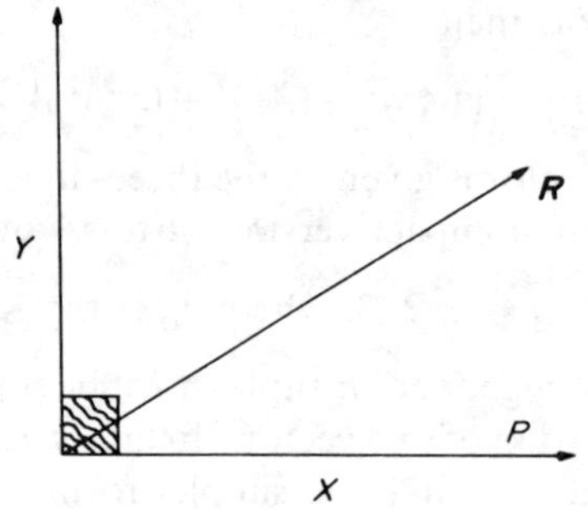

FIG. 1.5. Work = $\mathbf{F}\cdot\mathbf{OP}$

We have a block located at the origin of the coordinates. We wish to measure the work required to move the block in the X-direction to the point P. The arrow gives the magnitude and direction of the force exerted on the object. The work will be the component of the force in the X-direction multiplied by the distance OP. But this is $|\mathbf{F}||\mathbf{OP}|\cos\theta$, which is just the inner product of the vectors $\mathbf{F}$ and $\mathbf{OP}$.

The other type of vector multiplication is called the *cross* product or *vector* product, since the resultant is a vector rather than a scalar. The vector product of $\mathbf{A}$ and $\mathbf{B}$ is symbolized by $\mathbf{A}\times\mathbf{B}$. The interpretation of the vector product is more subtle, both geometrically and physically, than that of the scalar product. Accordingly, we shall begin simply with a consideration of the mechanical nature of the operation. If $\mathbf{A}$ and $\mathbf{B}$ are two arbitrary vectors then $\mathbf{A}\times\mathbf{B} = |\mathbf{A}||\mathbf{B}|\sin\theta\,\hat{n}$, where θ is the angle between $\mathbf{A}$ and $\mathbf{B}$ and $\hat{n}$ is a unit vector, to be determined, which defines the direction of $\mathbf{A}\times\mathbf{B}$. The simplest way to calculate the vector product is by forming a determinant in which the first row has the three basis vectors as its elements, the second row is the components of $\mathbf{A}$ and the third row is the components of $\mathbf{B}$.

$$\mathbf{A}\times\mathbf{B} = \begin{vmatrix} \hat{x}^1 & \hat{x}^2 & \hat{x}^3 \\ A^1 & A^2 & A^3 \\ B^1 & B^2 & B^3 \end{vmatrix} = (A^2B^3 - A^3B^2)\hat{x}^1 - (A^1B^3 - A^3B^1)\hat{x}^2 + (A^1B^2 - A^2B^1)\hat{x}^3.$$

We may again exploit one of our more sophisticated notational devices to represent this vector product in the compact form $\varepsilon_{ijk}A^jB^k = C^i$.

We may ascertain the direction of $\mathbf{A}\times\mathbf{B}$ by choosing a coordinate system in which $\mathbf{A}$ and $\mathbf{B}$ lie on the coordinate plane $\hat{x}^1\hat{x}^2$. Note that this entails no loss of generality since any two vectors which are not colinear define some plane or other. In terms of our compact notation it simply means that since **A** and **B** have no component in the X^3-direction the dummy indices j and k will run over 1 and 2 instead of 1, 2 and 3. Whence,

$$\varepsilon_{ijk} A^j B^k = (A^1 B^2 - A^2 B^1)_i = C^i.$$

But this expression vanishes unless $i = 3$. Hence, it is a vector whose sole component is in the X^3-direction. In short, $\mathbf{A} \times \mathbf{B}$ is a vector which is perpendicular to both **A** and **B**. We express this by saying that **C** is *normal to the plane* of **A** and **B**.

We may pause to verify these matters in an elementary way in which we exploit the fact that the scalar product of two orthogonal vectors vanishes. Thus we may form the *vector* product of **A** and **B** and verify that the inner product of this quantity and either **A** or **B** is zero. Let $\mathbf{A} = 3\hat{x}^1 + 2\hat{x}^2 + 4\hat{x}^3$.

$$\mathbf{B} = 2\hat{x}^1 - 2\hat{x}^2 + 3\hat{x}^3$$

$$\mathbf{A} \times \mathbf{B} = \begin{vmatrix} \hat{x}^1 & \hat{x}^2 & \hat{x}^3 \\ 3 & 2 & 4 \\ 2 & -2 & 3 \end{vmatrix} = 14\hat{x}^1 - \hat{x}^2 - 10\hat{x}^3 = \mathbf{C}$$

$$\mathbf{A} \cdot \mathbf{C} = 3(14) + 2(-1) + 4(-10) = 0$$

$$\mathbf{B} \cdot \mathbf{C} = 2(14) - 2(-1) + 3(-10) = 0.$$

The reader should be struck by the almost magic harmony of all this! So we have the general theorem $\mathbf{A} \cdot (\mathbf{A} \times \mathbf{B}) = \mathbf{B} \cdot (\mathbf{A} \times \mathbf{B}) = 0$. Moreover, we may prove this general result in a most elegant fashion by employing our "advanced" notation.

$$\mathbf{A} \times \mathbf{B} = \varepsilon_{ijk} A^j B^k$$
$$\mathbf{A} \cdot (\mathbf{A} \times \mathbf{B}) = \delta_{il} \varepsilon_{ijk} A^l A^j B^k = \varepsilon_{ijk} A^i A^j B^k$$
$$= \varepsilon_{1jk} A^1 (A^j B^k) + \varepsilon_{2jk} A^2 (A^j B^k) + \varepsilon_{3jk} A^3 (A^j B^k) = 0.$$

Finally, we note that the vector product operation has the property of *anticommutativity*, i.e. $\mathbf{A} \times \mathbf{B} = -(\mathbf{B} \times \mathbf{A})$. This property, again, follows from the nature of ε_{ijk}, specifically that it is antisymmetric in the indices j and k. So that $\varepsilon_{ijk} A^j B^k = -\varepsilon_{ikj} A^j B^k$.

A possible physical interpretation of the vector product arises in the following situation. Suppose that you are driving a screw into the wall with the aid of a wrench. The applied force is in a direction which is parallel to the plane of the wall but the resultant is in the direction of the screw, i.e. normal to the wall. As you probably know, the longer the wrench, the greater will be the effect of the given force. If **r** is the vector from the point of application of the force to the screw, then the resultant along the axis of the screw is known as the *torque* which is defined by

$$\mathbf{L} = \mathbf{r} \times \mathbf{F}.$$

Now that the reader has a reasonable grasp of the basic operations of vector algebra, it is appropriate to offer a more refined definition of the vector concept itself. Previously, we identified a vector with the triple of numbers representing the location of the arrowhead. The obvious limitation of that account is that it ties the vector to a specific coordinate system, for which there is no geometric justification. A Cartesian coordinate system is determined by the position of its origin and the orientation of its axes. But there are infinitely many such systems at our disposal and no one of them seems to have a preferred role. Thus, when we describe a physical system with the aid of a coordinate system, that description will be partly determined by an arbitrary choice of coordinates. For example, I may say that my home is located 3 miles due south of town *A* or may prefer to say that it is 6 miles to the west of town *B*. In the one case, the location is associated with the number 3 and in the other with the number 6. Clearly, this does not imply that my home is in more than one location, but simply that it is possible to specify its location with respect to more than one reference point. By the very same token, the various components of a vector will depend on an arbitrary choice of coordinate system. Thus it is a mistake to identify a vector with three numbers. The three numbers merely *represent* the vector in a particular coordinate system. In another system, the very same vector will be associated with entirely different numbers. Thus we see that a vector is not a triple (*n*-tuple) but an infinite set of triples any one of which serves as a *representative* of the vector. However, the representative of the vector would not be a useful concept unless there were a rule or law which related all of the representatives in such a way that given one any or all of the others could be computed. Less abstractly, given a triple of numbers associated with a vector in a given coordinate system S, we require a definite law to tell us what its components will be in another arbitrarily chosen system $\overline{S}$. Such a law is known as a *transformation* law. In effect, it operates on a given vector representative to map its components in one coordinate system onto its components in another coordinate system. We shall begin with a fairly simple but important two-dimensional case.

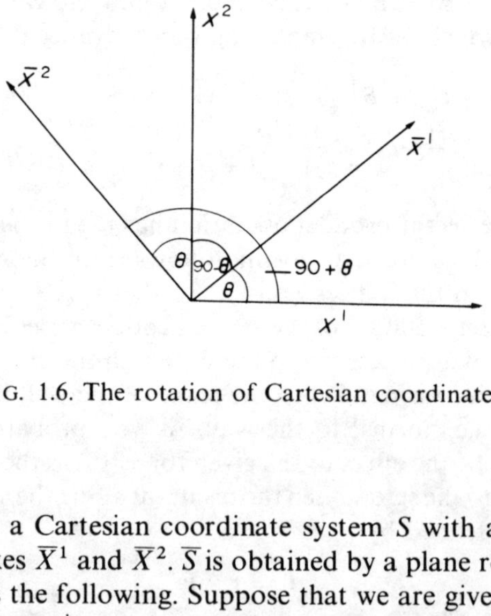

FIG. 1.6. The rotation of Cartesian coordinates

Figure 1.6 represents a Cartesian coordinate system S with axes X^1 and X^2, and a second system $\overline{S}$ with axes $\overline{X}^1$ and $\overline{X}^2$. $\overline{S}$ is obtained by a plane rotation of S through an angle θ. Our problem is the following. Suppose that we are given an arbitrary vector $\mathbf{V}$ whose components in S are V^i, how do we determine its components $\overline{V}^j$ with respect to $\overline{S}$? Clearly, we need a knowledge of the $\overline{X}^j$ in terms of the X^i. From our knowledge of elementary trigonometry, we deduce that $\overline{X}^1$ may be resolved into the sum $X^1 \cos\theta + X^2 \sin\theta = X^1 \cos\theta + X^2 \cos(90° - \theta)$. In other words, $\overline{X}^1$ is the product of X^1 and the cosine of the angle between $\overline{X}^1$ and X^1 *plus* the product of X^2 and the cosine of the angle between $\overline{X}^1$ and X^2. By somewhat similar reasoning, it may be shown that this result extends to $\overline{X}^2$. That is to say, that $\overline{X}^2$ may be expressed in terms of S in the form: $\overline{X}^2 = X^1 \cos(90° + \theta) + X^2 \cos\theta$. In summary, the transformation from S to $\overline{S}$ may be expressed by the transformation equations:

$$X^1 \to \bar{X}^1 = X^1 \cos\theta + X^2 \cos(90^\circ - \theta)$$
$$X^2 \to \bar{X}^2 = X^1 \cos(90^\circ + \theta) + X^2 \cos\theta.$$

We now make use of a notational convention which greatly simplifies the last result. We represent the cosine of the angle between the ith axis of $\bar{S}$ and the jth axis of S by C_{ij}. Thus, the C_{ij} are called the *direction cosines.* Then we express the transformation from S to $\bar{S}$ in the general form:

$$X^i \to \bar{X}^i = C_{ij}X^j.$$

Thus, the C_{ij} are the elements of what we call a *transformation matrix.*

$$C_{ij} = \begin{pmatrix} \cos\theta & \cos(90^\circ - \theta) \\ \cos(90^\circ + \theta) & \cos\theta \end{pmatrix} = \begin{pmatrix} \cos\theta & \sin\theta \\ -\sin\theta & \cos\theta \end{pmatrix}.$$

We note that Det $C_{ij} = |C_{ij}| = \cos^2\theta + \sin^2\theta = 1$. This is a sufficient condition for C_{ij} to be orthogonal; whence the inverse of C_{ij} is simply C_{ji}. The inverse of C_{ij} is the matrix of the transformation from $\bar{S}$ to S. It is obviously the case that the carrying out of two successive transformations, where the second is the inverse of the first, should be equivalent to carrying out no transformation or what is called the *identity* transformation. We may exploit this truism to verify that the inverse of C_{ij} is indeed C_{ji}. The result of two successive transformations is represented by the product matrix of the matrices of the individual transformations. Thus, in the case under consideration we have

$$\begin{pmatrix} \cos\theta & -\sin\theta \\ \sin\theta & \cos\theta \end{pmatrix} \begin{pmatrix} \cos\theta & \sin\theta \\ -\sin\theta & \cos\theta \end{pmatrix} = \begin{pmatrix} \cos^2\theta + \sin^2\theta & (\cos\theta\sin\theta - \sin\theta\cos\theta) \\ (\sin\theta\cos\theta - \cos\theta\sin\theta) & \cos^2\theta + \sin^2\theta \end{pmatrix}$$

$$= \begin{pmatrix} 1 & 0 \\ 0 & 1 \end{pmatrix} = \delta_{ij}.$$

Finally, we note that everything we have said about two-dimensional transformations holds precisely for three-dimensional transformations also.

Having determined the transformation equations for Cartesian coordinates, we may proceed to consider how a vector is affected by such a transformation. Thus, if the vector representative in S has the components V^i, we wish to determine the components $\bar{V}^i$ in the system $\bar{S}$. We shall not provide a rigorous proof of the result but simply explain the mechanical nature of the operation. However, we do provide a diagram of the situation which should be more valuable to the non-mathematical reader than a formal demonstration. Without significant loss of generality, we may again consider the two-dimensional case. Moreover, for computational ease, we choose the specific case where $\bar{S}$ is obtained from S by a rotation through an angle of 45°, since $\sin 45^\circ = \cos 45^\circ = 1/\sqrt{2}$.

Figure 1.7 illustrates the case of the two sets of axes S and $\bar{S}$. The bold arrow is a vector whose representative in S is $(3, 2) = 3\hat{x}^1 + 2\hat{x}^2$. The broken lines represent the projections of this vector on the axes of the two systems. Our problem is to determine the components of the vector in the new system on the basis of our knowledge of its components in the old system together with the transformation equation which relates the two systems. It turns out that the two sets of components are related (or transform) in precisely the same manner as the two sets of coordinates. Consequently, the new components may be computed by subjecting the old components to the same transformation which produced

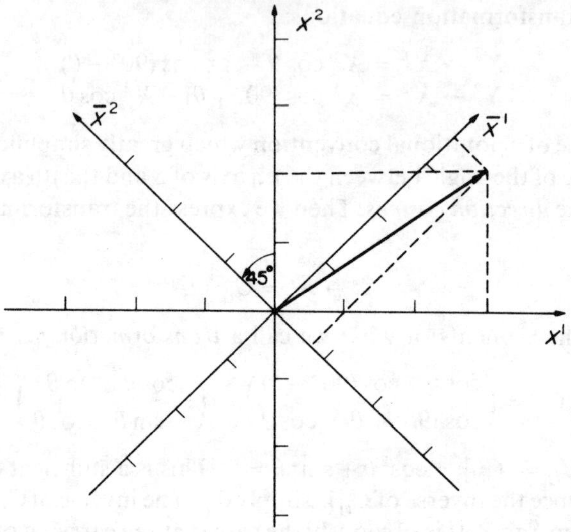

FIG. 1.7. Transformation of the components of a vector by a coordinate rotation

the new coordinates. This is a very simple exercise in matrix multiplication.

$$\begin{pmatrix} \cos 45^\circ \sin 45^\circ \\ -\sin 45^\circ \cos 45^\circ \end{pmatrix}\begin{pmatrix} 3 \\ 2 \end{pmatrix} = \begin{pmatrix} 1/\sqrt{2} & 1/\sqrt{2} \\ -1/\sqrt{2} & 1/\sqrt{2} \end{pmatrix}\begin{pmatrix} 3 \\ 2 \end{pmatrix} = \begin{pmatrix} 3/\sqrt{2}+2/\sqrt{2} \\ -3/\sqrt{2}+2/\sqrt{2} \end{pmatrix}$$
$$= \begin{pmatrix} 5/\sqrt{2} \\ -1/\sqrt{2} \end{pmatrix}.$$

Whence, $\overline{\mathbf{V}} = 5/\sqrt{2}\hat{\bar{\mathbf{x}}}^1 - 1/\sqrt{2}\hat{\bar{\mathbf{x}}}^2$.

The transformation law for a vector is therefore $\overline{V}^i = C_{ij}V^j$. Accordingly, we may define a vector as a set of n-tuples which are related by the foregoing law. However, later we shall find that our interpretation of this law stands in need of further generalization.

Before completing our unavoidably dry discussion of vectors, we shall offer our patient reader a momentary respite by pausing to consider some of the deeper ramifications of our discussion so far. We have discovered that a vector is a mathematical entity which is describable in terms of an arbitrarily chosen set of coordinates and which transforms according to a definite law under a change or transformation of coordinates. An entity with a definite law of transformation is called a *geometric object*. From an epistemological standpoint, the significance of a geometric object is its permanent or objective character. For example, a vector may be described or represented in terms of any one of infinitely many coordinate systems. Each such representation may be regarded, metaphorically speaking, as an appearance of the vector. But the vector itself transcends all such appearances, just as the "real table", dear to the hearts of philosophers, transcends the infinitely many appearances of the table.

Contrary to the archaic doctrine of positivism according to which the aim of physical theory is to construct compact summaries of appearances which are intended to serve as tools for the predicting of future appearances, the actual task of the physicist is to penetrate appearances, to ascertain the laws of the objective world—the nature of reality. A mathematical formalism which has the capability of expressing, in some degree, the objective features of the world is, therefore, an indispensable tool for such a venture. The vector formalism is a significant step in this direction.

One way of describing the spatio-temporal features of the world is to assign coordinate values to various physical magnitudes. However, the shortcoming of such a description is that it reveals a pattern which to some extent reflects not the world but the accidental features of the system of coordinates whose choice is always somewhat arbitrary. This arbitrariness may be significantly circumvented by describing the physical system of interest in vectorial form., i.e. in the form of vector equations. The terms which occur in such equations will be independent of any particular coordinate system and will accordingly possess an objective character. Expressed somewhat differently, a vector equation which is valid with respect to a given coordinate system will be valid with respect to every other such system. (This statement is not absolutely correct but the spirit of it is.) In short the vector formalism is a coordinate-free formalism.

To return to our metaphor, the vector formalism is akin to a method of describing the world which would be equally faithful to every individually perceived appearance of it and, moreover, would permit one to deduce the character of an individual appearance associated with a specific percipient subject. By the same token, a vectorial law will be consistent with the measurements whose results are expressed in terms of any particular coordinate system. Moreover, given the law in vectorial form, the numerical results of measurements expressed in terms of a specific coordinate system are predictable even though they vary from system to system. These numerical results correspond, of course, to the individual n-tuples associated with the vector.

When a geometric object composed of components—for example, a vector—transforms under a certain group of transformations so that the object is essentially unchanged even though the values of the components do undergo alteration, the object is said to be *covariant* under that group of transformations. Similarly, a physical law which is expressed by means of such geometric objects is said to be covariant under a given group of transformations. This means that such a law will preserve its form under any transformation which is an element of the group. Since all such transformations involve a change of coordinates, and since the choice of coordinates should have no more influence on the physical content of the law than, for example, the language in which the law is stated, it would appear to follow that the more inclusive the group of transformations under which the law is covariant the more objective the law will be. In fact, one of the principal philosophical motivations for the formulation of relativistic theories is precisely to write laws of such a form that their covariance group will be the largest one attainable. However, the precise meaning and significance of the role of covariance involves rather deep considerations and we shall discuss the matter in detail in a later chapter.

In the preceding paragraph, I alluded to a *group of transformations*. A *group* is not simply a synonym for a set but an algebraic concept which is of profound epistemological significance. Although it is technically possible to utilize the theory of relativity, for computational purposes, without recourse to the algebraic theory of groups, a mature conceptual grasp of the theory of relativity, which is the point of this book, demands at least a passing acquaintance with group theory. Thus we shall now embark on a painless introduction to the group concept.

Abstractly, a group is a set of elements $\{a_1, a_2, \ldots, a_n\}$ together with an operation for combining them, $a_i a_j$, such that the following axioms are satisfied:

(1) The composition of any two elements of the group is an element of the group, i.e. if $a_i \varepsilon \mathcal{G}$, $a_j \varepsilon \mathcal{G} \rightarrow a_i a_j = a_k \varepsilon \mathcal{G}$
(2) $a_i(a_j a_k) = (a_i a_j)a_k$. The group operation is associative.
(3) There is an *identity element*, I, such that for any $a_i \varepsilon \mathcal{G}$, $Ia_i = a_i I = a_i$.
(4) For every element $a_i \varepsilon \mathcal{G}$, there is a unique element a_i^{-1}, called the *inverse* of a_i, such that $a_i a_i^{-1} = a_i^{-1} a_i = I$.

A very familiar example of a group is the set of integers with the operation of addition. Obviously the sum of any two integers is an integer so that the first axiom of group theory is satisfied. Moreover, the identity element is zero. Every integer has an inverse which is simply the negative of the given integer. However, the elements of a group need not be numbers. They may be objects, actions, motions, mathematical operations, just in case the axioms are satisfied.

A case in point is the set of permutations of three playing cards. The total number of permutations of three objects is $3! = 3 \times 2 \times 1 = 6$. Let us denote the three cards by the symbols 1, 2, 3. Then the possible permutations are

a_1	a_2	a_3	a_4	a_5	a_6
$1 \rightarrow 1$	$1 \rightarrow 2$	$1 \rightarrow 3$	$1 \rightarrow 1$	$1 \rightarrow 2$	$1 \rightarrow 3$
$2 \rightarrow 2$	$2 \rightarrow 1$	$2 \rightarrow 2$	$2 \rightarrow 3$	$2 \rightarrow 3$	$2 \rightarrow 1$
$3 \rightarrow 3$	$3 \rightarrow 3$	$3 \rightarrow 1$	$3 \rightarrow 2$	$3 \rightarrow 1$	$3 \rightarrow 2$.

Each of the permutations may be regarded as an operation which *maps* the set $\{1, 2, 3\}$ onto itself. We take $\{a_1, \ldots a_6\}$ to be the elements of the group. The group operation is simply the performing of two mappings in succession. For example, $a_4 a_3$ is the permutation which results from performing a_4 on the result of a_3:

$$\begin{array}{l} \quad a_3 \quad a_4 \\ 1 \rightarrow 3 \rightarrow 2 \\ 2 \rightarrow 2 \rightarrow 3 = a_5 \\ 3 \rightarrow 1 \rightarrow 1. \end{array}$$

Note, however, that $a_3 a_4$, the result of performing a_3 on the outcome of a_4 is a_6. Thus, the group operation is not necessarily commutative. The reader should satisfy himself that the four axioms are indeed satisfied.

A *prima facie* instance of an altogether different type of group is the group of symmetries of an equilateral triangle. An element of this group is a motion which maps the equilateral triangle onto itself. For example, if we label the vertices A, B and C, then the result of a clockwise rotation in the plane through 120° will result in A occupying the position of B, B the position of C, and C the position of A. Another element will be a clockwise rotation through 240 degrees. The triangle may also be mapped onto itself by flipping it over while holding one of the vertices in a fixed position. For example, if one holds A in a fixed position and rotates the triangle 180° through space, A is mapped onto A, B onto C and C onto B. Clearly, two additional motions are allowed which are similar to the latter but which involve holding B and C, respectively, in a fixed position. Finally, the identity element would simply be no motion (or equivalently a plane rotation through

360°). It is easy to show that these six elements are the realization of a group—namely, the symmetry group of the equilateral triangle.

A consideration of the structure of this group leads us to an important mathematical notion—namely, the concept of *isomorphism*. Specifically, it turns out that when the six elements of this group are appropriately associated with the six elements of the group of permutations of three cards in a one-one fashion, then the product of two elements in the first group will be the element which corresponds to the product of the two corresponding elements of the second group. That is to say, that algebraically the two groups are concrete realizations of one and the same abstract group.

The reader is now in a position to grasp some of the significance in our earlier references to *groups of transformations*. It turns out that various types of coordinate transformations exhibit the group structure. This simply means that the product of two coordinate transformations of a given kind will be equivalent to a single coordinate transformation of the same kind; each transformation will have a corresponding inverse transformation, and so forth.

A fundamental property of groups in general is the existence of certain *invariants* which may serve to define a given group. Let S be a set of objects which are operated on by the elements of a group G. For example, a trio of playing cards operated on by the elements of the permutation group is a case in point. When an element $s \varepsilon S$ is operated on by an element $g \varepsilon G$, the resultant will be called the *image* of s under g, and may be denoted by s'. For example, $\langle 3, 2, 1 \rangle$ is the image of $\langle 1, 2, 3 \rangle$ under the transformation a_3. We may now define an invariant of a group G in the following way. Let s' be the image of $s \varepsilon S$ under an *arbitrary* element of G. Let F be a certain function of the $s_i \varepsilon S$. The function F is an *invariant of G* if and only if $F(s_i) = F(s_i')$.

An excellent example of one such invariant arises in a situation which we have already studied at some length. On page 22 we considered what happened to the vector representative (3, 2) under the coordinate rotation through 45°. In our present terminology, the image of (3, 2) under this particular element of the rotation group was found to be $(5/\sqrt{2}, -1/\sqrt{2})$. Now let us consider a certain function of the components (x, y) of a vector representative—namely, $f(x, y) = x^2 + y^2$. The reader will recall that f would adequately define the 'length' of the vector representative or arrow. Now we find the value of f for the representative (3, 2). We have

$$f(3, 2) = \sqrt{3^2 + 2^2} = \sqrt{13}.$$

Similarly,

$$f(5/\sqrt{2}, -1/\sqrt{2}) = \sqrt{25/2 + 1/2} = \sqrt{13} \text{ (or } 13^{1/2}\text{)}.$$

In general, the length or "norm" of a vector is an invariant of the group of coordinate rotations. This suggests that whereas the components of a vector are superficial aspects which are related to an arbitrarily chosen coordinate system, its length is an objective or real property. Thus we have arrived at a result which was previously developed on the basis of the Pythagorean theorem. However, we have now reached the result not by way of elementary geometry but via abstract algebra. Here we have another instance of the beautiful harmony to be found in mathematical topics.

It should be fairly clear that the invariants which are associated with the various groups of transformations are of distinct epistemological significance. It is intuitively obvious that any quantity that refers to an invariant property of the world should be independent of the choice of coordinates. A coordinate system is, figuratively speaking, a language employed for the space-time description of a physical system. Just as nature is indifferent to the choice of natural language which happens to be employed to describe it, so it should also be indifferent to the physicist's choice of technical language, e.g. coordinate system. Now if a property of an object or system is independent of the coordinatization with respect to which it is expressed, it is *a fortiori* invariant under a group of coordinate transformations. We often refer to such invariant properties as the symmetries of the object. We conclude that the discovery of the symmetries or invariant properties of a physical system, e.g. the total electric charge, has a fundamental significance in the theoretical descriptions of objective reality.

Many individuals who are confronted with the theory of relativity for the first time are apt to regard it as a piece of scientific conjuring which seeks to prove that black is white. In particular, its treatment of time seems counterintuitive and even inconsonant with "rational insight". However, the attentive student will soon learn that the theory of relativity was generated by the single-minded pursuit of the objective features of reality—the quest for objective knowledge. In this respect, it is philosophy in the tradition of Plato. If certain philosophers should contend that it fails to make sense, it is because they are bound to the world of appearances for want of those mathematical tools which are indispensable to the conceptual representation of reality. The philosopher who professes to be concerned with the nature of reality but is too indifferent to acquire the appropriate mathematical tools to uncover it is at best a dilettante and at worst a charlatan but, in any event, a prisoner.

We shall now complete our discussion of the concept of a vector. You will recall that we have repeatedly stressed the key significance of the fact that a vector is a geometric object which is independent of the choice of coordinate system. However, our account to this point has failed fully to divorce vectors from coordinate systems. Specifically, we have treated each vector representative or arrow as an object which is defined with reference to the origin of a coordinate system. We have shown that the essential properties of a vector are preserved under a coordinate rotation. However, there is another type of coordinate transformation which we have disregarded—namely, an arbitrary displacement or translation of the origin. That is to say, that the second or barred system is obtained from the unbarred system by a displacement which is such that the axes of the new system are held parallel to the axes of the old system. The algebraic representation of such a transformation is simply

$$X^i \to \bar{X}^i = X^i - X^i_0$$

where the X^i_0 are the coordinates of the origin of the barred system with respect to the unbarred system. A diagram of the two-dimensional case should fully clarify the situation. Let P be a point with coordinates X^1, X^2. It is obvious from Fig. 1.8 that its coordinates in the barred system depend on both the original coordinates and the way in which $\bar{O}$ is displaced with respect to O. The coordinate values in the second system are obtained by subtracting from the original values the displacements of $\bar{O}$ with respect to O in the X^1 and X^2-directions.

On the other hand, however, if X^1 and X^2 were the components of a vector with respect to S, then $\bar{X}^1, \bar{X}^2$ would not be the components of the same vector with respect to $\bar{S}$. The source of the difficulty is simply that we made the unwarranted restriction that a vector

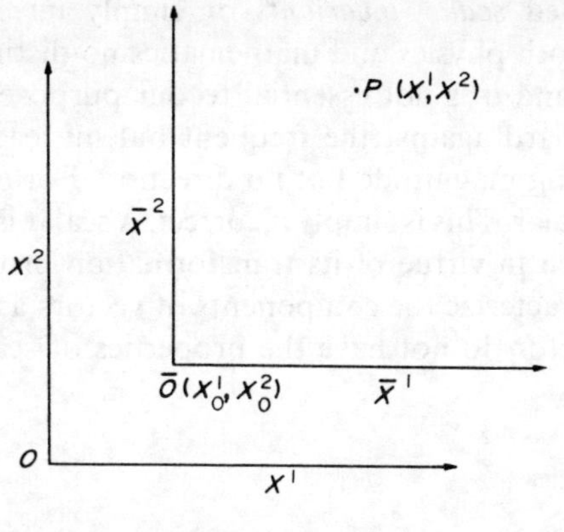

FIG. 1.8. Coordinate transformation by the displacement of the origin

arrow should have its tail at the origin. In fact, the arrow may originate at any well-defined point. Its components are therefore not coordinates but coordinate *differences.* Specifically, let V^i be a vector arrow at the point X_0^i, not necessarily the origin, and let its tip be at the point X^i. Then the components of V^i are $(X^i - X_0^i)$. We have simply hitherto been considering the special case in which $X_0^i = 0$, so that the components were numerically equal to the coordinates of the tip of the vector arrow. In fact, the majority of treatises in physics persist in referring to the special case $X_0^i = 0$ as the *position* or *radius vector*. This is obviously an elementary, albeit generally innocuous, error since a radius vector in one coordinate system may not be a radius vector in another coordinate system, which is simply to say that there is no such object as a radius vector although there are radius arrows.

In summary, we have determined that a vector is a geometric object which is covariant with respect to the group of rotations. Moreover, it is also covariant with respect to the group of translations. It should be intuitively clear that it is also covariant with respect to any coordinate transformation which involves both a translation of the origin as well as a rotation of the axes. The totality of these two types of transformations also forms a group, which we call the inhomogeneous rotation group. Thus, we may provisionally define a vector as a covariant quantity whose law of transformation is

$$V^i \rightarrow \overline{V}^i = a_{ij} V^j - C^i.$$

We introduce one more concept to complete this phase of the discussion. A *scalar* is a one-component object whose value is independent of any coordinate transformation. For example, the temperature at any point in a room is obviously independent of the coordinate representation of its location. Hence, the transformation law of a scalar is

$$\Phi(X^i) = \overline{\Phi}(\overline{X}^i).$$

Certain scalar quantities have the additional property of having the same algebraic form in all systems of coordinates. One such scalar is the magnitude of a vector which is of the form

$$\delta_{ij} V^i V^j = \delta_{ij} \overline{V}^i \overline{V}^j.$$

Such quantities are called *scalar invariants* or simply *invariants*. However, in most elementary treatises in both physics and mathematics no distinction is drawn between a scalar and an invariant and it is not essential to our purposes. On the other hand, the student should be on guard against the frequent but misleading characterization of a scalar as a quantity having magnitude but no direction. For example, a pure number is sometimes treated as a scalar. This is simply incorrect. A scalar is a type of geometric object and only qualifies as such in virtue of its transformation properties. In particular, it is simply erroneous to characterize the components of vectors as scalar quantities. Clearly, the components of a vector do not have the properties of scalars. In fact, they are not geometric objects.

The Calculus

Of all the branches of mathematics, calculus is the one which is pre-eminent in the applied realm. This is only to be expected in so far as science, especially physics, is concerned with change or process and calculus is very much the mathematical theory of change. Unfortunately, calculus has been traditionally regarded as part of "higher mathematics". Accordingly, it has tended to be treated with resigned awe by the non-mathematician. It is the key to most contemporary science, which has, therefore, been off limits to those who lack a mathematical education. While it is true that a rigorous treatment of calculus requires considerable mathematical acumen, the basic concepts are more unfamiliar than difficult and a working knowledge requires more practice than mathematical insight. The purpose of what follows is simply to enable the reader to understand the language of calculus so that the significance of a formula expressing a law of nature may be intuitively grasped. There are many books available for those who are desirous of attaining computational facility in the subject through a few weeks of concentrated effort.

The reader will recall that a function is an association of values of a given variable called the independent variable with unique values of a second variable called the dependent variable. A central, if not *the* central, problem of the calculus is to determine the rate at which the dependent variable changes in value with respect to the independent variable. For example, in the case of $y = 5x$ it is obvious that y is increasing five times more rapidly than x. Figuratively speaking, if we think of x as defining the "time" in which the function travels, and y as defining the "distance" covered by the function in that time, then the "speed" of the function, the rapidity of its change, would be given by $y/x = 5x/x = 5$.

In the foregoing example determination of the rate of change of y with respect to x offered no problem largely in virtue of the linearity of the chosen function. However, the appropriate answer would have been less obvious in the case of such a function as $y = 3x^2 + 7$, and still less so in the case of $y = \tan x$. The problem is analogous to that of determining the velocity of a moving vehicle at any given time. As everyone knows, if a vehicle covers a distance of 5 miles in 1 hour, then its *average* velocity over the entire trip is 5 miles per hour. However, its actual velocity at any given time during the trip may have been different from 5 miles an hour. One could, of course, assume a constant velocity, i.e. that the distance traversed is a *linear* function of the time. While such a stratagem might be appropriate in dealing with the behaviour of trains attempting to run on schedule, it would clearly be unrealistic in cases where one is attempting to express or to draw inferences from an accurate law of nature. For example, Galileo's law of falling bodies is expressed in the form $s = (\frac{1}{2})gt^2$, where s is the distance, t is the time, and g is a constant determined by the strength of the earth's gravitational attraction. Clearly, it would make no sense to treat this law as roughly linear, that is to suppose that the velocity of a falling stone is reasonably constant.

How, then, is one to grapple with the problem of non-linear change? While the averaging approach has been ruled out, it is still suggestive. To return to the case of the moving vehicle, we have seen that its average velocity over the entire trip is almost totally uninformative of its velocity at any specific time. However, if we considered the distance travelled during a relatively small interval of the entire period which we denote by Δt, the ratio or fraction $\Delta s/\Delta t$ would give a better clue as to the "instantaneous velocity" than s/t. Moreover, if we make Δt sufficiently small, Δs will also diminish accordingly, and the ratio $\Delta s/\Delta t$ will provide a reasonably accurate estimate of the velocity at a specific time. Clearly, as Δt gets progressively closer to zero $\Delta s/\Delta t$ will approach a specific limiting value which is the sought-after velocity. Instead of writing the cumbersome expression

$$\lim_{t \to o} \Delta s/\Delta t = v$$

we have a special notation which turns out to be even more fruitful than the index notation, which we introduced in connection with vectors and matrices. We write $ds/dt = v$, where s is the dependent variable of the function and t is the independent variable. ds/dt is said to be the *derivative* of the function $s(t)$ with respect to time. Although ds/dt is not a fraction, it turns out that for many purposes it may be treated as a fraction in which both numerator and denominator are infinitesimally small. A more accurate interpretation of ds/dt is that it is the limit or "last" member of an infinite series. However, it would be beyond the scope of this brief treatment to develop the idea of limits.

We may now proceed directly to the problem of the evaluation of the derivative of a given function. This is the central problem of what is called *differential calculus*. We have already noted the obvious result that the derivative of a function such as $f(x) = y = 5x$ is $dy/dx = 5$. Let us now consider the simplest non-linear function, $f(x) = y = x^2$. It is obvious that when the value of x is increased by a small increment Δx, there will be a corresponding increase in the value of y to $y + \Delta y$. Hence, we may write

$$y + \Delta y = (x + \Delta x)^2 = x^2 + 2x\Delta x + (\Delta x)^2.$$

Since we are given that $y = x^2$, we deduce by subtraction

$$\Delta y = 2x\Delta x + (\Delta x)^2.$$

Whence

$$\Delta y/\Delta x = 2x + \Delta x.$$

We now consider the question as to the behaviour of $\Delta y/\Delta x$ as Δx approaches zero. By inspection it is apparent that in the limit we have

$$dy/dx = 2x$$

The reader may experiment with $y = x^3$ to assure himself of the general rule when $y = x^n$, $dy/dx = nx^{n-1}$. Consequently, without going through the algebraic manipulation we may derive $d/dx(x^3) = 3x^2$, $d/dx(x^4) = 4x^3$, and so forth. It turns out that the general rule also holds for fractional and negative values of n. Consequently, we may write

$$\frac{d}{dx}(\sqrt{x}) = \frac{d}{dx}(x^{\frac{1}{2}}) = \frac{1}{2}x^{-\frac{1}{2}} = \frac{1}{2\sqrt{x}}$$

$$\frac{d}{dx}\frac{1}{x^2} = \frac{d}{dx}(x^{-2}) = -2x^{-3} = \frac{-2}{x^3}, \text{ and so forth.}$$

It should also be noted that when the expression which is to be *differentiated* has a constant coefficient, e.g. $y = 5x$, that the rate of change of the function will be increased by whatever factor is represented by that coefficient. That is to say, that $y = 5x^2$ is changing five times more rapidly that $y = x^2$. Accordingly, $d/dx\ (ax^n) = nax^{n-1}$.

We now state without proof that the derivative of a sum of terms is simply the sum of the derivatives of the individual terms. For example, if $y = 3x^2 + 7x + 2$, then $dy/dx = 6x + 7$. Another general result is that the derivative of the product of two terms is equal to the product of the first term and the derivative of the second plus the product of the derivative of the first term and the second term. For example,

$$y = (3x+2)(4x^3+7)$$

$$\frac{dy}{dx} = (3x+2)\frac{d}{dx}(4x^3+7) + \frac{d}{dx}(3x+2)(4x^3+7)$$

$$(3x+2)(12x^2) + 3(4x^3+7) = 48x^3 + 24x^2 + 21.$$

Next, we need to know how to deal with quotients. Thus, supposing that u and v are both functions of x, what is the derivative of u/v? It is quite easy to derive the general rule for differentiating a quotient from first principles. However, once again, we shall state the general rule without proof:

$$\frac{d}{dx}\left(\frac{u}{v}\right) = \frac{v\frac{du}{dx} - u\frac{dv}{dx}}{v^2}.$$

A simple example should suffice to illustrate the application of the quotient rule. Let $u = x + 3$, $v = 2x^2 + 5$.

$$\frac{d}{dx}\left(\frac{x+3}{2x^2+5}\right) \text{ is determined as follows:}$$

$$\frac{d}{dx}(x+3) = 1; \frac{d}{dx}(2x^2+5) = 4x$$

$$\frac{d}{dx}\left(\frac{u}{v}\right) = \frac{(2x^2+5)-(x+3)(4x)}{(2x^2+5)^2} = \frac{(2x^2+5)-(4x^2+12x)}{(2x^2+5)^2}$$

$$= -\frac{(2x^2+12x-5)}{(2x^2+5)^2}.$$

The reader may be puzzled by the fact that the derivative is negative. However, this is quite normal. An inspection of the original function will reveal that as x increases y will diminish. In other words, the function displays "negative growth" or decelerates. It is this characteristic which is revealed by the fact that the derivative of the function is negative.

The last of the basic rules of differentiation is that which provides a general method for differentiating a function of a function. This rule is known as the chain rule and illustrates the utility of interpreting the derivative as a fraction. It is possible to prove that if y is a function of u and u is a function of x then the following equation is identically true.

$$\frac{dy}{dx} = \frac{dy}{du}\frac{du}{dx}.$$

Intuitively, the quantity du cancels out.

The chain rule may be applied to the problem of finding the derivative of such a function as $y = (x^3 + 1)^2$. This is an example of what we have just referred to as a function of a function. In this case, $u = x^3 + 1$ and $y = u^2$. Following the chain rule, we first differentiate y with respect to u which yields $2u$. Then we differentiate u with respect to x which yields $3x^2$. Finally, we take the product:

$$2u\ .3x^2 = 2(x^3 + 1)(3x^2) = 6x^2(x^3 + 1) = 6x^5 + 6x^2 = \frac{dy}{dx}.$$

Before proceeding to the next phase of our discussion of the differential calculus, we should pause to consider the geometric interpretation of the derivative. Firstly, let us consider the simple linear function: $y = 2x$.

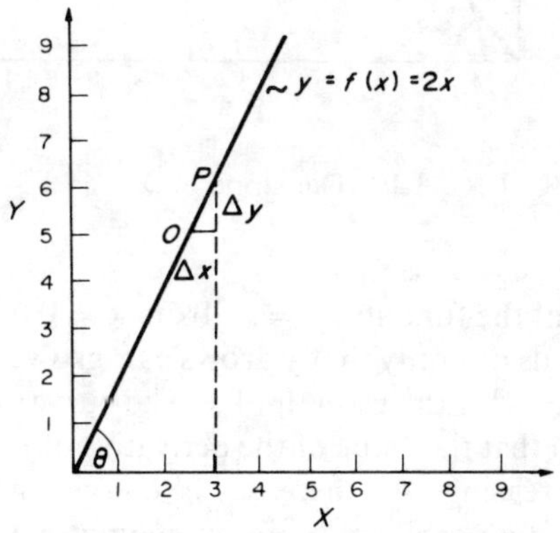

FIG. 1.9. The slope of $y = 2x$.

Figure 1.9 is the graph of the given function. The slope of the linear curve represents the steepness or gradient of the curve. We may think in terms of the slope of an actual hill or road. This one is quite steep indeed. In effect, if one were driving along such a road, for each hundred yards that one progressed in the horizontal direction, one would increase one's altitude above sea-level by two hundred yards. In other words, the gradient of f is the ratio y/x. But the reader will recall from our review of trigonometry that y/x is $\tan\theta$. Hence $\tan\theta$ is the gradient or rate of change of the function. Now consider the small triangle with hypotenuse OP, base Δx and height Δy. It is obvious that it is similar to the much larger triangle. Now imagine that the point P is continuously approaching O. Δy and Δx will shrink accordingly. But one will continue to have a triangle similar to the large one, so that the angle between the base and the hypotenuse will continue to be θ. No matter how small the segment OP, the gradient continues to be $\tan\theta$. Finally, *in the limit*, when O and P coincide, the gradient or slope at the point O will still be $\tan\theta$. Thus, it is possible to attach mathematical significance to the notion of the slope of a curve at a point. In this case, of course, the slope or derivative of f is the same for all points, i.e. for all values of the function, in virtue of its linearity. We now proceed to apply these notions to non-linear functions.

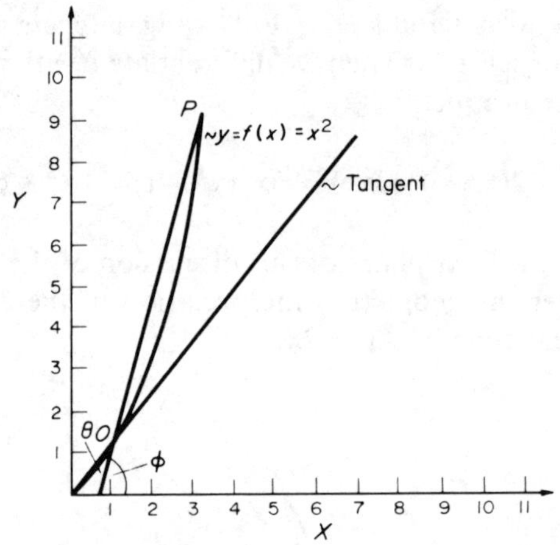

FIG. 1.10. The slope of $y = x^2$

Figure 1.10 is the graph of the function $y = x^2$ from $x = 0$ to $x = 3$. An examination of the shape of the graph reveals not only that y grows as x grows but moreover that the *rate of growth* of y also increases with the growth of x. The curve becomes increasingly steep. Accordingly, we may expect that the value of the derivative of f will be undergoing continuous change and will be increasing as x increases. We now wish to capture the notion of the value of the derivative at an arbitrary point in geometric terms. Take a fixed point O and a variable point P on the given curve. The chord OP makes the angle ϕ with the X-axis. Now let the point P approach O in a continuous manner. As P approaches O, the chord OP rotates in the direction of diminishing ϕ. Finally, in the limit, as P approaches O, OP coincides with the line which makes the angle θ with the X-axis. In the limit, the chord shrinks to a point located at O. Nevertheless, one may continue to associate an angle with the 'lost chord'. The line which OP approaches as P approaches O bears a particular relation to the curve. Namely, it is the only line through O which coincides with the curve at one and only one point without intersecting it. Such a line is said to be the *tangent* to the curve at the point O. The tangent of the angle which it makes with the X-axis corresponds to the value of the derivative at O. Clearly, for every point on the curve, there will correspond a unique tangent associated with the derivative at that point.

We now turn to the topic of successive differentiation, which is a simple extension of the concepts that have already been mastered. It is clear that the derivative of a function is also a function in its own right. For example, the derivative of $y = x^2$ is $dy/dx = 2x$, which is obviously a function. Thus we may take the derivative of a derivative. The latter is called a *second derivative*. Thus, the first derivative of $y = x^2$ is $2x$, while the second derivative is 2. The second derivative of y with respect to x is symbolized d^2y/dx^2. The superscripts must on no account be confused with exponents. Rather, d/dx should be regarded as an operation symbol, namely the operation of taking the first derivative, and d^2/dx^2 is likewise the symbol for the operation of taking the second derivative and so forth to d^n/dx^n.

The notion of higher order derivatives is fairly easy to grasp intuitively. Let s be the distance traversed by a particle. We have already learned that its velocity is given by ds/dt. Now if the particle happens to be continuously accelerating, i.e. its velocity is changing, then the second derivative of s with respect to t will exist and will be given by d^2s/dt^2. By stretching one's imagination somewhat, one may conceive of an accelerating particle such that the rate of acceleration is also continuously changing. Then the third derivative of s would exist and would be given by d^3s/dt^3.

So far we have restricted the discussion to the derivatives of simple algebraic functions. However, many other types of function exist, and their derivatives have important physical interpretations. We shall restrict our attention to the functions $y = \sin\theta$ and $y = \cos\theta$, both of which will recur frequently in later chapters. The procedure for differentiating $y = \sin\theta$ with respect to θ is exactly the same as for an algebraic function. We simply begin by considering the equation $y + \Delta y = \sin(\theta + \Delta\theta)$. However, the actual derivation goes slightly beyond the scope of the trigonometry which we have covered. Suffice it to state the result, which is a particularly pretty one—namely, $d/d\theta(\sin\theta) = \cos\theta$. An equally delightful result obtains for $d/d\theta(\cos\theta)$, which is found to be $-\sin\theta$. The reader should note that both of these functions are differentiable to any order. For example:

$$y = \sin\theta,$$

$$\frac{dy}{d\theta} = \cos\theta,$$

$$\frac{d^2y}{d\theta^2} = -\sin\theta,$$

$$\frac{d^2y}{d\theta^3} = -\cos\theta, \text{ and so on.}$$

It is perhaps appropriate to remark at this point that not all functions are differentiable. That is to say, that certain curves are either discontinuous or so shaped that the tangent to them cannot be well defined. Similarly, a function may be such that although it is differentiable, its derivative is not. A function such that only its first derivative exists is said to be first-order differentiable or of class C^1. If the derivatives exist up to the nth, it is of class C^n. A function such as sine or cosine which is C^∞ is said to be an *analytic* function. Intuitively, the class of a function is an indication of its 'smoothness'.

The final topic which we shall need to understand in differential calculus is that of *partial differentiation*. So far, our discussion has been limited to the problem of finding the derivative of a function of a single independent variable. However, many physical laws are concerned with processes which involve two or more independent variables. The mathematical analysis of such situations involves the partial differentiation of a function.

Although this is sometimes regarded as a slightly advanced topic, which may possibly not even be covered in a first calculus course, the conceptual difficulty is only a step beyond that of ordinary differentiation. An example of a function of two independent variables is that of the area of a rectangle which depends on both its length and breadth, neither of which depends on the other. Another simple example is that of a container which is being filled with water from two independent sources. The rate at which the container is being filled will then be a function of two independent variables.

Unfortunately, it is not possible to proceed directly to the mechanics of this topic. It will first be necessary to introduce the notion of the *differential.* You will recall that it was pointed out that although dy/dx is not be construed as a fraction, it may be treated as one for many purposes. In consequence of this, dx and dy, called the differentials of x and y respectively, must have a significance as independent quantities. We shall now attempt to determine what it is.

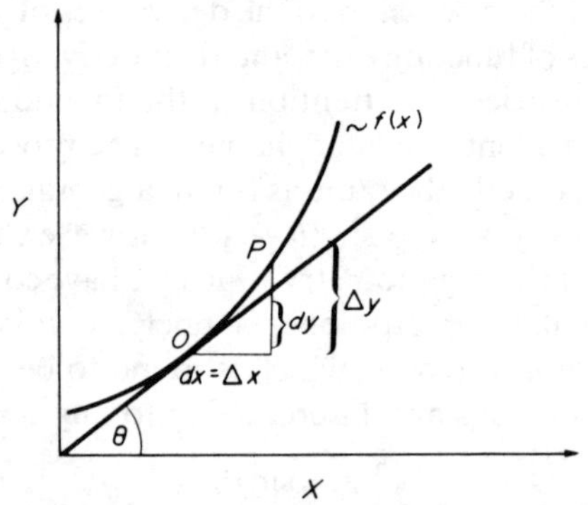

FIG. 1.11. Graphical representation of the differential

Figure 1.11 is a graph of an arbitrary function $f(x)$. We have already learned that the derivative of f at O is $dy/dx = \tan\theta$. We also know that $\tan\theta$ is the limiting value of the true fraction $\Delta y/\Delta x$ as $x \to O$. However, inspection of the diagram reveals that there is an asymmetry in the relationship between Δx and dx on the one hand and Δy and dy on the other. Specifically, no matter what the distance between O and P, dx is actually identical with Δx. Thus, we may define $dx = \Delta x$. On the other hand, when the derivative of f is treated as a ratio between two sides of an actual triangle dy is not the same as Δy, even though the difference between them will be negligible for very small triangles. In fact, whereas dx is an arbitrarily chosen finite increment of x, dy is a quantity which is determined by the choice of dx. Specifically, since $y = f(x)$, $dy/dx = df/dx$ and $dy = df \cdot dx$. For example, if $y = 3x^2$, then $dy/dx = 6x$, and $dy = 6xdx$. In this context, dx is a finite quantity so that $dy = 6xdx$ is also a finite quantity and dy/dx is the ratio between them. I realize that the reader may find all of this somewhat puzzling. On the one hand he has been told that the derivative is the instantaneous rate of change of a function which is the limiting value of $\Delta y/\Delta x$ as $x \to O$, while on the other hand he is now told that the derivative is a fraction or ratio involving two finite quantities. For obvious reasons, it is not possible to resolve the apparent contradiction within the scope of the present rapid survey. A good deal must simply be accepted on faith. However, we may at least dissolve any hint of paradox by reminding the reader that we have already seen that the derivative is itself a finite quantity, and that a finite quantity may be represented as a fraction with either a small numerator and denominator or a large numerator and denominator. Thus, $\frac{1}{2}$ = 1,000,000/2,000,000 = 0.000001/0.000002. From the standpoint of an instantaneous rate of change, the derivative involves *infinitesimal* quantities. From the standpoint of the tangent of an angle, the derivative involves the sides of a triangle which are *finite* quantities.

Let us now turn directly to the mechanics of partial differentiation. Consider the function $f(x, y) = z = x^3 + 2xy + 3y^2$. Our problem is to determine the rate of change of z with respect to simultaneous changes in both x and y. Since x and y are assumed to be independent, it is possible to conceptualize the rate of change of z with respect to x while y is constant. This quantity is called the *partial derivative* of z with respect to x. It is symbolized by $\partial z/\partial x$. In this case we have:

$$\frac{\partial z}{\partial x} = 3x^2 + 2y$$

Similarly,

$$\frac{\partial z}{\partial y} = 2x + 6y.$$

We are now led directly to the notion of the *partial differential*. In the case we have just considered, the two partial differentials of z would be:

$$dz_x = (3x^2 + 2y)\,dx$$

$$dz_y = (2x + 6y)\,dy$$

We may then form the total differential simply by adding the various partial differentials. In this case, it would be

$$dz = (3x^2 + 2y)dx + (2x + 6y)\,dy$$

In general, for any function $f(x, y, \ldots)$:

$$df = \frac{\partial f}{\partial x}dx + \frac{\partial f}{\partial y}dy + \ldots$$

However, we should add that the foregoing expression only makes sense because the various terms on the right-hand side have the same dimensionality. Clearly, one cannot add x miles/hour to y miles/gallon. Accordingly, the matter of the *total derivative* as distinct from the total differential presents a special problem. Thus, we are not entitled to write the total derivative as $\partial f/\partial x + \partial f/\partial y + \ldots$.

Nevertheless, there is a significant class of cases in which the total derivative may be expressed in terms of the partial derivatives. Although the independent variables are independent of each other, it may be the case that both are functions of a third variable. For example, the intensity of a certain field may depend on position, so that the coordinates will play the role of independent variables. However, it may be possible to express the coordinates x and y as functions of a single variable, as for example, time. In this context time is called a *parameter*. When this is the case, from the total differential,

$df = \partial f/\partial x dx + \partial f/\partial y dy$, we may derive the total derivative of f with respect to time in the form

$$\frac{df}{dt} = \frac{\partial f}{\partial x}\frac{dx}{dt} + \frac{\partial f}{\partial y}\frac{dy}{dt}$$

Let us work out an example to get the feel of things. Suppose we are once more presented with the function $f(x, y) = z = x^3 + 2xy + 3y^2$. But suppose that in addition we are given $x(t) = 3t + 6$ and $y(t) = 2t + 7$, which are the *parametric equations* for x and y. We have already computed $\partial f/\partial x = 3x^2 + 2y$ and $\partial f/\partial y = 2x + 6y$. We now compute $dx/dt = 3$ and $dy/dt = 2$. Whence:

$$df/dt = (3x^2 + 2y).3 + (2x + 6y).2 = 9x^2 + 4x + 18y.$$

In effect, we have replaced information concerning the partial rates of change of f in the directions of the X- and Y-axes with information concerning the total rate of change of the function in the direction of t. Hence, df/dt may be called the *directional derivative* of f in the direction of t.

We may now proceed to the other main branch of calculus, which is called *integral calculus*. Although the techniques associated with integral calculus tend to be considerably more difficult than those of the differential calculus, we shall devote rather less time to our exposition of this topic. A crude, partially true account of the relation between the two branches of calculus is that it is the business of the differential calculus to formulate physical laws in the form of differential equations and the business of the integral calculus to solve those equations. That is to say, that theories tend to be expressed in differential form and integral calculus provides the tools for deriving specific predictions from the given theory. Accordingly, since our concern is to attain conceptual understanding of physical theories we shall more frequently encounter applications of the differential calculus than of the integral calculus. However, that is not to say that it is possible to ignore the latter. To understand the full significance of a law, one must be aware of the nature of the information which it is able to provide. Put differently, to understand a problem one must know something about its solution and how such a solution may be obtained.

Mechanically speaking, the basic process of integration is simply the inverse of differentiation. Whereas the latter leads from a given function $f(x)$ to its derivative $f'(x)$, so the former leads from the derivative $f'(x)$ to the original function $f(x)$. However, before commenting on the mechanics of the operation, we shall attempt a geometrical interpretation of it. To anticipate the conclusion, we shall, in fact, find that just as the derivative could be geometrically interpreted as the tangent of a certain angle, so the integral may be interpreted as the area under a curve.

Figure 1.12 once more represents the curve $y = x^2$ from $x = 0$ to $x = 3$. Let us suppose that we wished to compute the area under this curve. You will note that this area is very crudely approximated in the diagram by three rectangular strips of width Δx and height $y + \Delta y$. The area of any one of these strips would be given by $\Delta A = \Delta x.(y + \Delta y)$. It would obviously be a very simple matter to compute the area of each strip and then add the three results. However, the overall result would obviously exceed the true area under the curve by an unacceptable amount. A more accurate approximation to A would be obtained by taking Δx much smaller. Thus, if the area were divided into one hundred strips instead of three, the resulting figure would be a fair approximation to A for many practical purposes. Specifically, Fig. 1.12 reveals that the area of each strip exceeds the desired amount by

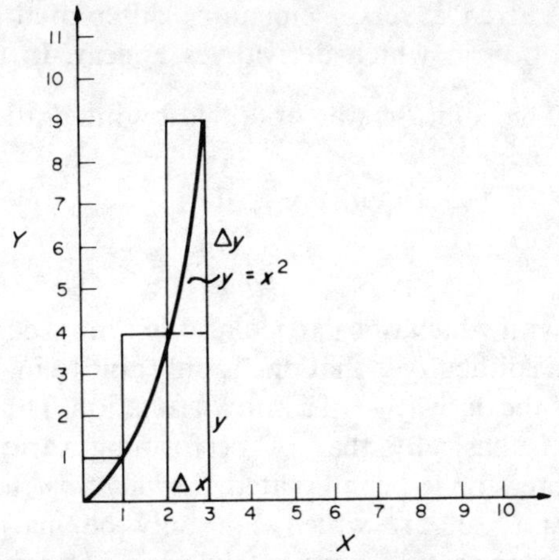

FIG. 1.12. Graphical representation of the integral of $y = x^2$

approximately $\frac{1}{2}\Delta x . \Delta y$. However, since Δy diminishes as Δx diminishes, by taking a sufficiently small Δx, the excess factor may be reduced to any desired extent. The perceptive reader will by now have surmised that a limiting process is implied by the foregoing discussion. Thus,

$$\Delta A = \Delta x(y + \Delta y).$$

Dividing by Δx, we have

$$\frac{\Delta A}{\Delta x} = y + \Delta y.$$

In the limit, as Δx approaches 0, Δy approaches 0, and this expression becomes,

$$\frac{dA}{dx} = y = x^2,$$

which is an *exact* result. It is intuitively obvious that A is a function of x. More specifically, we have discovered that the derivative of A with respect to x is equal to x^2. Thus, the problem of finding the area under the given curve has been translated into the problem of finding the function A of which x^2 is the derivative. Such a problem is characterized as finding the *integral* of x^2. This problem or the operation of solving it is symbolized by $\int x^2 dx$, which is read simply as "the integral of x^2". Thus $\int$——dx may be regarded as a single symbol like d/dx.

To integrate x^2 is to find the function of which x^2 is the derivative. Since we are familiar with the basic techniques of differentiation, the present problem should not prove to difficult. Considered as a derivative, x^2 was obtained from a function f by subtracting 1 from the exponent n and multiplying by that exponent. The result was x^2. Thus, we may determine f by adding one to the exponent of x^2 and dividing by the new exponent. Thus,

$$\int x^2\, dx = \tfrac{1}{3}x^3.$$

In effect, we have just solved a very elementary differential equation. A differential equation is just an equation in which derivatives appear. In this case the differential equation was $\frac{dA}{dx} = x^2$. The solution was obtained by integration as follows:

$$dA = x^2\,dx$$

$$A = \int x^2\,dx = \tfrac{1}{3}x^3.$$

Unfortunately, the ease with which this particular differential equation was solved is very far from typical. The basic difficulty is that one is only able to integrate a given expression when it is recognizable as the derivative of a known function. Thus, the problem of solving a differential equation is basically that of performing various algebraic and other manipulations on the expression to be integrated, the *integrand*, until it assumes some such recognizable form. This is an art which can only be mastered and sustained by considerable practice. However, this, fortunately, need not concern us.

Our characterization of integration as the inverse of differentiation such that the former operation undoes the effect of the latter is deficient in one very important respect. The derivative of a function is not affected by an additive constant of that function. Thus, for example, the derivative of x^2 is identical to the derivative of x^2+5. Accordingly, the integral of a function is only defined up to an additive constant but is not uniquely determined. Such integrals are called *indefinite integrals*. In fact, the indefinite integral of x^2 is not $\frac{1}{3}x^3$ but rather $\frac{1}{3}x^3 + C$, where C is an arbitrary constant called the constant of integration. In order to evaluate an indefinite integral, it is necessary to ascertain the value of this constant. In physical contexts, the values of the constants of integration are called initial conditions. The values of the initial conditions may be determined empirically. More generally, the value of C may be ascertained when one knows the value of y corresponding to just one value of x. Thus, in our illustrative case, we note the obvious fact that when $x = 0$, $A = 0$ so that C must be zero.

However, there is another type of integral for which this problem does not arise. It is called the *definite integral*. The indefinite integral is an expression for the general area under the curve for any value of x. However, there are many applications of the calculus in which one needs to determine the area under a discrete part of the curve, e.g. the part which extends from $x = 2$ to $x = 5$. It is this type of case which is associated with the definite integral. In such a case, in effect, one determines the area under the entire curve up to $x = 5$. Then one determines the area under the part up to $x = 2$ and subtracts the latter from the former. The result is clearly the value of that part of the total area which we wished to determine. The definite integral which we have just described is symbolized by $\int_2^5 x^2\,dx$. The result of integrating between the limits of 2 and 5 is represented by $\frac{1}{3}x^3\big|_2^5$. This expression is then evaluated by substituting the two limiting values of x, and subtracting the one from the other. That is: $\frac{1}{3}(5)^3 - \frac{1}{3}(2)^3 = 39$. It is important to note that the constant of integration may be disregarded when one evaluates a definite integral. Since it disappears in virtue of the subtraction which is performed.

It was explained earlier that it is possible to differentiate a derivative of a function to obtain a higher order derivative. By the same token, it is also possible to integrate the integral of a function. This will be required, for example, in order to solve a second order differential equation, i.e. an equation in which there occurs a second order derivative. The mechanics of successive integration presents no special problem. For example, suppose that we are given the equation: $d^2y/dx^2 = 2x$, and wish to obtain the general solution. We would proceed as follows. The double integral of $2x$ is written in the form: $\int\int 2x dx dx$. We integrate once to obtain

$$\frac{dy}{dx} = x^2 + C_1.$$

We then integrate the first derivative to obtain

$$y = \frac{x^3}{3} + C_1 x + C_2.$$

We note that there are now two constants of integration corresponding to two initial conditions both of which would have to be determined to obtain a particular solution.

A rather different type of successive integration is required in order to solve an equation which involves partial derivatives. In fact, this operation is called *partial integration.* Suppose that we are required to find the function z which satisfies

$$z = \int\int (x^2 + 2y)\, dy dx.$$

The reader will note that the occurrence of $dydx$ in the integrand indicates that a partial integration as distinct from an ordinary successive integration is to be carried out. One proceeds precisely as in the case of partial differentiation—namely by treating one of the independent variables as a constant. Thus, holding x constant one integrates with respect to y to obtain $x^2y + y^2 + \phi_1(x)$. It should be stressed that instead of having a constant of integration appear we have the more general expression $\phi_1(x)$, i.e. an arbitrary function in place of an arbitrary constant. The reason for this is that x has merely been treated as a constant when in fact it is an independent variable on which ϕ_1 may be dependent. We now perform the operation indicated by $\int (x^2y + y^2 + \phi_1(x))\, dx$, in which case y is now treated as a constant. This yields the final result:

$$z = \frac{x^3 y}{3} + xy^2 + \phi_1(x) + \phi_2(y).$$

The occurrence of ϕ_2 requires no further comment. However, it should be noted that ϕ_1 does not have the same significance as before since it is now a function of x which has just been integrated. The determination of the functions ϕ_1 and ϕ_2 is generally a difficult matter involving both physical insight as well as mathematical ability. Whereas the values of the constants of integration are referred to as initial conditions, the possible functions which are values of the ϕ_i are referred to as *boundary conditions.* Typically, they involve physical assumptions about the nature of a force field at infinity. However, as in the case of the C_i, the ϕ_i disappear when one integrates between definite limits.

We must now add a final remark concerning the nature of integration. When explaining the concept of the derivative, we gave it a geometrical interpretation as the slope or gradient of a curve, expressed as the tangent of an angle. However, the tangent of an angle is a dimensionless number, whereas the derivative of a given function may be a velocity, a momentum, a force or whatever. The only connection between the physical derivative and the tangent is mere numerical identity. Similarly, we interpreted basic integration geometrically in terms of the area under a curve which has the dimension of length squared. But physical integrals may be in terms of volume, length, work, electric charge, etc., depending on the nature of the problem. The geometrical interpretation as area gives the correct numerical result but is only one of many possible interpretations.

The reader will be immensely gratified to learn that our treatment of the mathematical groundwork has now come to an end. That is not to say that no further mathematics will be introduced. However, the reader who has devoted reasonable care and attention to the material which has so far been presented will have a solid enough foundation in the concepts of applied mathematics to understand the various extensions of these ideas which will be introduced as they are needed in the following chapters. You may have found the path up to this point somewhat arduous. However, you may be amply rewarded by at last having at your disposal the key to the most refined and precise area of all human knowledge—theoretical physics. The price is surely not too high for any philosopher who is genuinely concerned with both the nature of the world and our knowledge of it.

Bibliography for Chapter 1

ABBOT, P., *Teach Yourself Calculus*, English Universities Press, London, 1952. A conventional approach to elementary calculus which covers the essentials in a manner that may be appreciated by a beginner.

KAPLAN, W., *Advanced Calculus*, Addison-Wesley, Reading, Mass., 1952. Do not be deterred by the title. An excellent coverage of basic applied mathematics which is particularly suitable for the student of physics.

KLEPPNER, D., and N. RAMSEY, *Quick Calculus*, Wiley, New York, 1965. This is a programmed text which the reader should be able to cover in a few hours. It presents the absolute minimum that one needs to apply calculus to elementary physics and even includes some essentials of pre-calculus mathematical analysis.

LEECH, J. W., and D. J. NEWMAN, *How to Use Groups*, Methuen, London, 1969. A brief survey which pays attention to the notion of symmetry and indicates some of the physical applications of group theory.

MARION, J. B., *Principles of Vector Analysis*, Academic Press, New York, 1965. A rather sophisticated but quite readable treatment. It provides a good foundation for those who will proceed to study tensor analysis.

OWEN, G. E., *Fundamentals of Scientific Mathematics*, Harper & Row, New York, 1964. A readable survey designed for bright secondary level students. It is particularly good for its elementary introduction to vectors and matrices.

REICHMAN, W. J., *Calculus Explained*, Van Nostrand, Princeton, 1964. A very leisurely approach to elementary calculus, which stresses intuitive understanding rather than rigorous analysis.

SUPPES, P., *Introduction to Logic*, Van Nostrand, Princeton, 1957. A widely used introduction for university students. The second part contains a readable coverage of the rudiments of set theory, which is adequate to the needs of readers of this book.

2

Relativity and Newtonian Mechanics

The Concept of Relativity

THE reader will probably be aware of the fact that historically Einstein's theory of relativity superseded the classical mechanics of Newton. Why, then, do we choose to open our account of physical theory with an examination of a point of view which is now universally regarded as false? There are, in fact, several reasons for this approach. In the first place, although the name of Einstein is intimately associated with the idea of relativity, it happens to be the case that a principle of relativity is incorporated in Newtonian physics. Moreover, the Newtonian principle of relativity appears, prima facie, to be much more natural in the sense of being closer to our common-sense intuitions about the nature of space and time than the principle of special relativity of Einstein. Thus, in order to justify our preference for the latter, we are obliged to understand the defects in the former. Secondly, one of the livelier issues in recent philosophy of science is the problem of intertheory relations. In order to deal with this issue in terms of the relationship between Newtonian physics and relativity physics it is obviously necessary to be acquainted with both. Thirdly, the very status of the principle of special relativity is ambiguous. For example, there is some difference of opinion as to whether it is a statement about physical reality or about the nature of our knowledge of physical reality or about the nature of physical theories. Accordingly, it is most helpful to have two principles of relativity at our disposal for purposes of comparison. Specifically, this will put us in a better position to determine whether Einstein's principle is preferred to Newton's on the basis of purely factual considerations, or logical ones, or epistemological ones, or some combination of these. Finally, it is just a fact that Newtonian physics is intrinsically easier than relativity physics and so provides a useful introduction to it. In fact, despite the undeniable revolutionary aspects of Einstein's theories, one should be mindful that they are more a refinement than a repudiation of their Newtonian antecedent.

Unfortunately, it is possible to interpret the notion of relativity as it occurs in physical theories from two related but, nevertheless, different standpoints. One or the other of these tends to be emphasized according to one's philosophical predilections. While both are of significance, it often happens that philosophers stress the one while physicists stress the other. Let me attempt to explain this. Most of us are at least implicitly aware of the fact that some physical concepts have an absolute significance while others have a relative significance. Perhaps the most obvious example of a relativistic concept is that of position. Every time that I enter my office I find that my desk is in the same position as on previous occasions. Let us label the position of the desk at some arbitrary time t_1 by P_1. Let the position of the desk at a later time t_2 be P_2. Then we may express the claim that the desk is in the same position at t_2 as it was at t_1 by the formal statement: $P_1(t) = P_2(t)$. Although this equation seems both unambiguous and true, it is perfectly conceivable that someone might challenge it on the ground that during the interval of time $t_2 - t_1$ the desk has actually been moving in a circular trajectory in virtue of the earth's rotation. Moreover, the axis of rotation of the earth has itself been travelling at a high velocity in an elliptical orbit about the sun, and so forth. Hence, it is clearly not the case that $P_1(t) = P_2(t)$.

One might be inclined to dismiss this challenge as a mere quibble, on the ground that whenever one refers to the position of an object one has in mind the earth as a *frame of reference*, so that the concept of position invariably includes a tacit reference to the surface of the earth. However, not only would this retort seem excessively provincial to any hypothetical, extra-terrestrial bystander but moreover breaks down even in the human, terrestrial context. For example, suppose that a game of chess is in progress and that the pieces are in some arbitrary configuration at t_1. At this point in time the board is picked up and moved to another room and once more set down at t_2. We then consider the question whether the black queen is in the same position at t_2 as at t_1. Provided that the configuration has not been altered, one would probably be inclined to answer in the affirmative, even though the position of the queen is obviously different in terms of its relationship to the surface of the earth. In short, position now means with respect to the chess board as the frame of reference. Such examples could be multiplied indefinitely. Thus, consider what is meant when a mother on a moving train tells her child to stay in one place.

Such considerations oblige us to abandon the naive Aristotelian notion of absolute position and to replace it with that of relative position, i.e. the position of an object *relative to a specific frame of reference*. From the more sophisticated standpoint, the question as to whether an object at t_2 occupies the same position as it did at t_1 is seen to be indeterminate. Many empiricists, in fact, would make the stronger charge that it is meaningless. In any event, two points clearly emerge from our discussion. The first is that a statement about the position of an object is clearly elliptical in the absence of an unambiguous specification of a reference frame. The second is that two statements, in virtue of the aforesaid ellipsis, may appear to be logically incompatible when in fact they are not. Specifically, one may claim of a given object both $P_1 = P_2$ and $P_1 \neq P_2$. Both claims may be simultaneously true but with respect to different reference frames. The last logico-semantical point, albeit seemingly trivial, is essential to the conceptual grasp of Einstein's theory.

Once the door has been opened to relativistic or relational ideas, one quickly discovers that many other physical concepts stand in need of adjustment. Sometimes, the modification is an obvious one which the layman will accept the moment he is alerted to it, whereas at other times it will seem peculiar and counterintuitive. One of the simplest adjustments to be made is in the key concept of velocity. Everything that was said about the relativity of position applies, *mutatis mutandis*, to the concept of velocity. Suppose that one is running along the corridor of a moving train. Clearly, this is another case in which it would be elliptical to speak simply of one's velocity. The velocity with which one is moving relative to the train would be quite different from one's velocity with respect to the surface of the earth, and so on. In short, one cannot speak of the "true" or "real" velocity of an object in the sense of absolute velocity, but only of the relative velocity of an object with respect to this or that frame of reference.

Less evidently relativistic or relational is the concept of rectilinear or straight-line motion. Most people would think that the claim that the trajectory of an object is rectilinear must have an absolute significance. Suppose that one is in a room having normal, straight walls and floor. Suppose also that an object moves along the floor in such a way that its path is a line which is parallel to one side of the room. This surely constitutes a sufficient condition for rectilinear motion. However, if it should turn out that the room is actually rotating with respect to the containing building, it would follow that the path of the object would be a curvilinear one relative to the building. Thus, even the geometrical description of motion turns out to be relativistic. As a final example, consider the concept of weight. The ordinary man, as distinct from the physicist, would almost certainly suppose that the weight of an object is an intrinsic property that has no relational significance. In fact, most people suppose that weight is just the popular name for mass, which in turn tends to be identified with the quantity of matter in a given object. In fact, however, the weight of an object is a quantity that depends on both its mass and the gravitational force which it experiences. Now it happens that gravitational fields are not homogeneous but vary in strength from place to place. For example, the gravitational field of the earth is slightly stronger in the polar regions than in the equatorial zone. Consequently, an object of a given mass will weigh slightly more at the North Pole than it does at the equator. In the absence of an effective gravitational force an object will be weightless, although it continues to have the same mass. Accordingly, one cannot speak of the absolute weight of a body, but only of its weight in relation to its distance from a given gravitational source.

So far our discussion of the relativization of physical concepts has reflected only one of the two standpoints to which we earlier alluded. It is this standpoint on which philosophers have tended to focus their attention, when reflecting on the impact of relativistic physics. However, it is curious that when a physicist refers to a principle of relativity, he tends to emphasize the other side of the coin. Let us attempt to characterize the distinction. The former standpoint is to the effect that a great many of the concepts which we employ in our description of the physical world turn out to be "merely" relativistic. Consequently, a given description of the world, to the extent that it involves such concepts, is not to be regarded as the uniquely correct one. Different observers may be attached to different frames of reference. Accordingly, what is a rapidly-moving object to one observer is a motionless object relative to another. But since the one frame of reference is just as good as the other, the two descriptions are equally legitimate. Accordingly, one's description of the world is context-dependent and subjective. Physical description ceases to be regarded as an unambiguous source of objective knowledge.

The vast majority of physicists tend to interpret relativity from a very different perspective. You will recall the example of the configuration of chessmen in connection with the relativity of position. We saw that it is impossible to assert absolutely whether the black queen is in the same position at two different times. From the subjectivist standpoint, which we outlined in the preceding paragraph, it would seem that since no absolute answer can be given, one is bound to abandon any attempt at objective description of the facts and fall back on considerations of taste, standpoint or other personal predilections in one's preference for one description over another. The objectivist response to the problem is quite the opposite. The relativity of position, far from being an obstacle to objective knowledge, is a fact which may be exploited precisely for the purpose of framing a conceptual account of the situation of interest which is all the more objective. Specifically, since position is relativistic rather than absolute, it follows that for many important purposes it may be disregarded. If the position of an object in the universe had absolute significance, one would have to take that position into account in describing a particular state of affairs. However, if position has only relative significance one may quite profitably ignore it. Thus, it simply doesn't matter whether the chessboard is moved from the living-room to the bedroom. In either case, the fact that the black queen is attacking the white king remains constant. Thus, one may describe the configuration of the game in a manner which would have precisely the same significance were it taking place in London, Paris or Rome. Since position is relative, it follows that natural laws may be formulated in a manner which ignores position. Such formulations will thereby possess greater universality and corresponding objectivity.

The relativity of position enables one to formulate a limited principle of relativity for chess: The state of any game of chess is independent of its location. An alternative way of expressing this principle of relativity would be: From the state of a game of chess, it is impossible to deduce the location of that game. The operative factor is that when the queen is moved all of the other pieces are moved in a manner which precisely compensates the motion or change of position of the queen. Thus, yet another formulation of our relativity principle for chess might be: The configuration of a chess game is *invariant* with respect to arbitrary changes of position of the board. If position were of absolute significance, the well-known chess notation for describing the course of a game might have to be abandoned, since it would then be conceivable that a move made in London would have a different significance from a similar move made in Paris.

Both of the foregoing interpretations seem plausible. But surely they cannot both be correct. In fact, they are not as incompatible as they may first appear. I have suggested that they are two sides of one and the same coin. They correspond in a crude and metaphorical way to the venerable philosophical distinction between appearance and reality. Our immediate apprehension of the world is in terms of appearances. But appearances vary from one individual to another. However, it is fallacious to infer that a knowledge which transcends appearances is therefore impossible to attain. On the contrary, the high degree of variation in appearances should be taken to indicate that appearances, *per se*, are of no great importance but stand in need of intellectual penetration, so that one may come to grips with the unchanging, invariant properties of the real world. Contrary to some popular misconceptions, the province of this activity belongs as much to science as to metaphysics. In conclusion, each standpoint is correct but limited. To stress the one without giving due consideration to the other is philosophically defective. The scientist is concerned with invariant, objective patterns. However, all of the experiments which he performs to test his hypotheses concerning the structure of such patterns are tied to laboratories, which are just frames of reference. Thus, the scientific method involves an interplay between appearance and reality. That which appears is not necessarily devoid of objectivity.

Elementary Foundations of Newtonian Mechanics

In order to present Newtonian mechanics in its most perspicuous form, we need an additional mathematical notion—namely the derivative of a vector. This presents no particular difficulty. Suppose that **R** is an arbitrary vector with components R^i in some Cartesian coordinate system S. Then the derivative of **R** with respect to time is given in S by the vector representative dR^i/dt, i.e. $(dR^1/dt, dR^2/dt, dR^3/dt)$. In short, the derivative of a vector is simply the vector formed from the derivatives of its individual components. For example, let x^i be the spatial coordinates of a moving particle. Then dx^i/dt is the velocity vector (arrow) of the moving particle.

The basis of Newtonian mechanics is contained in three fundamental postulates. They may be expressed as follows:

1. The Law of Inertia

> "A body on which no external force is acting will persevere in a state of rest or uniform rectilinear motion."

This is simply to say that in the absence of external forces, a body will have zero acceleration. Accordingly, the equation of motion of such a body will be: $m(d^2x^i/dt^2) = 0$, where m is the mass of the body. The concept of mass or inertia is simply the capacity which is inherent in every material object to resist a change of motion. That is to say, that the greater the mass of the object, the greater will be the force required to produce a given change of motion or acceleration. The quantity of motion or ***momentum*** is defined by $m(dx^i/dt) = m\mathrm{V} = \mathbf{P}$. Thus, an alternative formulation of Newton's first law is that in the absence of external forces, the momentum is constant or conserved, i.e. $d\mathbf{P}/dt = 0$. However, on the assumption that the mass of the particle of interest is constant, the two formulations are mathematically equivalent.

2. The Law of Motion

> "The change of motion of a body is equal to the force which acts on it."

It is this law which is regarded as the central hypothesis of Newtonian mechanics. Mathematically, it may be expressed as

$$\mathbf{F} = m\frac{d^2x^i}{dt^2} = \frac{d\mathbf{P}}{dt}$$

The substance of this law is that *force* is that which is required to produce a change of motion. More precisely, the change of motion of a body is proportional to the force which acts on it. The status of Newton's second law has been the subject of considerable controversy. On the one hand, it has sometimes been declared to be a mere definition; it is charged that force is merely defined as the product of mass and acceleration. Without undertaking a lengthy analysis of this claim, we may declare it to be clearly unfounded on the ground that it is possible to deduce from the second law matters of fact which could not be predicted without it. On the other hand, a definition has no factual content. It is a convention to the effect that a given expression, the *definiendum*, may be introduced by another semantically equivalent expression, the *definiens*. Since definitions are merely concerned with the relationship between words they are non-creative, in the sense that they add nothing factual to the theories to which they are adjoined. In short, a definition is theoretically dispensable. But without the second law, Newtonian mechanics would be nothing. It is possible that the plausibility of the claim that $\mathbf{F} = m\mathbf{a}$ is a definition derives from the tendency to neglect the fact that the second law must be supplemented with specific information concerning the nature of the force in question, i.e. the so-called force-function. For example, we shall learn in a later chapter that the gravitational force-function is given by the expression $\mathbf{F} = Gm_1m_2\hat{r}/r^2$. When this explicit expression for the force is inserted in the equation of motion, the claim that the latter is simply a definition seems far less plausible.

On the other hand, some philosophers once argued, perhaps less reprehensibly, that the second law is a synthetic *a priori* principle; that it does have factual content but is universally necessary or incorrigible. The general line of thought on which this claim is based is that the law of motion follows from the principle of sufficient reason, which states that nothing happens without an adequate cause. On this view, a body which is moving in a certain direction at a given speed is bound to persist in that state of motion unless something causes it to undergo a change. Hence, a force is required to explain the changes of motion or accelerations of bodies. This argument is easily disposed of. The fallacy of the reasoning is revealed by a consideration of Aristotelian physics. According to Aristotle, the natural state of a body is one of rest. A body which moves with a uniform speed requires a constant force to sustain that motion. Thus, an Aristotelian might just as well argue, on the basis of sufficient reason, that a body will remain in the same position in the absence of anything which causes it to undergo a change of position. Thus, one must have recourse to force to explain change of position or velocity. For the Newtonian, the law of motion is bound to be a second order differential equation, whereas for the Aristotelian it is just as clearly bound to be of the first order. The arguments on both sides are equally compelling. The correct account is that both are factual claims about the physical world, which have the status of physical hypotheses. It is quite conceivable that the Aristotelian claim might have triumphed. It is a fact, however, that refined methods of measurement and ingeniously devised experiments supported the Newtonian position. For example, if Aristotelian dynamics were correct, one would expect a falling stone to have a uniform velocity. In fact, it was found to undergo a uniform acceleration, within the limits of the existing means for measuring that quantity.

3. *The Law of the Equality of Action and Reaction*

"When two bodies interact, for the force exerted by the first body on the second there will be an equal and opposite force exerted by the second body on the first."

Letting $\mathbf{F}_{ij}$ symbolize the force exerted by the ith particle on the jth particle the third law may be given mathematical expression in the form: $\mathbf{F}_{12} = -\mathbf{F}_{21}$. For a system of many interacting particles, this may be generalized to $\mathbf{F}_{ij} = -\mathbf{F}_{ji}$.

The foregoing laws together with the various theorems which are deducible from them constitute an instance of what philosophers of science call a physical model. This particular one is a very simple or idealized model, in so far as it is restricted to the behaviour of bodies which are treated as idealized points, having mass but no spatial extension. However, it is our intention to keep the physics in our discussion as simple as possible in so far as this is consistent with the conceptual understanding of the theory of relativity.

The most striking feature of this model is that its three basic hypotheses are in the form of *vector equations*. The reader will recall that the principal implication of this is that such laws are coordinate free. It needs to be stressed that the choice of vectors for the representation of the theory is not a freely chosen device but an assumption about the world which could be falsified. For example, since force is represented as a vector, it follows that the resultant of two forces $\mathbf{F}_1$ and $\mathbf{F}_2$ acting on a particle may be determined simply by computing the vector sum $\mathbf{F}_1 + \mathbf{F}_2 = \mathbf{F}_3$. The fact that a particle under the influence of two independent forces behaves as though it were subject to a single force which is the vector sum of the independent forces is by no means a mathematical necessity. This simple fact that is presented in secondary schools as the parallelogram law for the addition of forces is not mathematical but physical. It is always essential when performing mathematical operations in the context of scientific theories to be aware of the underlying physical assumptions.

From the vectorial character of Newton's laws one may infer certain features of the nature of the space in which they are maintained to hold. In the first place, a vector is independent of position. Accordingly, whether a mechanical experiment is performed at $P(X_1^i)$ or at $P(X_2^i)$, the outcome should be unaffected. The implication of this is that from a mechanical standpoint space is *homogeneous*. All points in space are equivalent, i.e. mechanically indistinguishable. In the second place, a vector is independent of direction. Mathematically, as you already know, it is covariant with respect to a rotation of the coordinate system. The implication of this is that the action of a force is independent of the direction in which it is acting. Hence, at least from the standpoint of mechanics, space is *isotropic*, the same in all directions. We may summarize these two features of Newtonian space by saying that the laws of Newtonian mechanics are covariant with respect to the inhomogeneous rotation group, which may be labelled O_{i3}, where the numerical subscript refers to the dimensionality of the space.

The reader will recall that a group of transformations has associated with it certain invariants. It is easily shown that the principal invariant of O_{i3} is length or distance. That is to say, that Newtonian space contains a length concept or implies a definition of length which is invariant with respect to the transformations O_{i3}, which reflect the symmetry of Newtonian space. Such an invariant definition of distance is called a *metric*. The metric of Newtonian space, which we now offer in differential form is given by

$$(ds)^2 = (dX^1)^2 + (dX^2)^2 + (dX^3)^2 = \delta_{ij} dX^i dX^j (i, j = 1, 2, 3)$$

The δ_{ij}, the familiar Kronecker symbol, are called the *metric coefficients*.

To summarize, Newton postulated a set of basic laws which presupposed a physical space which is homogeneous and isotropic and which has a definite metric or distance function defined on it. It should be noted that in the absence of such a spatial background, the laws of mechanics would appear to be devoid of significance. The law of inertia refers to uniform, rectilinear motion. But, as we have seen, such a trajectory may only be defined when we have a frame of reference which, in turn, is associated with a coordinate system. For Newton, such a frame of reference is *absolute space*. A suitable coordinate system would be any system of Cartesian coordinates which is defined up to an inhomogeneous rotation. This means that while there is a unique frame of reference, there is no unique coordinatization of that frame. Given one adequate coordinate system, any other which may be obtained from the first by an element of the inhomogeneous rotation group will be equally adequate.

It is instructive to consider Newton's own account of the basic frame of reference which served as the backdrop to the totality of mechanical phenomena. "Absolute space, in its own nature, without regard to any thing external, remains always similar and immoveable." It is clear that Newton regarded space as an absolute substance or object. It could have, as we shall see, an effect on certain physical phenomena but was not subject to any modification by something else.

In addition to the doctrine of space, Newton also required a definite doctrine of time. Just as the basic laws of physics reveal their independence of location and direction, so they also reveal their independence of temporal location. Time never appears *explicitly* in the laws of mechanics. When a force acts on a particle, the actual time of the occurrence makes no difference to the outcome. All instants of time are mechanically equivalent. Time, like space, is homogeneous. Although moments of time are mechanically indistinguishable, that is not to say that time is irrelevant to mechanical processes. On the contrary, the concept of time is manifest in the first and second laws in which it occurs as the independent variable. These laws refer to rates of change with respect to time. Such a rate of change would be but vaguely defined in the absence of a strict concept of time. In particular, we require that the differential operator d/dt have a fixed significance. Moreover, when we come to integrate a differential equation to determine, for example, the total distance or length of trajectory of a particle which has been constantly accelerated by a given force, we perform the integration between definite temporal limits, i.e. for a fixed period of time. Thus, we require that just as there is a definite concept of spatial distance, so there will be a definite concept of temporal distance or duration. In other words, we require that the integral operator $\int_{t_1}^{t_2} \text{——} \, dt$ have a fixed significance.

Specifically, we require, or rather the laws of mechanics require, that a time interval be independent of the choice of coordinate system. In short, time is supposed to be a *scalar* quantity. In virtue of the homogeneity of time, we may for computational purposes select any instant of time as the arbitrary zero point, just as we may select any point in space as the origin of a spatial coordinate system. If time is truly homogeneous, then this fact should be reflected in our physical laws. Specifically, it should be possible to select any other instant of time as the zero point without influencing the essential form of those laws. That is to say, that in addition to being covariant with respect to O_{i3}, the laws of mechanics should be covariant with respect to the group of time translations T. In sum, the covariance group of Newtonian mechanics is the direct product of $O_{i3} \times T$. By the *direct product* of two groups is meant simply the group whose elements are composed of an element of each group, where it is assumed that any two such elements commute.

Although time does not occur manifestly in the third law, it is surely one of its conceptual presuppositions. In particular, the third law asserts that if a particle at P_1 has an influence on a particle in some distant region P_2, then *simultaneously* the particle at P_2 will have an equal but opposite influence on the first particle. Thus, simultaneity or the notion of two events having different spatial locations but the very same temporal coordinate is presupposed by the third law. Once again, there is no better way of capturing the flavour of the Newtonian picture of reality than by considering his own words: "Absolute, True, and Mathematical Time, of itself, and from its own nature flows equably without regard to any thing external, and by another name is called Duration. . . ." It is obvious that time, like space, has the status of an absolute object or substance.

The Principle of Classical Relativity

Although Newton was satisfied that his doctrine of absolute space and time was conceptually adequate, he was, nonetheless, aware of a certain practical problem which it posed. The fundamental goal of science is to provide conceptual representations of objective physical laws. The scientist assumes that nature has an objective, nomic structure that is constant throughout space and time. Consequently, the theoretical laws which depict this structure should have global rather than mere local significance. One implication of this requirement is that the formulation of a law should not be dependent on any local factors; ideally it should be free of contextual reference.

We have noted that this ideal of objectivity is partially realized by writing laws in vectorial form. A vectorial law has precisely the same significance with respect to infinitely many coordinate systems and is objective at least to that extent. However, while there are inifinitely many suitable systems of coordinates, there are infinitely many more which are unsuitable. From what was said in the previous section, it follows that if we have one suitable system at our disposal, then we may obtain the rest by a transformation which is an element of $O_{i3} \times T$. The assumption here is that any such element leads to an admissible coordinatization of absolute space and time, the receptacle of physical processes.

Clearly, then, one task of theoretical mechanics is to specify a suitable coordinate system. The reason for this is the following. Let us suppose that a scientist wishes to carry out an experimental test of Newton's second law. Suppose, also, that he sets up the experiment in some moving vehicle, say a train. This is not too far-fetched, since the earth is a type of moving vehicle. The scientist, suitably equipped with instruments for measuring off equal distances and durations, proceeds to associate a coordinate system with the train. For example, he might take his own position as the origin, and the longitudinal axis passing through this point as X^1. Then it is a simple matter to choose X^2 and X^3, so that all X^i will be mutually perpendicular or orthogonal. Now he is ready to perform his mechanical experiment and to record the results in terms of the coordinates X^i. Now, as you know, if the velocity of the train is not constant, if the trajectory of the train is curved or jerky, the results of the experiment will be affected. The processes which take place in the train are influenced by forces of a special kind which we call *inertial forces*. To take a simple case, let us suppose that the experiment is performed while the train is undergoing a constant acceleration $\mathbf{a}_t$. It will then be found that the experiment is not in conformity with the basic law $\mathbf{F} = m\mathbf{a}$ but rather with $F = m\mathbf{a} - m\mathbf{a}_t$. This result is clearly unsatisfactory, not because it is inconsistent with Newton's law of motion, which is, after all, a falsifiable physical hypothesis, but because the second term on the right-hand side refers specifically to the coordinate system or the frame of reference, the moving train, to which it is attached. Thus, the result of the experiment has local rather than global significance. It is dependent on a particular vehicle. Philosophers of science have frequently made a similar point in a more abstract way by formulating the requirement that a law of nature (or rather its conceptual representation) should contain no essential occurrence of a proper name. In this case, the defective formulation contains, as it were, the name of the frame of reference.

The situation is not vastly improved by selecting a coordinate system which is attached to the earth. The earth's rotation gives rise to inertial forces which influence the outcome of mechanical experiments. Our task is to find a point which is literally fixed or at rest in space, to which all mechanical phenomena may be referred. But how may one find such a point or, indeed, recognize it for what it is? Our difficulty stems from the fact that apart from mechanical considerations, there always seems to be an inescapable *kinematic* relativity. By *kinematics* is meant the geometric description of motion apart from all consideration of forces. The principle of kinematic relativity is to the effect that any description of moving objects is dependent on a completely *arbitrary* choice of one object which is defined to be at rest. For example, it is possible in principle that the moon is a truly fixed body in space. But even if it were, one could not recognize it as such since the moon appears to be in motion with respect to the earth. It may well be that the relative motion between the earth and the moon is wholly due to the 'real' motion of the earth but it is a matter which is impossible to decide.

Newton was able to overcome these difficulties in virtue of a particular feature of the fundamental laws of his theory. In the first place, the problem posed by kinematic relativity may be partially overcome on the basis of dynamic considerations. Visually, one cannot distinguish between the motion of the moon with respect to the earth and that of the earth with respect to the moon. However, mechanical experiments performed on the earth and the moon will give some indication of the actual state of affairs in virtue of the extent to which the results of such experiments deviate from the Newtonian laws, thus indicating the presence of inertial forces. In the second place, astronomical observations give a sufficiently clear indication that the laws of mechanics, especially in the case of gravitational forces, hold rigorously when referred to the system of apparently fixed stars. That is not to say that the distant stars are "truly" at rest with respect to absolute space, but that, at least, a coordinate system which is attached to them will be adequate in the sense that no frame-dependent, inertial-force terms will occur in a description of mechanical phenomena which is referred to such a system. Moreover, and this is the *crucial* point, any coordinate system which is attached to a frame of reference which is in a state of uniform rectilinear, i.e. unaccelerated, motion with respect to the fixed stars will be equally adequate. That is to say, that the basic laws of mechanics will hold rigorously with respect to any such system.

It emerges, therefore, that in addition to the basic uniformity or symmetry of space and time which is associated with $O_{i3} \times T$, there is another uniformity. Not only are mechanical laws indifferent with respect to position, direction and time but also with respect to velocities or unaccelerated motions. Physically, this means that it would be impossible to devise any mechanical experiment to ascertain the absolute velocity of the vehicle or frame of reference in which the experiment is performed. For example, while one may ascertain the *relative* velocity of the earth with respect to the sun, one cannot determine its *absolute* velocity with respect to absolute space. Thus, to a limited extent, the foregoing considerations give support to the principle of kinematic relativity.

At first sight, it might appear that the relativity of velocity deprives mechanics of some of its objectivity. However, in line with certain of our earlier remarks, the very opposite claim may be made. It has been emphasized that the fact that the laws of mechanics are covariant under $O_{i3} \times T$ adds to their global significance. It now turns out that there are vastly many more adequate coordinate systems at our disposal than previously appeared. It is intuitively obvious that a single, physical frame of reference may be associated with infinitely many distinct coordinate systems. But we now know that there are in principle infinitely many potential frames of reference which are adequate to Newtonian mechanics. Any such frame is called an *inertial* reference frame. In virtue of the potential infinitude of inertial frames of reference, it follows that the claims of Newtonian mechanics to global significance are given much added support. The laws of mechanics continue to be valid, regardless of the velocity of the vehicle—that is, regardless of an accidental, local circumstance.

This additional uniformity is given mathematical expression in the form of a most important set of transformation laws, which collectively are called the *Galilean transformation.* Without loss of generality we assume that two inertial frames of reference have been so coordinatized that at an arbitrary zero point of time the origins of S and $\overline{S}$ coincide and that $\overline{S}$ is moving with velocity $\mathbf{v}$ with respect to S in the X^1-direction only and that the corresponding coordinate axes are all parallel. This gives rise to what we may call the special, homogeneous Galilean transformation. This is now expressed as follows:

$$\begin{aligned} X^1 &\rightarrow \overline{X}^1 = X^1 - \mathbf{v}t \\ X^2 &\rightarrow \overline{X}^2 = X^2 \\ X^3 &\rightarrow \overline{X}^3 = X^3 \\ t &\rightarrow t \quad\;\, = t. \end{aligned}$$

The reader should assure himself, at least intuitively, that the totality of such transformations are the realization of a group, which we label G_s. Thus, G_s is a symmetry of Newtonian mechanics over and above $O_{i3} \times T$. The principle of classical relativity may be expressed in the statement: The laws of mechanics are covariant under G_s. A somewhat more elementary version of this is: The laws of mechanics are the same for all frames of reference which are in a state of mutual uniform motion.

The reader will note that we have once again spoken of the *covariance* of physical laws. However, it is absolutely fundamental to the point of view to be developed in this book that a distinction be drawn between two senses of covariance. Covariance is primarily a geometrical concept. That a vectorial equation is covariant under O_{i3} is true by definition of the geometric concept of a 3-vector. Thus, we may write any arbitrary 3-vector equation in complete assurance that it is covariant under O_{i3}. This is not the case for G_s. The Galilean group has nothing directly to do with the vector concept. Specifically, as you may recall, the definition of a vector contains a reference to the covariance group whose elements relate the various representatives of a given vector. But a vector is not such with respect to the Galilean group.

This is a point which is so frequently blurred in the literature of physics that it is well to dwell on it. There is a very widespread, although not universal, interpretation of relativity principles to the effect that since their mathematical expression is in the form of a principle of covariance, it follows that such principles are metatheoretical statements or, more precisely, metatheorems of mathematics, hence devoid of factual significance. This is a rather deep issue, which we shall discuss in a later chapter. At the moment our intention is simply to clarify certain of the facts. In particular, we wish to point out that the Galilei covariance of Newtonian mechanics does not follow from the fact that its basic laws may be written in vectorial form. Rather, the Galilei covariance of these laws is a consequence of their physical content.

That Galilei covariance is not purely geometrical may be easily deomonstrated by the obvious fact that a velocity vector may be "transformed away" by a Galilean transformation. Suppose that a particle has the velocity $\mathbf{v}$ with respect to S. Then one may perform a Galilean transformation to $\overline{S}$ which coordinatizes the rest-frame of the particle. That is to say, that the velocity has been made to vanish with respect to the new frame. A transformation of this kind destroys the objective character of the vector. In particular, the reader will recall that every vector has a definite magnitude which is independent of the coordinate system. A transformation which may alter the magnitude of a vector, which may even make it vanish, is not allowable. (Alternatively, it may be said that velocity is not a geometric object with respect to the Galilean transformation.)

The implication of this would appear curious or even paradoxical. We have said that the Galilean transformation is not "allowable". Yet physicists do employ it as a matter of course. The point is that "allowable" is here being employed in a particular mathematical sense. It has been said that rules are made to be broken. The obvious proviso is that one should be aware of the rule one is prepared to break and only break it for some useful purpose. Every student of physics will know that the Galilean transformation is a very useful device, particularly for simplifying the physics of otherwise complex problems. How is it, then, that one is able to get away with it? The answer, to repeat, lies in the physical content of Newton's laws.

In the first place, it happens to be the case that the basic law of motion is a second order equation. It is a matter of fact that forces are proportional to accelerations rather than to velocities. When one takes the second derivative with respect to time, velocity vanishes. Thus, when one is analyzing the motion of a particle, one will be concerned, in particular with the law of change of its velocity, not with the velocity itself. It requires the very same force to accelerate a particle from 5 miles/sec. to 6 miles/sec. in a given time interval, as to accelerate it from 1000 miles/sec. to 1001 miles/sec. in the same interval. The velocity doesn't make any difference. This is obviously just another way of expressing the principle of classical relativity. Accordingly, the laws of motion are insensitive to Galilean transformations.

In the second place, notwithstanding what has just been said, while velocities do not occur in the basic axioms of mechanics, they do occur in certain high-level theorems. In spite of this, the Galilean transformation is still applicable to those situations in which such theorems are of significance. How is that possible? Let us consider the case of the fundamental theorem of conservation of linear momentum. Take a system of n particles, where i is the ith particle of the system. Assuming a fixed system of coordinates, the position of the ith particle may be symbolized by $\mathbf{r}_i = X^j_i$. Then the momentum of the ith particle is $m_i dr_i/dt$. (It should be noted that the subscripts in this expression do not refer to components of vectors, but to individual particles. Accordingly, no summation is to be performed over a repeated index. In such contexts, we employ the standard summation symbol, Σ, when a summation is intended.) Then the theorem asserts that in the absence of external forces, i.e. provided that the system is isolated from the exterior, the total momentum of the system will be conserved. In mathematical form:

$$\sum_{i}{}_1^n m_i \frac{dr^i}{dt} = \text{constant}.$$

Regardless of how the particles collide, assuming they are elastic, the overall momentum will be neither increased nor diminished.

Consider the very simple example of two elastic spheres, m_1 and m_2 which are approaching each other with velocities $\mathbf{v}_1$ and $\mathbf{v}_2$ from the $+X$ and $-X$ directions, respectively. After they collide, they recede from each other with velocities $\mathbf{v}'_1$ and $\mathbf{v}'_2$.

According to the law of conservation of momentum, we may deduce the following result:

$$m_1 v_1 + m_2 v_2 = m_2 v'_2 + m_1 v'_1$$
$$m_1(v_1 - v'_1) = m_2(v'_2 - v_2).$$

It should be obvious that if we referred this result to a second coordinate system associated with a frame of reference having an arbitrary velocity **u** with respect to S, we should find that although the transformed velocities, $\bar{v}_1, \bar{v}'_1$, etc., would be different, the law would be preserved. The reason for this is simply that although the magnitude of a velocity will be affected by a Galilean transformation, the difference of two velocities will not be.

We have frequently alluded to the distinction between frames of reference and coordinate systems. This is a distinction which is unfortunately too often neglected by many physicists. To do so is to sow the seed of misunderstanding of the various principles of relativity. Accordingly, we shall briefly consider the matter to ensure that the reader has it clear. A coordinate system is a conceptual construct. It has no existence other than a mathematical one. One cannot ascribe physical properties to it. Not only are there infinitely many such systems, as we have seen, but also infinitely many types, albeit only in principle. So far we have restricted ourselves to the most common kind, the orthogonal Cartesian system. In addition there are oblique Cartesian systems, in which the axes are rectilinear but not mutually perpendicular. Also, there are curvilinear systems, orthogonal or otherwise, in which the axes, and coordinate lines generally, are curves of various kinds.

By contrast, a frame of reference is a physical object of some kind, such as the earth, the sun, the fixed stars, a laboratory, a radar station, and so forth. We have seen the necessity for specifying reference systems in order to speak about positions, velocities, trajectories, and so on. However, to be of use in its intended role, the reference frame must be associated with a coordinate system which provides a numerical label for every ideal physical point of the former, as well as an identification of the points in space as they are situated in relation to that frame.

Having distinguished frames of reference from coordinate systems, we may now proceed to distinguish between frame transformations and coordinate transformations. A coordinatization of space is a one-one correspondence between triples of R^3, where R^3 is the third Cartesian product of the real numbers, and points of space. A coordinate transformation is simply a continuous one-one transformation from R^3 to R^3. That is to say, that every triple of R^3 is mapped onto another not necessarily distinct triple of R^3. More succinctly, it is a mapping of R^3 onto itself. The most common motivation for this mathematical operation is that of enhancing the description of a physical system from the standpoint of perspicuity and mathematical tractability. For example, most of us are aware of the desirability of using spherical coordinates, lines of latitude and longitude, for specifying locations on the surface of the earth. They are more convenient than Cartesian coordinates simply because the earth exhibits a high degree of spherical symmetry. A similar comment applies to the earth's gravitational field. The differential equation which describes that field is more readily solved when it is expressed in terms of spherical coordinates.

A frame transformation is rather more than this. Take two frames of reference K and $\bar{K}$ such that the latter is in some arbitrary state of motion with respect to the former. Coordinatize K by S and $\bar{K}$ by $\bar{S}$. Then the transformations relating S and $\bar{S}$ are the laws which model the kinematical relations between the two frames. Thus, we see that every frame transformation is a coordinate transformation but not conversely. In general, the mark of a physically significant frame transformation is that the coefficients of the transformation equations involve the variable of time, e.g. as in the Galilean transformation.

It is most important to note that such time-dependent frame transformations do have physical content. They are statements about the world. This may be illustrated in a direct and simple way. Let us suppose that a frame $\bar{K}$ has a velocity **v** with respect to a frame K. Furthermore, we suppose that a particle has a trajectory $x^i(t)$ with respect to $\bar{K}$. We now wish to determine the velocity of the particle with respect to K. The answer to this question is directly deducible from the Galilean transformation in the following manner. We first write down the inverse Galilean transformation relating the coordinates of the particle in the K-frame to those in the $\bar{K}$-frame:

$$X^i = \bar{X}^i + v\bar{t}.$$

Differentiating, we derive:

$$\frac{dX^i}{dt} = \frac{d\bar{X}^i}{dt} + \frac{dv}{dt}\bar{t} + v\frac{d\bar{t}}{dt} = \frac{d\bar{X}^i}{dt} + v.$$

We may rewrite this result in vectorial form as

$$\mathbf{w} = \bar{\mathbf{u}} + \mathbf{v}$$

This is the Newtonian law for the composition of velocities. Assuming that all the vectors are parallel, we find that the composition of velocities is arithmetic. Hence, we have derived a definite physical law from the Galilean transformation, proving that the latter is factual or, more technically, that it is a statement in the *object language* of Newtonian particle mechanics.

All of this may strike the reader as rather dubious. In effect, I have been arguing that if a car is heading due north with a velocity of 50 m.p.h. and a rifle with a muzzle velocity of 150 m.p.h. is fired from the car in the direction of its motion, then Newtonian mechanics predicts that the bullet will travel with an initial velocity of 150 + 50 m.p.h. = 200 m.p.h. with respect to the earth. Moreover, it has been argued that this is not a law of arithmetic, given *a priori*, but a putative fact of nature. But surely, the reader may say, any reasonably intelligent person would be able to compute this result solely on the basis of his knowledge of arithmetic, and in complete ignorance of Newtonian mechanics.

To this we would reply that one has learned the law of velocity composition through experience. That what one has learned happens to be in agreement with a certain physical theory is not altogether surprising when it is recalled that experience is also a criterion for the veracity of physical theories. Few people would suppose that the result of mixing two liquids, each having a temperature of 50 °F., will be a mixture with a temperature of 100 °F. Yet arithmetic cannot tell us when the physical composition of two magnitudes is arithmetically additive and when it is not. Every philosopher should know how frequent is the error of mistaking the commonplace for the *a priori*.

We have already discussed the fact that Newton presupposed the concepts of absolute space and absolute time in formulating his theory of mechanics. Both of these presuppositions are reflected in part in the Galilean transformation: absolute space in so far as distance $\delta_{ij}dX^idX^j$ is invariant under a Galilean transformation, absolute time in so far as t is invariant. However, there is an obvious difference. Even though distance is invariant, it is not the case that the coordinates themselves are invariant. Thus it was not necessary from a mathematical standpoint to retain so strong a presupposition. Methodologically, the Newtonian scheme would appear to involve the relativity of position. However, we shall see that despite the implication of the Galilean transformation, Newton had excellent reasons for clinging to his position about space. The case of time is somewhat different. Methodologically, time is through and through absolute. The temporal coordinate of an event is the same for every frame of reference. This is equivalent to saying that when we speak of the time of an event there is no need to specify a frame of reference. Thus, time would appear to be totally devoid of any relativistic characteristics.

In order that we may hope eventually to understand the Einsteinian theory of time, it is necessary to be clear concerning the significance of the Newtonian theory. This is a matter which has been unfortunately obscured by empiricism, especially in its operationalist form. Too many physicists and philosophers of physics have made the claim that one's theory of time is an hypothesis about the behaviour of clocks. This, I regret to say, is misleading, if not sheer nonsense. It is not the case that clocks determine the properties of time but, on the contrary, that one selects or constructs acceptable clocks on the basis of the properties of time. Clocks measure time but do not define it. The absolute character of Newtonian time has the status of a physical postulate which is testable with the aid of clocks but is not an inductive generalization about clocks. By the same token, the properties of Newtonian space are postulates and not generalizations about the behaviour of measuring rods. In the absence of such postulates, one would be unable to make the choice between a metal rod and an elastic band as an appropriate instrument for the measurement of spatial distance. The concept is prior to the instrument which determines its magnitude.

While Newton would appear to have had excellent grounds for postulating absolute time, one wonders why he found it desirable to postulate absolute space, given the equivalence of a potential infinity of spatial frames of reference. Does not the Galilean transformation and the closely related principle of classical relativity imply that space is a relativistic concept? Basically, what is at issue is whether there exist absolute motions as well as relative ones. According to Newton there are, in fact, ample instances of absolute motions. One of his examples is much more frequently cited than any other and is usually referred to as the "bucket experiment".

We take a bucket of water which is suspended by a string. The string is strongly twisted and then the bucket is released and allowed to rotate. Initially, there is considerable *relative* motion between the bucket and the water. During this period of maximum relative motion the surface of the water remains in its original plane state. Gradually, however, motion is communicated to the water. That is to say, that the rotation of the bucket induces a rotation of the water. Thus the relative motion between the two objects gradually diminishes. During this period the surface of the water becomes concave as the water tends to recede from the axis of rotation in virtue of a centrifugal force. At last, the bucket and the water have no relative motion with respect to each other. *Kinematically*, they are in a similar state to that which obtained before the bucket rotated. However, there is the essential difference that now the surface of the water is in its state of maximum concavity. It clearly follows that what caused the deformation of the surface was not the original relative motion of the water with respect to the bucket but its eventual absolute motion with respect to space.

The point of Newton's argument is readily perceived if we regard the bucket as a frame of reference. Newton's point is that when the water has its greatest motion with respect to this frame of reference there is no dynamic effect, but when the motion with respect to the frame of reference has vanished there is the maximum dynamic effect. The latter is, therefore, not the result of motion with respect to the frame of reference but with respect to something else. But what else is there? According to Newton, the only possible answer is absolute space.

You will recall that we originally suggested that the weight of the claim to the absolute character of space was reduced in virtue of the principle of relativity which implies that velocity has no dynamic effects. The force of Newton's argument is that although velocities do not produce dynamic effects, accelerations do. Moreover, the bucket experiment is designed to show that such effects are not the product of relative acceleration but of absolute or "real" acceleration. Thus, while it is experimentally impossible to distinguish absolute velocity from relative velocity, it is definitely possible to distinguish absolute acceleration from relative acceleration. Moreover, it follows that since there are real accelerations with respect to absolute space, there must also be real velocities, even though the laws of mechanics are such that these real velocities are unobservable. Hence, although Newton incorporated a principle of relativity in his mechanical theory, he refrained from attaching much importance to it. It was more the result of a practical contingency than the reflection of a deep theoretical feature of the world.

We shall conclude this section with an attempt to formulate a reasonably precise statement of the principle of classical relativity. In the first place, we must broaden the notion of a Galilean transformation to include transformations between coordinate systems whose axes are not parallel, so that all three barred coordinate variables will be changing with respect to the unbarred ones in time and, moreover, the origins are not arbitrarily assumed to coincide at the zero point of time. However, it should be stressed that the generalization is virtually trivial since any two frames of reference which are in relative uniform motion may be so coordinatized that the restricted Galilean transformation may be employed. In fact, the general, inhomogeneous Galilean transformation is simply the group $O_{i3} \times T \times G_s$.

The correct formulation of the principle of classical relativity is not as uncontroversial a matter as some would suppose. In fact, any formulation will inevitably reflect some measure of philosophical and methodological bias. The best one may hope to achieve is an explicit awareness and justification of such biases.

The majority of formulations depend on the concept of an inertial frame of reference. This is unfortunate for two reasons. In the first place, there is less than complete agreement as to the meaning of "inertial frame". Some writers define it as a frame in which Newton's first law is valid. Others define it as a frame in which *all* of Newton's basic laws are valid, and still others as a frame which is in a state of uniform motion (with respect to what?). Moreover, it is by no means obvious that the concept of inertial frame which occurs in Newtonian physics is identical with that which occurs in relativistic physics. In the second place, there is a real danger that the inclusion of the concept in the formulation of the principle of relativity may render the latter circular. For example, certain formulations amount to the claim that "The laws of mechanics are valid in all inertial frames of reference". But since an inertial frame may be defined as one in which the laws of mechanics are valid, such a formulation runs a clear risk of being analytically true.

Writers who dispense with the concept of inertial frame tend to express the principle of relativity in such a manner as to make it appear to be mathematical rather than physical.

For example, "The laws of mechanics are Galilei covariant". It should be noted that this formulation is metatheoretical. That is to say, that it is not a statement that refers to the world but one which refers to Newton's theory. We hope that the reader has by now acquired some feeling for the physical or factual character of the principle of relativity. If so, he will recognize the metatheoretical version of it to be misleading. Moreover, if principles of relativity are properly construed as metatheoretical, then one is obliged to accept the implication that Einstein's principle of relativity is not incompatible with that of Newton. The one is a correct statement above the mathematical form of Newton's theory while the other is an equally correct statement about the mathematical form of Einstein's theory. However, the overwhelming weight of scientific opinion favours the view that the two principles are factually incompatible, and I do not believe that there occurs in this context sufficient ground for violating the orthodox viewpoint. Finally, we might mention that it is unclear as to whether the principle should be formulated as a universally general statement with a possibly empty reference class or as a numerically indefinite existential statement or as a numerically definite statement which postulates the existence of at least two frames of reference in which Newton's laws are found to hold and which are related by a Galilean transformation. In the last regard, it is clear that Newton certainly attached existential significance to the principle in so far as he believed in the existence of at least one acceptable frame of reference—namely, absolute space. Moreover, an axiomatic version of a scientific theory should always postulate non-empty reference classes for its concepts. Accordingly, an informal non-axiomatic version should strive to reflect this point. Electrons are not conceived as mere possibilities or conceptual constructs which would conform to certain laws if only they happen to exist. A theory of electrons postulates that electrons exist. Similarly, a theory involving frames of reference in which Newton's laws hold must, rightly or wrongly, postulate the existence of such frames.

With these various considerations in mind we may now attempt our own formulation of the principle of classical relativity. We first introduce the following predicates.

$$Fx = x \text{ is a frame of reference.}$$
$$Nx = \text{Newton's laws hold with respect to } x.$$
$$Gxy = x \text{ is related to } y \text{ by a Galilean transformation.}$$

The principle of classical relativity may be formulated as

$$(\exists x)\{(Fx \wedge Nx) \wedge (y)[(Fy \wedge Gyx) \rightarrow Ny]\}.$$

In words: There is at least one frame of reference with respect to which Newton's laws hold, and every frame of reference which is related to it by a Galilean transformation is also one with respect to which Newton's laws hold.

Several defensive comments are now appropriate. Certain methodologists would object to our formulation of the principle of relativity on the ground that an existentially quantified statement cannot enjoy the status of a law of nature. Specifically, any such statement is not, strictly speaking, falsifiable. For example, "Green rabbits exists" cannot be falsified and cannot, therefore, be a candidate for the status of scientific hypothesis. Sympathetic though we may be to the spirit of this objection, we believe it to be an oversimplification. First, the case of a mixed general statement, i.e. one containing both existential and universal quantifiers, stands in need of further analysis. For all that, it is true that one may derive the purely existential statement $(\exists x)(Fx \wedge Nx)$ from the principle as we have formulated it. But even in this case the matter is not so clear cut as an abstract logical analysis might suggest. The reason for this is that reference frames are not necessarily natural objects. The most suitable frames of reference tend to be artifacts. Thus the existential clause may be interpreted as a claim to the effect that it is possible to construct a frame of reference in which the results of experiments would be in conformity with the Newtonian predictions. Now it happens to be the case, as we shall eventually learn, that if Einstein's *general* theory of relativity is correct, then no such frame can be constructed. Such frames are ruled out on theoretical grounds. Consequently, any direct confirmation of general relativity counts strongly, albeit not decisively, against the principle of classical relativity. Thus, while it may not be decisively falsified, its correctness may be so much in doubt that it is bound to be discarded. We believe this to be quite adequate as a criterion for a statement's status as a scientific hypothesis.

There is another side to this coin. While there are virtually overwhelming theoretical grounds for doubting the possibility of the existence of inertial frames of the kind envisaged by Newton or, for that matter, of the kind envisaged by Einstein in his earlier theory of *special* relativity, it is, nevertheless, possible to construct a very good approximation to such a frame. Without going into the technical aspects which are handled in the later discussion of general relativity, we may present the situation in an intuitive manner which should serve temporarily to dispel the reader's bewilderment. According to Newton, there are frames of reference with respect to which all material bodies in the universe will obey the laws of mechanics. In other words, it is possible to coordinatize the entire universe with a single suitable coordinate system. According to general relativity, a suitable coordinate system, i.e. one for which Newton's laws might be valid, cannot be extended globally, but has a purely local status. Mathematically, "local" means in the infinitesimal neighbourhood of a point. In practice, however, the neighbourhood may be extended a matter of feet or even miles under certain circumstances. Thus a practical counterpart of an inertial reference frame is constructible. One form of such a frame would be an artificial satellite orbiting the earth. An object in such a frame would remain at rest wherever it might be placed, and if subjected to a force would move in accordance with Newton's second law. Hence, a type of confirmation of the principle of classical relativity would be possible. Although such confirmation would be imperfect, it would serve, for example, as a basis for deciding whether *locally* classical relativity or *special* relativity is the better theoretical principle.

A second possible methodological objection must be faced. As has already been stated, this book is written from the standpoint that the various principles of relativity are factual hypotheses which are capable of empirical falsification. More technically, it is my contention that they belong to the *object* language of physics as distinct from the *metalanguage* of physical theory. Now one mark of a metatheoretical statement is that it contains names and predicates which refer to the object theory. Thus, in the metalanguage of the theory under discussion, there would occur statements in which Newton's laws are mentioned or named rather than used. Now the accusation could be levelled that we have employed a strategem to eliminate the name of Newton's laws, e.g. N1, N2, etc., by introducing the predicate Nx in a manner reminiscent of Quine.

My answer to that possible objection is that what is really at issue is whether the "name" of Newton's laws occurs essentially in the formula of interest. I would claim that it does not. On the contrary, the predicate Nx was only employed for purposes of brevity and reasonable perspicuity. Nx may be eliminated in a completely natural way by writing a conjunction of formulas, one for each of the basic laws of the theory, in which we make use of the predicate $K^i x = x$ is a set of frame coordinates. A typical conjunct, in this case for the second law, would be

$$(\exists x)(K^i x \wedge F^i = ma^i) \wedge (y)(\overline{K}^i y = G_{ij} K^j x \rightarrow \overline{F}^i = m\bar{a}^i),$$

where G_{ij} is the Galilean transformation matrix. The first part of this formula, for example, asserts that there is a frame of reference and that the behaviour of a particle under the influence of a force in this frame is given by $F^i = ma^i$, where the superscripts refer to the components of the force and acceleration along the axes of the reference frame. All of this clearly belongs to the object language, especially when one is mindful of the important fact that a frame of reference is a part of the physical world.

Our final comment about the formula is that it deliberately leaves open the question of the existence of absolute space. A corollary of any relativity principle is that there is no uniquely preferred frame of reference. Newton, himself, believed that of all inertial frames there is one that enjoys a preeminent status—namely, absolute space. The laws of mechanics apparently made it impossible to identify that unique frame. On the other hand, such considerations as the bucket experiment led Newton to suppose that there were independent grounds for postulating its existence. Thus, Newton was not completely faithful to his own principle of relativity. However, the philosopher-physicist Ernst Mach was to point out that there was an ungrounded assumption in Newton's reasoning. According to Newton, there would be an esential difference between the case of a bucket's rotating with respect to the fixed stars, and that of the fixed stars' rotating with respect to the bucket, even though the two are kinematically equivalent. Thus, it was inconceivable to Newton that the rotation of the fixed stars with respect to the bucket could possibly be relevant to the concavity of the surface of the water. Mach pointed out that this was an unwarranted assumption. In so doing, he dimly presaged, at least in a heuristic way, the principle of general relativity. In any event, we are interested not in Newton but in his theory, and a formulation of the principle of relativity which does not involve the hypostasisation of space is methodologically more faithful to the theory.

I shall now summarize the most essential implications of this chapter. The existence of a principle of relativity may be interpreted as expressing not the subjective features of the world but its objective symmetries. The laws of classical or Newtonian mechanics may be expressed in vectorial form. Although the notion of a vector is a geometric concept the possibility of expressing laws of nature in vectorial form has factual implications or presuppositions—namely, that space is both homogeneous and isotropic. Furthermore, spatial and temporal intervals are invariant. In particular, the zero temporal interval or *simultaneity* has absolute significance. Over and above the symmetries of the classical laws which are reflected in their vectorial character, there exists the group of Galilean symmetries. The factual status of Galilean relativity has been demonstrated. The principle of classical relativity was interpreted as a factual statement concerning the existence of, and relationships among, the set of reference frames with respect to which the laws of mechanics may be expressed in their fundamental, global form.

Bibiliography for Chapter 2

BUNGE, M., *Foundations of Physics*, Springer-Verlag, Berlin, 1967. A mine of information and insights into all the fundamental branches of physics, which emphasizes logical and philosophical foundations. Some parts are definitely not for the beginner but there is something for almost everyone.

FEYNMAN, R. P., R. B. LEIGHTON, and M. SANDS, *The Feynman Lectures on Physics*, vol. 1, Addison-Wesley, Reading, Mass., 1963. A readable yet sophisticated introduction to mechanics, which stresses concepts much more than computational techniques. Highly recommended for students of philosophy.

GOLDSTEIN, H., *Classical Mechanics*, Addison-Wesley, Reading, Mass., 1950. An outstanding textbook for those who wish to pursue the more advanced aspects of mechanics. Considerable attention is given to transformation properties.

KITTEL, C., W. D. KNIGHT, and M. A. RUDERMAN, *Mechanics*, McGraw-Hill, New York, 1965. This book provides the reader with a real working knowledge of mechanics at an introductory level. Emphasis is placed on invariance and relativity at an early stage.

3

The Principle of Special Relativity

The Breakdown of Classical Relativity

WE have seen that Newton postulated as a background for his theory of mechanics an absolute space and an absolute time. Although we found ourselves obliged to question the necessity for the introduction of absolute space, it is indeed true that his theory required that an absolute significance be given to the concept of length or distance. For example, the force that chiefly interested Newton is that of gravity. The gravitational force is inversely proportional to the square of the distance between the gravitationally interacting bodies. Now if distance were a frame or coordinate dependent quantity, the law of gravitation would be ambiguous. Since the acceleration produced by a gravitational field is the same in all inertial frames of reference, it follows that, assuming constancy of mass, the distance between two bodies at a given time must be the same in all inertial frames. In fact, one is entitled to make the stronger mathematical claim that distance is invariant with respect to *all* frames of reference, since the definition $s^2 = \delta_{ij}X^iX^j$ is independent of time, and hence of velocity and acceleration.

In the mid-nineteenth century, a new development in physical theory arose which would eventually place the presupposition of Newtonian mechanics in crucial jeopardy. We refer to the theory of electrodynamics of Maxwell. The crisis was only satisfactorily resolved in 1905 in the first of Einstein's papers on special relativity, entitled *On the Electrodynamics of Moving Bodies*. The title of this paper deserves to be emphasized since philosophical discussions of Einstein's first theory of relativity have too frequently conveyed the impression that it is simply a new theory of kinematics, whereas, in fact, it is as intimitately connected with electrodynamics as the principle of classical relativity is connected with classical mechanics. Thus, the significant physical content of the principle is sometimes apt to be understated. However, for paedagogical reasons, it is extremely useful to approach the theory from the kinematic standpoint. Although we shall follow this simple route, the reader should try to keep in mind that electrodynamics is the physical foundation of special relativity just as Newtonian mechanics provides the foundation of classical relativity.

It was observed in the previous chapter that the principle of relativity implies that there is no preferred frame of reference. This has the effect of dividing the laws of physics into fundamental and non-fundamental laws. A fundamental law is one in which the frame of reference plays no role. A non-fundamental law, however, will be tied to a particular reference frame. For example, the laws of sound propagation depend in a significant way on the medium of propagation. Accordingly, the appropriate frame of reference for the formulation of such laws is the uniquely determined frame with respect to which that medium is at rest. By the same token, if one were describing the propagation of ripples produced by a stone's being dropped into a pond, one would do so employing coordinates attached to the water, in which case the mathematical description would manifest the appropriate radial symmetry which we perceive in the concentric ripples. Thus, non-fundamental laws are associated with privileged or preferred frames of reference, which may be specific media. Although such laws are non-relativistic, it must be pointed out that they are not incompatible with the principle of relativity, since they do not pick out a *globally preferred* reference frame, but only one which is preferred for a specific local situation.

In spite of its striking mathematical elegance, Maxwell's electrodynamics was apparently non-fundamental in the foregoing sense. Its basic laws describe the interconnections of such quantities as electric charge and electric and magnetic field intensities. These quantities are not of a mechanical nature. However, the theory treated electromagnetic phenomena as the propagation of disturbances, or more precisely waves, through some medium. Thus, the theory was non-fundamental in the sense that ultimately one might expect to be able to deduce its laws from fundamental ones pertaining to the mechanical stresses and strains of the medium through which electromagnetic phenomena are propagated.

However, the actual situation was not that simple. We may grasp the situation by considering the phenomenon of light propagation which, according to the theory, is an instance of electromagnetic propagation. Not only is the wave character of light predicted on the basis of Maxwell's theory but moreover there is an abundance of evidence to support this view independently of the theory. Accordingly, our mechanical intuition demands that there exist a medium through which light is transmitted. Mechanically, sound waves and water waves are unthinkable in the absence of an appropriate medium. Likewise, optical waves are unthinkable in the absence of their appropriate medium. Accordingly, a medium called the *luminiferous aether* was postulated as the basis for optical phenomena and electromagnetic phenomena in general.

Although the aether appeared to be theoretically necessary it presented many problems. It is well known that light can be transmitted through a physical vacuum. Thus, the analogy with air and water breaks down. In the second place, while matter appeared to have an effect on the behaviour of the aether, no reciprocal effect was detectable. Thirdly, since light can be transmitted through transparent bodies, it followed that the aether was capable of permeating not only a vacuum but also solid objects. In short, the aether was a mechanical medium which possessed certain properties which were incompatible with mechanical theory.

On the other hand, the aether did offer one tantalizing theoretical possibility. According to Maxwell's theory, all electromagnetic phenomena, including light, are propagated in a vacuum with a constant velocity, $c \approx 186{,}000$ miles/sec. This velocity should not be dependent on the velocity of its source but only on the characteristics of the stationary medium. The apparent significance of c, the constant which occurs in the laws of electrodynamics, is as a quantity which has a fixed value with respect to a uniquely preferred frame of reference—namely, the aether-frame. This provided the prospect of determining absolute velocities by optical or electromagnetic means. For example, there was definite physical evidence to the effect that the earth moved freely through the aether without dragging the latter along with it. Accordingly, the velocity of light should have different apparent values in different directions. Suppose that the speed of the earth's orbital motion is v with respect to the aether. Then the observed speed of light in the direction of the earth's motion would be $c-v$, and $c+v$ in the opposite direction. Hence, by measuring the difference between these two observed values, one easily obtains the absolute velocity of the earth!

It should be noted that this reasoning is predicated on the implicit assumption that electrodynamics is not a relativistic theory. Thus, since light has a constant velocity which is the same in all directions in the stationary aether, if one considers a light-pulse emitted from a source which is located at the origin of a suitably chosen coordinate system, the wave-front which is propagated in all directions will be given by the equation $(x^1)^2+(x^2)^2+(x^3)^2=c^2t^2$. That is to say, that the optical wave-front will be a spherical surface centred at the origin. However, it is obvious that this equation is not Galilei covariant. Optical phenomena should, therefore, be influenced by uniform motions of observers. Because optical processes are not relativistic, an optical experiment, unlike a mechanical one, may be performed to ascertain the velocity of the reference frame in which it is carried out.

Such an experiment was performed by the American physicists Michelson and Morley in 1887. They employed an instrument of Michelson's invention called an interferometer, the details of which are no longer important, which was capable of measuring differences in the velocity of light in different directions to the second order of accuracy, that is to the order of v^2/c^2, where v is the velocity of the rest-frame of the interferometer, i.e. the earth. The outcome of the experiment, however, was that to accuracy of the second order, the velocity of light is the same in all directions, independent of the velocity of the reference frame.

The situation in physics at the end of the nineteenth century was one of profound conceptual crisis. On the one hand, classical mechanics had been developed to the point of near perfection. The principle of relativity which was incorporated in this theoretical framework implied that absolute velocity is not observable. On the other hand, the new theory of electrodynamics was of such a mathematical form as to appear to imply that it is observable. Yet every experiment that had been performed failed to reveal any absolute effect of velocity. Electrodynamics should have been incompatible with the principle of relativity, yet experiment failed to reveal the effects of the incompatibility which theory demanded. Many piecemeal attempts were made to patch up the conceptual scheme of physics but none was wholly satisfactory although one such attempt by the Dutch physicist, Lorentz, was a work of great brilliance. The reader will find a solid but elementary account of this period in Max Born (1962).

Relativistic Kinematics

In the aforementioned paper of 1905, Einstein provided the method of reconciling mechanics, electrodynamics and the principle of relativity. Despite its forbidding reputation, the reader of average ability in theoretical physics is struck by the exquisite simplicity of this paper. Although the central ideas have rich physical implications, they are not based on new physical assumptions. All of the physical hypotheses necessary for Einstein's new theory had been available in the nineteenth century. What was essentially new was Einstein's profound revision of the classical concepts of space and time. In many respects, therefore, this paper may be properly regarded as belonging to philosophy, although the philosophy of one who is informed, a philosophy which must remain ever obscure to one whose principal occupation is the exploration of linguistic behaviour.

The theory is based on two fundamental postulates, philosophically speaking. The first of these, the principle of relativity, is that in conformity with experience the fundamental laws of nature, including those of electrodynamics, are unaffected by uniform, rectilinear motions. Einstein was convinced that the concept of absolute velocity was empty. What mechanics had laid to rest, electrodynamics could not resurrect. For Einstein it was inconceivable that a substance, absolute space or aether, on which matter could have no effect, could have an active effect on the behaviour of material bodies and processes. Thus, classical relativity is extended by Einstein to electrodynamics. That is to say, that Einstein's postulate of *restricted* or *special* relativity asserts the existence of a class of inertial frames which play a fundamental role for the formulation of both mechanical and electrodynamic laws. Thus, the principle of special relativity confirms the impossibility of devising a method for measuring an absolute velocity or, what is the same thing, of ascertaining the absolute velocity of one's own reference frame.

The second fundamental postulate is that the velocity of light, in conformity with Maxwell's laws, is a universal constant which is independent of the velocity of its source. This postulate has an immediate and important consequence—namely, that c is independent of the velocity of the frame of reference. Suppose that light is emitted from a source which has the velocity v with respect to a 'stationary' reference frame K. The postulate of constancy asserts that the velocity of propagation will be c, independent of v. If we now consider the rest-frame of the light source, $\overline{K}$, the postulate asserts that the light will be propagated with the velocity c with respect to its source, i.e. $\overline{K}$. Thus, the law of light propagation is treated as a basic, frame-independent law of nature.

It will occur to the perceptive reader that it was precisely these two claims that were declared to be incompatible in the previous section, where we mentioned the conceptual problems of late-nineteenth-century physics. The great philosophical thrust of Einstein's 1905 theory was that in fact there is no formal, logical incompatibility between them. Clearly, however, it is a mathematical fact that the law of light propagation is not Galilei covariant. Accordingly, one must modify the procedures for carrying out frame transformations between electromagnetic inertial frames, i.e. reference frames with respect to which Maxwell's laws hold in their basic form.

Let us then attempt to determine what the appropriate transformation laws must be. What follows is essentially a simplified account of Einstein's own procedure. We require that the optical wave front equation $(x^1)^2 + (x^2)^2 + (x^3)^2 = c^2t^2$ be covariant when referred to a second frame which has a uniform velocity with respect to the original one. More compactly, we require $\delta_{ij}X^iX^j - c^2t^2 = 0$ to have the form $\delta_{ij}\overline{X}^i\overline{X}^j - c^2\overline{t}^2 = 0$ in the barred frame.

For ease of interpretation, we shall resort to the coordinate variables X, Y, Z. Furthermore, without serious loss of generality, we shall assume, as in the previous chapter, that K and $\overline{K}$ are coordinatized by S and $\overline{S}$ so that corresponding axes are always parallel and the respective origins coincide at $t = 0$. With respect to K, the law of interest is of the form

$$x^2 + y^2 + z^2 = c^2t^2. \tag{3.1}$$

$\overline{K}$ is, *ex hypothesi*, also an inertial frame. Accordingly, the form of the wave-front equation must, with respect to $\bar{K}$, be

$$\bar{x}^2 + \bar{y}^2 + \bar{z}^2 = c^2\bar{t}^2. \tag{3.2}$$

We first assure the reader of the truth of our claim that (3.1) is not Galilei covariant, a fact that we have hitherto claimed on intuitive grounds. Performing the special, homogeneous Galilean transformation on (3.1), we get

$$(x - vt)^2 + y^2 + z^2 = x^2 - 2xvt + v^2t^2 + y^2 + z^2 = c^2t^2. \tag{3.3}$$

It is obvious that (3.3) is inconsistent with (3.1) or that the wave-front equation is not Galilei covariant.

To determine the appropriate group of transformations for (3.1) we shall forsake complete mathematical rigour in favour of certain simplifying assumptions which happen, we assure the possibly sceptical reader, to be correct. Firstly, considerations of symmetry, as well as of relativity, require that in our special or restricted case, we must have the mappings:

$$\begin{aligned} y &\to \bar{y} = y \\ z &\to \bar{z} = z. \end{aligned} \tag{3.4}$$

In the second place, considerations of relativity require that the uniformity of expansion of the wave-front in one inertial frame entail its uniformity in every inertial frame. The alternative would be to abandon the concept of inertial frame. The preservation of uniformity provides the valuable clue that the transformation be linear. Combining the property of linearity with the natural requirement of dimensional homogeneity, i.e. that distances be transformed into distances and times into times, it follows that there can be no relative acceleration between the two frames. It is of some methodological significance that the absence of mutual acceleration need not be employed in the definition of an inertial frame, although most writers do so, but may be deduced as a consequence of the two frames being inertial. From all of these qualitative considerations which could have been provided with rigorous mathematical justification, we may write the general form of the required transformation as:

$$\begin{aligned} x &\to \bar{x} = f(x) - g(vt) \\ y &\to \bar{y} = y \\ z &\to \bar{z} = z \\ t &\to \bar{t} = j(t) + h(x/v). \end{aligned} \tag{3.5}$$

The reader should note that (3.5) simply appears as a generalization of the common sense special Galilean transformation. The first of (3.5) asserts that the new coordinate is functionally dependent on the old coordinate, the velocity and the time. The last of (3.5) asserts that the value of time is functionally dependent on the original time, the original coordinate and the velocity. In the classical, Galilean case, we have g, f, and j as *identity* functions while h is the constant function, h: $x/v \to 0$. At the moment, the only restriction placed on the four functions is that they be linear in the coordinates and the time.

We have now to apply the transformations (3.5) to (3.1) in order to discover the explicit form of the four unknown functions. This yields

$$\begin{aligned} &f^2(x) - 2g(vt)f(x) + g^2(vt) + y^2 + z^2 \\ &= c^2j^2(t) + 2c^2j(t)h(x/v) + c^2h^2(x/v). \end{aligned} \tag{3.6}$$

Now we must choose f, g, j, h such that (3.6) reduces identically to the form of (3.1). Immediate inspection suggests

$$f(x) = x;\ j(t) = t;\ g(vt) = vt. \tag{3.7}$$

The substitution of (3.7) into (3.6) yields

$$\begin{aligned} &x^2 - 2xvt + v^2t^2 + y^2 + z^2 \\ &= c^2t^2 + 2c^2th(x/v) + c^2h^2(x/v). \end{aligned} \tag{3.8}$$

If we now set $h(x/v) = -vx/c^2$, (3.8) reduces to

$$\begin{aligned} x^2 - 2xvt + v^2t^2 + y^2 + z^2 &= c^2t^2 + 2c^2t(-vx/c^2) + v^2x^2/c^2 \\ x^2 - x^2v^2/c^2 + y^2 + z^2 &= c^2t^2 - v^2t^2 \\ x^2(1 - v^2/c^2) + y^2 + z^2 &= c^2t^2(1 - v^2/c^2). \end{aligned} \tag{3.9}$$

(3.9) is close to what we need apart from a constant multiplicative factor. Accordingly, we must divide by the factor $(1 - v^2/c^2)^{\frac{1}{2}}$, and our required mappings are:

$$\begin{aligned} x &\to \bar{x} = (x - vt)/(1 - v^2/c^2)^{\frac{1}{2}} \\ y &\to \bar{y} = y \\ z &\to \bar{z} = z \\ t &\to \bar{t} = (t - vx/c^2)/(1 - v^2/c^2)^{\frac{1}{2}}. \end{aligned} \tag{3.10}$$

(3.10) constitute the special, homogeneous *Lorentz* transformation. It turns out to be applicable not only to the optical wave-front equation but also to the more general basic equations of Maxwell's electrodynamics. Before proceeding, I would like to take a moment to offer my congratulations to every philosophy student who has made the effort to follow and understand the foregoing derivation. It is the fundamental result of Einstein's theory of special relativity and constitutes the key to all the ramifications of that theory!

The upshot of the derivation of (3.10) is to show that there is no contradiction between the universality of c and the principle of relativity. However, the two claims are only congruent *provided* that the relationship between the coordinates of two inertial reference frames whose relative velocity is $\mathbf{v}$ is given by (3.10) and not by the intuitively acceptable Galilean transformation. In other words, their compatibility requires that electrodynamics be Lorentz covariant rather than Galilei covariant.

Before dealing with the dramatic implications of the Lorentz transformation, we wish to focus on its origin. Partly as a result of Einstein's own manner of presentation, two generations of philosophers have been misled into supposing that the basis of the theory consists in hypotheses about the behaviour of clocks and measuring rods. Not only is this interpretation incorrect from a methodological and philosophical standpoint, but it is also unpaedagogical in so far as it gives the theory an arbitrary appearance. On this interpretation, it would seem that the laws of relativistic physics would be different if only we had different measuring devices at our disposal. As we shall see, it actually gave rise to a thoroughly idealistic interpretation, according to which the theory is not about the world but about our knowledge of the world, as it is constrained by our methods of observation.

This, however, is to put the cart before the horse. It is not that the theory is generated by our method of carrying out measurements or choice of instruments, but rather that our decision as to what constitute adequate instruments is partly determined by the theory. As every scientist knows, when the outcome of a measurement fails to accord with one's theory, the adequacy of the measuring device is immediately under suspicion. Not that this is a decisive consideration, but it is surely suggestive. What the reader should be mindful of, therefore, is that (3.10) was derived without any reference to the mode of making space and time measurements or to the kinds of devices used for such measurements. That is not to say that (3.10) fails to have implications in these areas. We shall now proceed to examine the most obvious of them.

The most startling aspect of the Lorentz transformation is what is known as the relativization of time. In classical physics, time was regarded as absolute. This is reflected in the mapping: $t \to \bar{t} = t$. In relativistic physics, given the time of an event in a frame K, it is not the case that it will have the same value of time with respect to $\bar{K}$. Rather, its $\bar{K}$-time will be $\bar{t} = (t - vx/c^2)/(1 - v^2/c^2)^{\frac{1}{2}}$. As Einstein has argued, fundamental to all our temporal determinations is the concept of *simultaneity*. In classical or Newtonian physics, it was taken as axiomatic that the relationship of simultaneity between events was absolute and unambiguous. To say that E_1 and E_2 occurred simultaneously was to make a claim that had nothing to do with the frame of reference. Despite its centrality, the notion was never analyzed or explicated in classical physics. Its status was too evident. The Lorentz transformation of time, however, obliges one to re-evaluate one's notion of simultaneity.

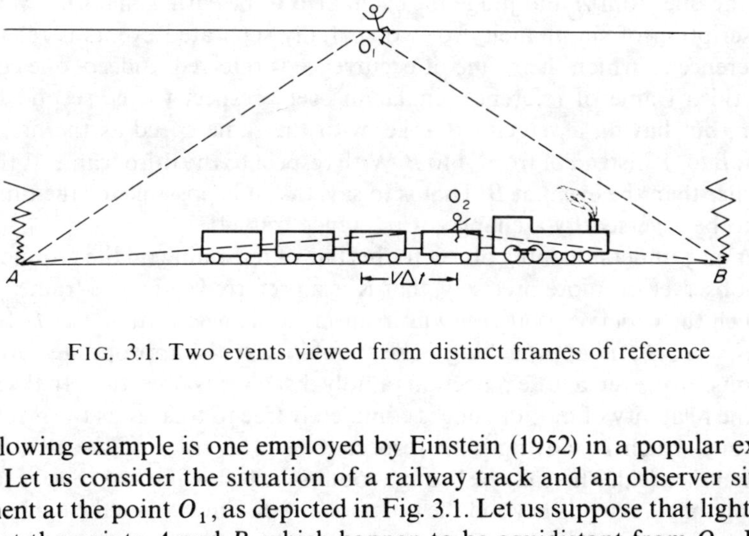

FIG. 3.1. Two events viewed from distinct frames of reference

The following example is one employed by Einstein (1952) in a popular exposition of relativity. Let us consider the situation of a railway track and an observer sitting on the embankment at the point O_1, as depicted in Fig. 3.1. Let us suppose that lightning strikes the track at the points A and B, which happen to be equidistant from O_1. Under what circumstances would O_1 declare the two strokes to have occurred simultaneously? Suppose that he is equipped with a measuring device which records the precise time of reception of a light-signal. Then O_1 would find the events to have occurred simultaneously if his apparatus records the light-signals from the two events as reaching O_1 at the same time. Since the velocity of light is the same in all directions, and since the distance traversed by each signal is the same, it follows that the two signals arrive simultaneously at O_1 if and only if they departed simultaneously from A and B, respectively.

It must be remarked that Einstein's recourse to light-signals in judgments of simultaneity is one of the most frequently misinterpreted aspects of his theory. It is widely supposed that since c plays a fundamental role in the Lorentz transformation, its role in the measurement of time is equally fundamental. This, however, is not the case. The use of light-signals for the determination of simultaneity in no way precludes other methods, such as the use of sound-signals. The choice of light-signals is based on the obvious consideration that light, unlike sound, has a precise theoretically determined velocity which is independent of the characteristics of any medium. (This is not completely accurate. It is propagated at less than c through translucent media, when its actual velocity is dependent on the refractive index of the given medium. However, it doesn't actually require any medium at all, unlike sound, etc.) This makes it the most reliable signal but not the only one possible. The ascription of a unique role to light-signals is mitigated by one consideration. One of the implications of the Lorentz transformation, which we have still to explore, is that under no theoretically possible circumstance could any signal play a *superior* role to that of light, but, again, that is not to say that it is unique.

Returning to our illustration, let us now suppose a second observer O_2 who is on a train moving with the velocity $\mathbf{v}$ from A to B. The reader will recognize this as merely a picturesque way of specifying a second frame of reference $\bar{K}$. We suppose that at the time of signal emission from A and B, O_2 is also equidistant from the two points. Moreover, he is equipped with a similar apparatus to that of O_1. During the interval of time $\Delta t = AO_2/c$ that the light would require to reach him, were he at rest like O_1, he will have travelled the distance $v\Delta t$ away from A and closer to B. Accordingly, he will receive the signal from B earlier than the one from A and judge the event at B to be earlier than the event at A. In short, the ascription of simultaneity to two spatially separated events is relative to the frame of reference to which their time of occurrence is referred. Indeed, one could easily introduce a third frame of reference similar in every respect to the second frame, the moving train, but having a velocity $-\mathbf{v}$, i.e. with the same speed as the first train but moving from B to A, instead of from A to B. With respect to the third frame $\bar{\bar{K}}$, the event at A will be earlier than the event at B. That is to say, that it is possible for the time order of two events to be reversed by a change of reference frame!

The reader may understandably object to this line of reasoning on the ground that O_1 is a privileged observer or, more precisely, that K is a preferred reference frame. This is the frame in which the observer is at rest with respect to the events at A and B. Since O_2 is moving away from A and approaching B, he should make the appropriate corrections in his calculations. However, a little reflection rapidly dispels this objection. In the first place, in virtue of the relativity of motion, one is completely free to treat $\bar{K}$ as the rest frame and K as the frame which is moving with respect to it. The roles of the two frames are completely symmetrical. This may be brought out in the following way. Suppose that O_2's train is so long that it extends beyond the points A and B. Thus, when the lightning strikes the track at A and B, it also strikes the train at A_T and B_T. Since A and A_T coincide, both observers would agree that the events at A and A_T are simultaneous. Similar considerations apply to the events at B and B_T. There is no problem connected with the ascription of simultaneity to coincident events. However, if O_1 could regard himself as privileged to judge the simultaneity of the events at A and B from which he is constantly equidistant, then O_2 could regard himself as privileged to judge the events at A_T and B_T from which he is constantly equidistant. However, O_2 will regard his pair of events as successive, while O_1 will regard his as simultaneous. But, in fact, they are making judgments about what is, for practical purposes, the identical pair of events. Clearly, there is no way of distinguishing a uniquely privileged frame.

Perhaps, it may be suggested, there is an alternative method of determining the time intervals between distant events that does not involve the transmission of a signal. The most obvious stratagem would be to employ two clocks which are first synchronized in the same location, say at the origin O of S. Then one of the clocks is taken to the point x_1. Each clock is equipped with a device for printing the time of occurrence of an event of a particular type, say a stroke of lightning. Suppose that lightning strokes occur at O and at x_1, and each clock prints the time of the local event as it is recorded. Then any observer may compare the two printed records at his leisure and have the basis for an unambiguous and objective judgment.

The defect in this suggestion is that it is based on the unwarranted assumption that the running of a clock is unaffected by its being transported from O to x_1. The method presupposes that the clocks continue to be synchronized when they are no longer in the same region. But this can only be confirmed by having a signal transmitted between the clocks after their separation. Moreover, it is known empirically that different clocks behave differently when subjected to the same accelerations. Undoubtedly, the best method for assigning time values to different events will involve only a single basic clock. Let us then pursue this course, the one followed by Einstein.

A standard reference clock is located at the origin of S. A light-signal is transmitted at time t_1 and arrives x_1 at t_2. The clock at x_1 is adjusted so that $t_2 = t_1 + x_1/c$. The signal is immediately reflected back to the origin, where it arrives at t_3 as recorded by the basic reference clock. In virtue of the constancy of the velocity of light, we shall have: *duration of outgoing trip* = *duration of return trip* or $t_2 - t_1 = t_3 - t_2$. Equivalently, t_2 must be the exact half-way mark in the time of the total trip, i.e. $t_2 = \frac{1}{2}(t_3 - t_1)$. According to Einstein, provided that such relations hold, the two clocks may be regarded as synchronized and, therefore, as good a basis as any for the determination of simultaneity of the events at O and x_1.

Unfortunately, Einstein's proposal has been clouded by much philosophical dust. In particular, one of the most respected expositors of his theory, Hans Reichenbach, misread the text and supposed that Einstein intended that the fraction $\frac{1}{2}$ was conventional and replaceable by any other fraction between 0 and 1. Thus, Reichenbach suggests the formula: $t_2 = \varepsilon(t_3 - t_1), 0 < \varepsilon < 1$. According to Reichenbach, any of the allowed values of ε would provide a standard of synchronization, a method of extending time through space, that would be just as consistent as the one determined by the choice of $\varepsilon = \frac{1}{2}$. While our purpose is definitely not that of scholarly exegesis, since we are concerned with the theory rather than the opinions of its author, so much time has been devoted to the *pros* and *contras* of this issue, in my view needlessly, that it would be to the point to cite the critical passage. Einstein (1905) makes the point that one may establish a local time at point A and another local time at point B, by placing a clock at each location. He then proceeds to write: "We have so far defined only an 'A time' and a 'B time'. We have not defined a common 'time' for A and B, for the latter can not be defined at all unless we establish *by definition that* the 'time' required by light to travel from A to B equals the 'time' it requires to travel from B to A" (p. 40).

The choice of $\varepsilon = \frac{1}{2}$ is clearly equivalent to the supposition that light has a velocity which is independent of direction. The obvious point which Einstein wished to make was simply that in order to establish the velocity of light, it is necessary to have putatively synchronous clocks at different spatial locations. In other words, one cannot specify the equality of velocities in the absence of a standard for measuring time. Accordingly, one can only establish such a standard with the aid of light-signals at the cost of circularity. Recognizing this, Einstein proposed to *define* the two signal velocities as equal, and to proceed from there. It would have been better had he used the verb *to postulate* in preference to *define*, for clearly this was his intention. The principle of constancy is not a matter of definition but a physical hypothesis, which follows from Maxwell's equations. Thus, while Einstein spoke of "defining" the two signal velocities as equal, he clearly did not intend to imply that there is any viable alternative. In fact, the significance of the Lorentz transformation is destroyed on any other interpretation. This matter will be discussed in detail in Chapter 5.

Do we, then, have an objective and absolute standard of time after all? Objective yes, absolute no! So far, our discussion of the relativity of time has been largely qualitative. Let us now make it more precise. We presuppose a frame K equipped with two clocks, C_1 and C_2, which are arbitrarily situated on the X-axis and synchronized by the light-signal method, and a frame $\overline{K}$ moving with velocity $\mathbf{v}$ with respect to K in the $+X$-direction, and equipped with a single clock $\overline{C}$, situated at $\overline{x}_1$. At a given time, $\overline{C}$ coincides with C_1 in K. C_1 records the time of the coincidence as t_1, while $\overline{C}$ records $\overline{t}_1$. Subsequently, $\overline{C}$ coincides with the second K clock C_2. C_2 records the time of the coincidence as t_2, while $\overline{C}$ records $\overline{t}_2$. The time which elapses between the two events in $\overline{K}$ is $\overline{t}_2 - \overline{t}_1$. Similarly, in K it is $t_2 - t_1$. Now, by means of the Lorentz time transformation equation, we deduce the following result:

$$\Delta t = t_2 - t_1 = \left(\overline{t}_2 + \frac{v\overline{x}_1}{c^2}\right)\Big/\sqrt{1-\frac{v^2}{c^2}} - \left(\overline{t}_1 + \frac{v\overline{x}_1}{c^2}\right)\Big/\sqrt{1-\frac{v^2}{c^2}}$$

$$= \frac{\overline{t}_2 - \overline{t}_1}{\sqrt{1-\dfrac{v^2}{c^2}}}. \qquad (3.11)$$

It is important to be aware of the distinction between the two quantities $t_2 - t_1$ and $\overline{t}_2 - \overline{t}_1$. Clearly, the former exceeds the latter by a factor which is a function of the relative velocity of the two frames. The systematic difference between them is that the unbarred quantity represents the result of two spatially separated clock readings on instruments which have been synchronized according to the previously discussed 'standard' method. The barred quantity, on the other hand, represents the result of the reading of a single clock which is *present at both events.* A mathematical check of (3.11) reveals the fact that the latter is a Lorentz invariant quantity. It is the minimum time interval between two events and its Lorentz invariance implies that it has the same value in all inertial frames of reference. This is precisely what one would expect on the basis of intuition. Let us label two arbitrary events $E_1(t_1)$ and $E_2(t_2)$. If a clock in the immediate vicinity of E_1 records t_1 and the same clock records t_2 in the immediate vicinity of E_2, then all observers, i.e. with respect to all reference frames, will agree as to the time interval between the two events, as recorded by that clock. This is the feature of the situation which is reflected by the Lorentz invariance of the quantity which has been measured. Now it might appear that a similar argument would hold, *mutatis mutandis*, in the case of two spatially separated clocks, one located in the vicinity of E_1 and the other in the vicinity of E_2. Surely, it would be argued, all moving observers would agree that the clock at E_1 recorded t_1 and the clock at E_2 recorded t_2. While this is true, the various observers would contend that the two clocks in question were not properly synchronized. Moreover, they would be in disagreement as to which clock was running slow and by how much.

In view of the distinguished character of the Lorentz invariant time duration, we call it the *proper-time* interval between two events. The time interval that is determined by the "spreading of time through space", i.e. by spatially separated clocks which have been synchronized by means of a procedure which is appropriate to the particular reference frame in which the clocks are at rest, is called *coordinate time*. A coordinate-time interval is not an invariant. Symbolically, we label *proper-time* by τ and coordinate time by t. Intuitively, the nature of the distinction may be illustrated by the following example. Someone takes a trip by car from town A to town B. He notes the departure time on the town hall clock at A and the arrival time on the town hall clock at B. The two public clocks have been synchronized by the standard method, i.e. by means of a radio time signal based on a standard reference clock. The elapsed time measured by the traveller is the 'improper' coordinate-time interval $t_2 - t_1$. However, should the car be equipped with a clock of its own which will, of course, be transported along with the driver, which is to say that it is at rest in the driver's own rest-frame, it will measure an interval $\Delta\tau$ which is less than Δt by the factor $\sqrt{1 - v^2/c^2}$. Of course, if the velocity of the car is very small compared with the velocity of light, $v \ll c$, the Lorentz factor is indiscernible from unity and will be ignored, which is the situation of everyday experience. On the other hand, if v is a significant fraction of c, then the effect will be eminently remarkable. An onlooker would attribute the discrepancy between the two measurements to the fact that although the driver was careful to synchronize his own clock with the town hall clock at the beginning of the trip, the driver's clock ran slow with respect to the town hall clocks. It is sometimes said that moving clocks run slow with respect to stationary clocks, according to the theory of relativity. While this statement is in some sense true, one could hardly choose a more unfortunate or confusing way of referring to the phenomenon of *time dilation* (also called dilatation). In fact, it is precisely the aforementioned mode of speaking that has led to the so-called "clock paradox", which is so baffling to the newcomer to relativity theory and which has led many philosophers to misinterpret the significance of the Lorentz transformation.

The Lorentz transformation equations have an inverse which is obtained in the natural manner by reversing the signs of the velocities and interchanging barred and unbarred coordinates. (However, the reader should note that this has no effect on the sign of v^2.) The outcome of the existence of the inverse Lorentz transformation may be illustrated as follows. An observer in a given frame of reference will maintain that any clock which is in motion with respect to his frame is running slow with respect to a clock which is at rest in his frame. By the same token, however, a second observer who is at rest in the frame of the moving clock, which he regards as a *stationary* clock, will maintain that the first clock is running slow with respect to his own. More succinctly, if C and $\bar{C}$ are clocks at rest in K and $\bar{K}$, respectively, then C is running slow with respect to $\bar{C}$ and $\bar{C}$ is running slow with respect to C. In short, there is a perfect symmetry between the two frames. While this is required on the ground of the principle of relativity, it, nevertheless, appears to contravene the principles of logic. Some philosophers have attempted to resolve the paradox by the too facile argument that C *seems* to run slow with respect to $\bar{C}$, and $\bar{C}$ *seems* to run slow with respect to C. It is then pointed out that between the latter two claims there is no logical contradiction. However, the immediate corollary is that time dilation is merely an *apparent* effect resulting from a change of reference frame and that it is not to be regarded as real. It may also be said that the effect is purely kinematical. This is correct, but kinematical phenomena are not necessarily unreal phenomena.

A less misleading manner of speaking is to say that a proper-time interval is always less than a corresponding coordinate-time interval. In terms of the distinction between these two kinds of interval, one is able to analyze the situation in such a way that the paradox is resolved without the cost of depriving this aspect of special relativity of its real physical significance. It is to this matter that we shall now turn.

That our analysis may be as perspicuous as possible, we shall have recourse to a hypothetical device called a *light-clock*. The virtue of such a device is that its mechanism is thoroughly understood and it may be assumed to be at least as reliable as any physical clock. However, it is essential to recognize that the conclusion of the following analysis is independent of the nature of the instrument and could be reached, although less perspicuously, with any other acceptable device, e.g. a good chronometer or an atomic clock. However, I should warn the reader that what follows does require an effort of considerable concentration to be fully understood. The mathematics of special relativity is surprisingly easy, deceptively so. Too many make the mistake of supposing that having grasped the mathematics, they understand the physical basis of it. Therefore, I suggest that you study the following example several times until it makes sense intuitively. Then you will truly be on the path to thinking in a relativistically correct manner.

A light-clock is an imaginary device consisting of a long rigid platform of length L. At one end of the platform is a light source and a tape punch mechanism, and at the other end is located a mirror. A light pulse is sent from the source to the mirror from which it is reflected back to the source. When the signal returns to its origin, the mechanism is activated and punches a hole in a tape, signifying the passage of 1 unit of time. In effect, the clock is measuring the number of intervals between pairs of events—namely, the emission and reception of returning light-rays. Each interval or unit will be of duration $2L/c$. Now refer to the diagram of Fig. 3.2.

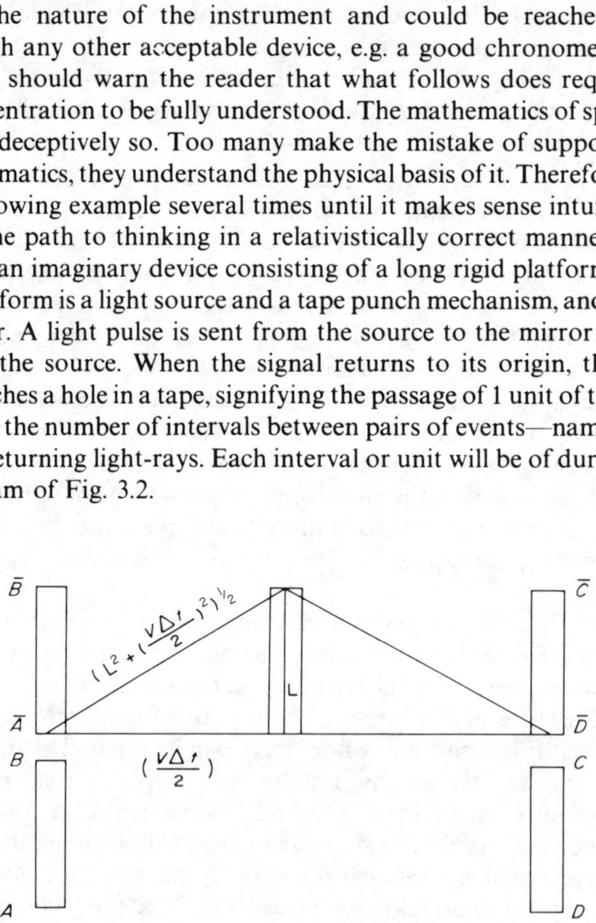

FIG. 3.2. The kinematics of two light-clocks

The upper part of this diagram is intended to represent a light-clock which is located at $\overline{AB}$ but is moving to the right with respect to a 'stationary' frame K, so that after an interval of time Δt, it coincides with $\overline{DC}$. The lower part of the diagram represents two light-clocks at rest in K at the points AB and DC. While the latter has been drawn below the former, for purposes of clarity, the reader should regard them as being superimposed, so that initially $\overline{AB}$ coincides with AB and finally, after Δt, with DC. The total distance traversed by the moving clock will obviously be $v\Delta t$, where v is the velocity with respect to K. We begin by considering the principle of time dilation according to which a moving clock runs slow with respect to a stationary clock. To simplify matters, we suppose that Δt is a single time unit. When the moving clock coincides with the first stationary clock, coincident light-pulses are sent from A and $\overline{A}$. When the light-ray from $\overline{A}$ returns to the point of origin, the mechanism of the moving clock punches a single hole, signifying the passage of one time unit. Let us now consider what transpired with respect to K. As the upper part of the diagram indicates, the light-ray does not make a simple out-and-return trip of length $2L$. Rather, its trajectory is a bent line whose total length, by an obvious application of the Pythagorean theorem, is

$$2\left(L^2+\left(\frac{v\Delta t}{2}\right)^2\right)^{\frac{1}{2}}.$$

But since light has the known velocity c, the length must be $c\Delta t$. Thus we may write the equation

$$c\Delta t = 2\left(L^2+\left(\frac{v\Delta t}{2}\right)^2\right)^{\frac{1}{2}}. \tag{3.12}$$

We may now solve (3.12) for Δt by elementary algebra.

$$c^2\Delta t^2 = 4\left(L^2+\left(\frac{v\Delta t}{2}\right)^2\right)$$

$$\Delta t^2 = \frac{4L^2}{c^2}+\frac{v^2\Delta t^2}{c^2}.$$

$$\Delta t^2 - \frac{v^2\Delta t^2}{c^2} = \frac{4L^2}{c^2}$$

$$\Delta t^2(1-v^2/c^2) = \frac{4L^2}{c^2}$$

$$\Delta t = \frac{2L}{c}\Big/(1-v^2/c^2)^{\frac{1}{2}}. \tag{3.13}$$

$2L/c$ is the proper-time between two events as measured by the moving clock. Δt is the improper time measured by the spatially separated coordinate clocks in K. The reader will note that Δt exceeds the proper-time interval by precisely the amount predicted by the Lorentz transformation.

It cannot be overemphasized that the difference between the two measurements is wholly objective. If the tapes of the moving clock and the second stationary clock were compared, we might, for example, find that whereas the first had one hole in it, the second had five holes. This is a clearly asymmetrical effect to which all possible observers would be bound to assent when comparing the tapes.

But given the objectivity of this asymmetrical effect, how are we to explain the clock paradox, which follows from the equivalence of all inertial frames as reflected in the

Lorentz transformation equations? Referring once more to Fig. 3.2, we now treat the clock situated along *DC* as moving to the left and the clock at $\overline{AB}$ as stationary. By the very same reasoning as before, when the clock which was initially at *DC* coincides with the clock at $\overline{AB}$, the former will record on its tape the elapse of one time unit, while the latter will record the elapse of five. Once again, the effect is strictly objective—the number of holes in a tape—and all observers would be obliged to agree. But have we not encountered a real contradiction? Are we not asserting the existence of two physically real effects which are mutually incompatible? In fact, we are not.

Let us use the characters *A*, $\bar{A}$, etc., to denote events, specifically the emission and reception of light-rays. In the first part of our analysis, what was revealed was that a moving clock assigned the interval of one unit to the pair of events $\overline{AA}$—that is, the emission of a light-ray from $\bar{A}$ and its return to $\bar{A}$. The stationary clock assigned the interval of five units to the pair of events *AD*. So far there is no contradiction. In the second part of the analysis, the moving clock assigned the interval of one unit to the pair of events *DD*, while the stationary clock assigned the interval of five units to $\overline{DA}$. Not only is there no contradiction between the two sets of measurements but after being subjected to a little reflection they begin to accord with our intuitions!

The principal philosophical message to emerge from the resolution of the clock paradox is that when speaking of the measurement of time, one is actually referring to the measurement of a temporal interval or separation between events. Indeed, many philosophers would interpret the relativization of time, i.e. its frame-dependent status, to vindicate the *relational* theory of time. According to this doctrine, time is nothing but, or may be reduced to, a relationship which holds among actual or possible events. Although I do see some merit in this metaphysical inference from special relativity, I suspect that it goes too far. It implies that in the absence of events there would be no time or duration. Duration without a sequence of events would be a meaningless abstraction. I certainly admit that it is difficult to *imagine* time divorced from the physical world. Moreover, the type of situation involved in the clock paradox may be rather neatly disposed of on the relational theory. However, it does seem that the notion of a *possible* event which the relationist is obliged to invoke is scarcely less abstract than that of physically eventless time. In any case, what is clearly implied by the relativity of time is that time, *per se*, lacks an intrinsic metric. Specifically, the duration or amount of time that is ascribed to the separation between two events involves a comparison between the given pair of events and a standard sequence of events registered by the ticking of a clock, the latter being the proper-time of the clock. The relativity of time simply means that there exist different procedures for implementing such a comparison, no one of which enjoys a globally preferred status. These various procedures simply correspond to the various choices of reference frames at one's disposal. Although the procedure corresponding to the measurement of a proper-time interval between a particular pair of events may seem to be preferred to any other, it fails to provide the determination of the proper-time for other such pairs. The attenuated relationalism to which I subscribe may be called the relational theory of the time *metric*. This constitutes a weaker or minimal inference from the theory of special relativity. However, I would contend that it is fully consonant with that theory, especially since the latter does recognize the existence of an absolute, i.e. Lorentz invariant, proper-time interval.

On the relational theory of time there would be a monumental difference between an imaginary empty universe and one which contains a single particle. The former would be timeless while the latter would be temporal. On my theory, the difference between the two hypothetical universes would be far less pronounced. It would consist only in there being no basis for the measurement of the temporal separation of events in the former whereas there would be in the latter, especially if the unique particle were an oscillating atom or other isochronous system. Nevertheless, I am mindful of the fact that the relationist would, of course, object to my hypothetical comparison on the ground that an empty universe is a contradiction in terms. This genuinely weighty subject will be scrutinized more fully in Chapter 11, in the context of the theory of general relativity. We shall now turn to the relativization of length or distance.

Let K be the frame in which an observer is at rest. A rigid rod moves past him in the $+X$-direction with velocity $\mathbf{v}$. We now wish to determine the length of the rod with respect to the 'stationary' observer or frame K. We select a frame $\bar{K}$ such that the rod is at rest along the $\bar{X}$-axis. Its rest-length will be given by $\bar{L} = \bar{x}_2 - \bar{x}_1$. By the first of (3.10), we compute its relativistic length L as follows:

$$\bar{x}_2 - \bar{x}_1 = \frac{x_2 - vt}{(1 - v^2/c^2)^{\frac{1}{2}}} - \frac{x_1 - vt}{(1 - v^2/c^2)^{\frac{1}{2}}} = \frac{x_2 - x_1}{(1 - v^2/c^2)^{\frac{1}{2}}} = \frac{L}{(1 - v^2/c^2)^{\frac{1}{2}}}$$

or $$L = \bar{L}\sqrt{1 - v^2/c^2} \qquad (3.14)$$

(3.14) signifies that the rest-length or *proper-length* of an object is its greatest length. Conversely, a rod in motion is contracted in the direction of the motion by the factor $\sqrt{1 - v^2/c^2}$.

It is once again easily verified that length contraction, like time dilation is a symmetrical phenomenon. That is to say, that a measuring rod of unit length in the rest frame K is contracted with respect to $\bar{K}$ by the Lorentz factor, while a rod of unit length at rest in $\bar{K}$ is likewise contracted by the Lorentz factor, with respect to K. This has given rise to another apparent paradox which precisely parallels the clock paradox. It is said that two inertial observers who are in relative motion will find that the standard measuring rod of the other is contracted with respect to his own. The paradox may again be resolved in the facile way by pointing out that since it is logically impossible for each of two rods to be shorter than the other, the contraction is not real but only apparent. Sometimes the analogy is made between the reciprocal relativistic length contraction and the case of two people who are situated some distance apart, each of whom perceives the other to be much shorter than himself. Thus, the Lorentz transformation is now incorrectly likened to an optical illusion. The correct approach, however, is much more in keeping with the spirit of scientific realism.

The most direct approach to the phenomenon of the relativity of distance is based on the recognition that the concept of distance is inextricably tied to the concept of time. Accordingly, we shall show that the relativity of distance is directly implied by the relativity of time. Consider the case of a frame K which is equipped with two clocks at widely separate points on the X-axis. A vehicle travels in the $+X$-direction with velocity $\mathbf{v}$ with respect to K. As it passes the first clock the time t_1 is recorded. As it passes the second clock the time t_2 is recorded. The total time $\Delta t = t_2 - t_1$ is an improper-time interval between the two events. The distance L between the two events will be given by $v\Delta t$. Let us now suppose that the vehicle is also equipped with a clock. The time interval with respect to the vehicle is $\Delta\bar{t}$. The latter is the proper-time interval between the events, i.e. as recorded by a single clock which is present at both events. By (3.11), we deduce

$$v\Delta t = \frac{v\Delta \bar{t}}{\sqrt{1-v^2/c^2}}$$

or

$$\bar{L} = L\sqrt{1-v^2/c^2} \qquad \text{(3.14, again)}$$

In other words, in the frame of reference with respect to which the two clocks are in motion, the distance between them is decreased. The attentive reader may object that we have not, in fact, derived (3.14) by a different route. It appears that the barred and unbarred coordinates have been interchanged in the two equations. Such an observation would be correct. However, it merely serves to indicate that one should not perform frame transformations in a completely mechanical fashion without analyzing the physics of the situation. In fact, the discrepancy is a purely notational one. In the first case L represented the rest-length, whereas in the present one it represents the contracted or relativistic length. All that this reveals is the symmetry of the Lorentz contraction with respect to two inertial frames, either of which may play the role of stationary or unbarred frame.

We may approach the matter from a somewhat different standpoint, which should provide the reader with the necessary insight to perceive that there is really nothing paradoxical in the reciprocity or symmetry of the Lorentz contraction. However, the reader should be reminded that the hypothetical length measurement experiment which we are about to discuss should not be interpreted as a method of "defining" length in the operational manner. Indeed, if magnitudes were defined by the method of measuring them, there could be no such thing as a poor or inaccurate measurement. Given a satisfactory theoretical system, our measurements should be in conformity with the concepts belonging to it, not the other way round. But, admittedly, Einstein sometimes failed to realize this in his earliest writings on relativity, which may be the source of the erroneous view that the theory of relativity is a theory of measurements. Like all scientific theories, it merely predicts the outcome of measurements.

We consider the obviously hypothetical case of a rest-frame K and a moving frame $\bar{K}$. In this case, $\bar{K}$ is nothing but a measuring rod with an observer seated at each end of it. At a certain prearranged time $\bar{t}$, each of the observers riding on $\bar{K}$ makes a chalk mark on the frame K as they pass it by. We label these marks $\bar{x}_2$ and $\bar{x}_1$, since they are $\bar{K}$ observations. The observers on $\bar{K}$ conclude that the length of their rod (vehicle) is $\bar{x}_2 - \bar{x}_1$. Moreover, this is a quantity which any observer could verify by extending a chain from one chalk mark to the other and counting the number of links. In this regard, the situation is analogous to that of the time dilation experiment with the number of links playing a similar role to that played by the number of punch holes. It is an objective result on which sensory illusion has no bearing. It is not a 'mere appearance'. However, the interpretation of the result is another matter. From the standpoint of an observer at rest in K, $\bar{t}$ is a function of position. The chalk marks were made at

$$t(\bar{x}_1) = \left(\bar{t} + \frac{v\bar{x}_1}{c^2}\right)\Big/\sqrt{1-v^2/c^2} \text{ and } t(\bar{x}_2) = \left(\bar{t} + \frac{v\bar{x}_2}{c^2}\right)\Big/\sqrt{1-v^2/c^2}.$$

In short, in K-time $t(x_1)$ is not simultaneous with but earlier than $t(x_2)$. Hence, $\bar{x}_2 - \bar{x}_1$ is too large, since the first chalk mark was made and then a period of time elapsed during which the rod moved a certain distance before the second chalk mark was made.

Now consider the situation from the reverse kinematic standpoint. $\bar{K}$ is regarded as being at rest while K is moving to the left with velocity $-\mathbf{v}$. Two observers at rest in K and a suitable distance apart simultaneously make chalk marks on $\bar{K}$ as they pass it by. At time t a chalk mark is made at x_4 and another at x_3. Suppose that these marks coincide precisely with the end-points of $\bar{K}$. Then the length of $\bar{K}$ with respect to K is $x_4 - x_3$. But with respect to $\bar{K}$, t is now a function of position. The mark at x_4 was made at

$$\bar{t}(x_4) = \left(t - \frac{vx_4}{c^2}\right) \Big/ \sqrt{1 - v^2/c^2} \text{ and } \bar{t}(x_3) = \left(t - \frac{vx_3}{c^2}\right) \Big/ \sqrt{1 - v^2/c^2}.$$

Since $x_4 > x_3, \bar{t}(x_4)$ is earlier than $\bar{t}(x_3)$, whence the result of the K-measurement was too long, even though all would agree as to the location of the chalk marks and the length of the chain extending between them.

The moral of the preceding analysis is that the determination of length cannot be divorced from that of time. One ascertains the length of an object by subtracting the values of the coordinates of the endpoints. However, such a method clearly depends for its validity on the coordinate values being determined simultaneously. It is not simply a matter of computing $x_2 - x_1$, but $x_2(t_2) - x_1(t_1)$ on the assumption that $t_2 = t_1$. In other words, what is involved in the determination of length is not simply two spatial points but two *events*. But as we have seen, the simultaneity of distant events is a frame-dependent phenomenon. It is this that accounts both for the Lorentz contraction and its reciprocity. No contradiction is involved, not because the effect is only apparent but because the analysis involved four distinct pairs of events.

We may now consider some other kinematical consequences of the Lorentz transformation which involve nothing more than simple arithmetic. In the first place, it should be noted that it places a theoretical limit on possible relative velocities. Specifically, in virtue of the ubiquity of the factor $\sqrt{1 - v^2/c^2}$, it follows that if $v > c$, lengths and durations would acquire imaginary values to which no physical interpretation could be given. In particular, the equations become singular or indeterminate when $v = c$. Thus, according to special relativity, the velocity of light is not only a universal constant, but moreover is the upper bound of all physical velocities as measured with respect to inertial frames of reference. Conversely, it is also noteworthy that for velocities which are very small compared with c, the velocities of everyday life, including even those of space travel, v^2/c^2 may be ignored and the Lorentz transformation becomes effectively identical with the classical Galilean transformation. This explains why relativistic effects went undetected for so long and continue to defy our intuition.

Another striking feature of special relativity kinematics pertains to the transformation law for velocities. Consider two frames K and $\bar{K}$ in the standard configuration. Suppose that a particle has the velocity $\bar{u}$ in the $+\bar{X}$-direction with respect to $\bar{K}$. What is its velocity with respect to K? The classical and intuitive answer, of course, is $\bar{u} + v$, which is the Newtonian law for the composition of velocities. On this hypothesis, the composition of velocities is an additive operation. We may derive the Einsteinian counterpart in a very elementary manner by employing the inverse Lorentz transformation:

$$x = \frac{\bar{x} + v\bar{t}}{\sqrt{1 - v^2/c^2}} \tag{3.15}$$

$$t = \frac{\bar{t} + \dfrac{v\bar{x}}{c^2}}{\sqrt{1 - v^2/c^2}}.$$

Dividing the first of (3.15) by the second, we obtain

$$u = \frac{x}{t} = (\bar{x} + v\bar{t}) \Big/ \left(\bar{t} + \frac{v\bar{x}}{c^2}\right). \tag{3.16}$$

Now divide and multiply (3.16) by $\bar{t}$ to obtain

$$u = (\bar{x}/\bar{t} + v) \Big/ \left(1 + \frac{v\bar{x}}{c^2\bar{t}}\right)$$

or

$$u = \frac{\bar{u} + v}{1 + \dfrac{\bar{u}v}{c^2}}. \tag{3.17}$$

In words, the resultant of the composition of two velocities is not a velocity which is the arithmetic sum of the given velocities but a quantity which is less than the latter by an amount which is a function of the given velocities.

To many, the last result would be ground for protesting that the theory of relativity is incompatible with the laws of arithmetic. This, of course, would be tantamount to declaring relativity theory to be logically inconsistent. The truth of the matter is that (3.17) is not a claim about the addition of numbers but about the combining of physical quantities—namely, velocities. Like its Newtonian counterpart, it is a falsifiable physical hypothesis. In a somewhat different vein, the physicist and founder of operationalism, Bridgman, expressed his own doubts about the propriety of this law on the ground that it involves the mixing of two different methods of determining velocity, one which involves the differential operator d/dt, and the other involving $d/d\bar{t}$. However, it would not be difficult to settle the matter by measuring the velocity of a particle by means of two radar devices, one attached to the earth and the other in a rapidly moving rocketship.

It should also be noted that (3.17) is fully consistent with the limiting role of c. In fact, it even provides added insight into this perplexing fact about the physical world. One would suppose that if an object were moving with a speed close to c, it would be a simple matter to select an inertial frame with respect to which the speed of the object would exceed c. However, an application of (3.17) reveals the impossibility of this. Suppose that we wish to compute the resultant of the composition of $(\frac{4}{5})c$ and $(\frac{4}{5})c$. Firstly, we may simplify the arithmetic in virtue of the fact that the numerical value of c depends on an arbitrary choice of units. So we follow the widespread stratagem in relativistic computations of setting c equal to unity. Then by (3.17) we have:

$$\frac{\frac{4}{5} + \frac{4}{5}}{1 + \frac{16}{25}} = \frac{40}{41},$$

which is still less than unity, the velocity of light. Finally, the reader should satisfy himself that the Lorentz velocity transformation (3.17) preserves the invariance of c.

No introductory account of relativistic kinematics would be complete without some discussion of the *twin paradox*. It is deeper and rather more baffling than the clock paradox and has generated a vast amount of literature both in physics and philosophy. In fact, many serious thinkers have supposed it to be symptomatic of a serious flaw in the logical or conceptual structure of special relativity. Although it may be formulated with reference to clocks, it is usually cast in a form which refers to human beings. A human being, in so far as he possesses a reasonably regular heart rate, may play the role of a biological clock. For purposes of the argument, his heart may be idealized as a perfectly isochronous system. Now since time dilation is a kinematic phenomenon which has nothing to do with the manner in which actual clocks are constructed, it follows that relativistic laws must apply to biological clocks to the same extent and in the same way that they are applicable to light-clocks, atomic clocks or whatever.

Suppose that twins somewhere on the surface of the earth have synchronous heart rates. One of the twins, whom we shall designate by T_m, embarks on a long space flight at a velocity which is a significant fraction of the velocity of light. Let us suppose that each possesses a device whereby he may measure the heart rate of the other. Special relativity predicts that each will find that the other's heart rate has slowed down in comparison with his own. So far, there is nothing peculiar in this situation from the perspective of one who has succeeded in resolving the clock paradox. However, there is more to the situation than this. The theory of special relativity actually entails that when T_m returns to earth, he will find, depending on the length of the round trip, that he is considerably younger than his earth-bound brother, whom we shall call T_r. Whereas T_m is still a vigorous young space traveller, T_r is now spending a good deal of time in his rocking chair while enjoying his retirement.

The mathematical account of this phenomenon is quite simple. We adopt the standard notational convention of labelling proper-time by τ. Then the relationship between a proper-time interval and a coordinate-time interval is given by

$$\Delta t = \frac{\Delta \tau}{\sqrt{1 - v^2/c^2}}. \tag{3.18}$$

Let us place T_r at the origin of S. T_m leaves the origin at t_0, moving in the $+X$-direction with velocity $\mathbf{v}$. He arrives at his destination $x_1 = vt_1$, which could be the location of a distant star, where t_1 is the coordinate time in S. The clock on the space ship will, however, record the time of arrival as the proper-time $t_1\sqrt{1 - v^2/c^2}$. T_m immediately sets off on the return leg of his journey in space, travelling at the velocity $-\mathbf{v}$ along the X-axis. When he reaches his original point of departure, the coordinate time, which is also a proper-time for T_r, is now $t_2 = 2t_1$. However, the space-ship clock will now register $2t_1\sqrt{1 - v^2/c^2}$. In short, T_m's clock is found to be lagging behind that of his twin brother when the two are brought together for direct comparison. *Mutatis mutandis*, the space traveller has aged far less than his twin brother. The reader will begin to appreciate the reasons for my insistence on the reality of kinematic phenomena.

So far, we have established one of the truly remarkable deductive consequences of the postulates of special relativity. It would be pointless to attempt to explain it away, since it corresponds to many highly confirmed results of experimental physics. However, this dramatic feature of physical reality is still not what is referred to when one speaks of the twin paradox. The unforeseen and peculiar is not thereby the paradoxical. The paradoxical aspect of the situation is the following. Since motion is relative, it follows that it should be possible to redescribe the situation with respect to the rest-frame of T_m, i.e. the space ship, in such a way that the circumstances of the brothers are reversed. That is to say, that one should be able to treat the space ship as a stationary frame and the earth as a moving one, so that T_r will initially recede from T_m in the $-X$-direction. The earth will subsequently stop and return to the stationary space ship where the twins once more encounter each other. But by parity of reasoning, since T_m has been stationary he should be much older than T_r. At last we are confronted not merely with a peculiar physical fact but with a genuine contradiction. It is this that is the essence of the twin paradox.

Proposals for resolving the paradox are legion. Perhaps the most radical of these is to suggest that since special relativity yields a contradiction, it is simply a false theory that one would do well to abandon. Scarcely less radical than this is the suggestion that the aether should be reinstated in the context of the relativistic formalism. The gist of this neo-Lorentzian viewpoint is that there must be a genuine asymmetry between the two brothers, which derives from their being in different states of motion relative to the aether. This amounts to both praising and violating the theory at the same time. A rather more reasonable approach is to point out that there is a real physical difference between the respective situations of the two brothers. In particular, T_m was subjected to a huge acceleration at the beginning of his trip, a second one in coming to rest at his stellar destination, a third when starting on the return journey and a fourth when finally coming to rest on the earth. In the meantime, T_r was subjected to no such accelerative forces. It is argued that the objective difference between T_m and his brother is simply the aforesaid accelerative forces which the one experiences and the other does not. This, it is held, should account for the different rates of aging. Furthermore, it is claimed that special relativity is restricted to the treatment of inertial or unaccelerated frames of reference and is simply incompetent to deal with accelerated motion, which is the province of general relativity.

Although the last proposal is more moderate and even to some extent on the right track, it is, nevertheless, seriously flawed. Firstly, it is an empirical fact that differently constructed clocks react differently to accelerative forces. In particular, it is generally agreed that an atomic oscillator behaves in appreciably the same way as an ideal clock, which is to say that any effects of acceleration on it are negligible. Indeed, we shall subsequently find that this is somewhat crucial within the theoretical framework of general relativity. Now the twin paradox may involve any sort of clock independently of the latter's construction. It is, therefore, a kinematic phenomenon. The invocation of accelerative *forces* is simply inappropriate in the context. That, of course, is not to say that acceleration may not be treated from a purely kinematic standpoint. However, even if one grants that acceleration may be treated kinematically there is a more serious difficulty to be faced. The discrepancy between the two clocks is a function of the total distance or total elapsed time. Hence, by choosing a sufficiently great distance one could render the putative effect of acceleration negligibly small. That is to say, that a trip to the sun and back at a given velocity would involve roughly the same accelerations as a trip involving that same velocity to a distant star and back, yet the theoretical prediction is that the discrepancy in aging would be very much greater in the latter case than in the former. Thirdly, the claim that special relativity is incompetent to handle accelerated motion is plainly false. Special relativity is no different from classical physics in this regard. Both are, in a sense, tied to non-accelerating frames but both routinely deal with accelerated motion. Admittedly, the kinematics of accelerated motion is considerably more complex in relativistic contexts; however, there is no significant difference in principle. Finally, I do believe it to be true that a conceptual resolution of the twin paradox is more readily attainable within the framework of general relativity. Nevertheless, within its own prescribed limits, special relativity is a complete theory, which is capable of handling all of the problems, including the twin paradox, which fall within its scope.

I shall now attempt to sketch the outline of the solution to this problem. The full resolution would require a rather detailed analysis of the relativistic Doppler effect.

However, I do not wish to swamp the reader with the details of physical theories. In any case, I believe that the following account should suffice to reveal the nature and direction of the solution, which is all that is needed to satisfy our present philosophic purposes.

I remarked that the last proposal is on the right track. Specifically, it is indeed true that a genuine asymmetry exists between T_r and T_m. Whereas T_r remains at rest in a *single* inertial frame during the entire period of the journey, T_m changes from one inertial frame to another once at the beginning, twice at the midpoint and once more at the conclusion of his trip. From the standpoint of space-time geometry, which will be discussed in the next section, T_m's path would look very different from that of T_r. The difference is an intrinsic one, which could not be effaced by a coordinate transformation. The most perspicuous way of explaining this difference is to compare the respective situations of T_m and T_r from the standpoint of the Lorentz invariant and hence neutral proper-time, which is the natural parameter for a space-time trajectory. The account may be simplified by making the following harmless assumptions. T_m starts his trip by jumping aboard the space ship, which is already passing the earth at full velocity. On reaching his destination, he jumps from his own vehicle to another which is already heading towards the earth at full velocity. Finally, he jumps to earth as the second space ship passes it by. In this way, we may limit our account to just three inertial frames: one for T_r and two for T_m.

The key to the analysis consists simply in the correct appreciation of the meaning of proper-time and of its relationship to coordinate time as expressed in (3.18). In particular, this formula serves to connect time intervals in just two frames of reference, both of which are inertial and one of which is endowed with a clock which is present at both end points of the measured interval. For T_r the total elapsed time is $\Delta\tau_r$. Since T_r is in a single inertial frame, using a single clock, namely himself, which is present at both events, $\Delta\tau_r$ is a proper-time interval. Moreover, T_r will assign an interval of $\frac{1}{2}\Delta\tau_r$ to the first leg of T_m's trip and an equal interval to the second leg. However, neither of the latter is a proper-time interval. Now we turn to T_m. The interval from earth to star is a proper-time interval for T_m. In order that we may compare τ_m with τ_r, we shall express the former in terms of the latter. Accordingly, we calculate that the proper-time for the first part of T_m's trip will be $\frac{1}{2}\Delta\tau_r\sqrt{1-v^2/c^2}$. By the same token, the proper-time for the second half of T_m's trip will also be $\frac{1}{2}\Delta\tau_r\sqrt{1-v^2/c^2}$. The only way in which T_m can possibly assign a duration to the round trip is by adding the durations of the two-one-way trips. Accordingly, he must assign $\Delta\tau_r\sqrt{1-v^2/c^2}$ to the total duration. Since all of the quantities which enter this calculation are proper-time intervals, T_r and T_m would agree as to their propriety. That is to say, that T_r and T_m would be bound to agree that T_r had experienced much more proper-time than T_m. They would be poignantly aware of the peculiarity of their relationship but recognize it as a self-consistent consequence of special relativity. The important point is that they would not be in disagreement as to who had actually aged less. The upshot of this entire discussion is that time dilation is a kinematic phenomenon which is sometimes symmetrical and sometimes asymmetrical but which is, in either case, entirely real. By that, I mean that philosophers may not brush it aside as a mere appearance.

Four-Vector Formalism and Space-Time Geometry

No new physics will be introduced in this section. Its purpose is rather to gather together the various kinematic results of the previous section and present them in a more integrated fashion in the context of four-dimensional space-time geometry. Although it would have been possible to have initiated the discussion of special relativity in this manner and, consequently, to have avoided our rather laborious and piecemeal approach to relativistic kinematics, such a procedure would have been paedagogically unsound. It would have had the virtues of brevity, elegance and considerably less repetition, but at the cost of serious concealment. The formalism which will be introduced in this section was deliberately designed to present relativity theory in an economical style. It achieves this end with so high a degree of success that many of the real conceptual problems which we were obliged to confront and resolve simply fail to be apparent. A natural formalism is desirable so long as those who employ it are aware of what lies behind it. In particular, it seems to me to be essential that the student of the philosophical foundations of the theory of relativity must first become aware of the counterintuitive and even paradoxical aspects of the theory, learn how to come to conceptual terms with them, and only then have recourse to a formalism which tends to suppress them. Indeed, only then can the real virtues of the formalism be appreciated.

In the context of contemporary relativity theory, Einstein's original paper seems almost classical and old-fashioned. In particular, Einstein did not drop the concepts of space and time but rather undertook their relativistic reinterpretation. The later approach derives from the contributions of his older contemporary and teacher, Hermann Minkowski. It was he who first clearly perceived that special relativity may be more adequately viewed against the background of four-dimensional spacetime. In short, he realized that the various relativistic phenomena should be coordinated with a new kind of geometry.

The initial psychological impact of special relativity is to suggest that the physical world is deprived of much of the absolute objectivity that is ascribed to it by classical mechanics. In particular, lengths and durations or space and time are now deprived of their absolute, frame-independent significance. However, while the invariants of the Galilean group are indeed lost, the set of Lorentz transformations is also the realization of a group, known as the *Lorentz group*. Accordingly, one should expect to discover new objective features of reality through a consideration of the invariants of the Lorentz group.

We have already noted that the wave-front equation is a Lorentz covariant expression. A somewhat similar but more general expression which also has the property of Lorentz covariance is

$$s^2 = x^2 + y^2 + z^2 - c^2t^2 \qquad (3.19)$$

This expression defines the *space-time interval* between two events. It is not merely the case that (3.19) is covariant under a Lorentz transformation. It also turns out that s is a scalar quantity, analogous to ordinary distance. Thus, although the spatial separation and temporal separation between events are frame-dependent, the space-time separation is absolute. Metaphorically, space and time may be split up in various ways according to the frame of reference which one selects. With respect to one frame of reference there will be more space and more time between events, while with respect to another there is less space and less time between them. The situation is somewhat analogous to that of two coordinate systems which are distinguished by a pure rotation. With respect to one such system there will be more X-distance and less Y-distance, while with respect to the other there will be less X-separation and more of Y. However, the total distance between the points will be the same in all cases.

By the same token, it emerges that one may regard a Lorentz transformation as a rotation in four-dimensional spacetime. With respect to one set of coordinates there is more spatial separation between events and more temporal separation, with respect to another set there is less, but in all cases the total space-time separation is invariant. Thus it would seem that the fundamental geometric reality is neither space nor time but a combination of the two. Such considerations occasioned the famous remark of Minkowski in a paper written in 1908 in which special relativity was given its four-dimensional formulation. "Henceforth space by itself, and time by itself, are doomed to fade away into mere shadows, and only a kind of union of the two will preserve an independent reality" (1908, p. 75).

Minkowski gave geometrical substance to this poetic comment by constructing a formalism in which (3.19) is automatically Lorentz covariant, much as a spatial vector is covariant with respect to the group of inhomogeneous rotations. In the Minkowski formalism, time is treated as a fourth independent coordinate in addition to the X^i. Specifically, he sets $X^4 = ict$, where i is the imaginary number $\sqrt{-1}$, so that $i^2 = -1$. Just as Latin indices range over the three spatial coordinates, in Minkowski's notation, Greek indices will range over the four space-time coordinates. The differential form of (3.19) may then be represented by the expression

$$(ds)^2 = \delta_{\mu\nu} dX^\mu dX^\nu \tag{3.20}$$

(3.20) is the expression for the space-time interval. It is the *metric form* of special relativity. We now define a *four-vector* A^μ to be any quantity which transforms like the dX^μ of (3.20). Thus a four-vector is a geometric object with respect to a four-dimensional, space-time transformation, which is formally analogous to the familiar three-vector with respect to a spatial rotation. However, it should not be forgotten that this formal similarity is achieved by making dX^4 an imaginary quantity. Thus, even in special relativity there is a sharp distinction between space and time. There are other significant features of a four-vector which distinguish it from a three-vector. The *norm* or length of a four-vector is given by

$$(A^\mu)^2 = (A^1)^2 + (A^2)^2 + (A^3)^2 - c^2(\dot{A}^4)^2 \tag{3.21}$$

It should be noted that unlike the case of a three-vector, the last expression may be zero, even though the individual components are non-vanishing. If the norm of a four-vector is greater than zero, i.e. dX^i is the dominant quantity, the vector is said to be *spacelike*. If it is negative, i.e. dX^4 is the dominant quantity, the vector is said to be *timelike*. Finally, if the norm is zero, then it is a *null* or *lightlike* vector, which corresponds to the condition: $(dX^i)^2 = c^2(dt)^2$, which is just the condition for light-propagation. In short, the path of a light-ray in this four-dimensional context is always represented by a null vector. However, it must be remembered that we are now referring to space-time paths and not ordinary three-dimensional trajectories. It should be noted that the spatial components (1, 2, 3) of a four-vector transform among themselves like those of a three-vector while the time-component of a four-vector transforms like a spatial scalar. When a four-vector is spacelike there will always exist a Lorentz transformation under which the time-component vanishes, whereas when a four-vector is timelike there will always exist a transformation under which the spatial components vanish. However, there is no allowable transformation under which a spacelike vector is transformed into a timelike vector or conversely.

These aspects of the formal behaviour of four-vectors reflect several of the fundamental ideas at which we arrived somewhat laboriously in the previous section. For example, it was explained how a distinction must be drawn between the coordinate-time separation between events $E_1(t_1)$ and $E_2(t_2)$ and their proper-time separation. From the Minkowski standpoint, there is a space-time separation ΔX^μ between them. ΔX^μ may be subjected to a Lorentz frame-transformation to a frame $\bar{K}$, in which the two events are located at the same point in space, which also happens to be the location of a clock which measures the proper-time interval $t_2 - t_1 = \Delta\tau$. This simply corresponds to a transformation of the space-time interval under which ΔX^i vanishes. Thus, the space-time separation must be a timelike separation. On the other hand, if the coordinate time interval between them had been so small and the spatial separation so great that it would have been impossible to have a single clock at both events without requiring it to exceed the velocity of light, then the separation would have been spacelike. The two events would then be simultaneous with respect to an appropriate reference frame, which could be reached via a Lorentz transformation. Some writers refer to the relationship between two such events as that of *topological simultaneity*.

Thus the immediate physical significance of the distinction between spacelike and timelike separations is as follows. When the separation is spacelike, there is a frame of reference in which the events are contemporaneous and, therefore, causally independent. When the separation is timelike, there is a Lorentz invariant proper-time interval between the two events so that they are absolutely successive and, therefore, possibly stand in some causal relationship. Certain philosophical critics of special relativity have argued that the relativistic treatment of time is incompatible with the principle of causality in so far as it permits the reversal of the temporal order of two events so that there would be certain observers in the universe who would perceive an effect's preceding its cause. This idea has even been exploited by science-fiction writers to suggest that if one travels with a sufficiently great velocity one may travel backwards in time. Such suggestions have been generated by insufficient attention to the mathematical details of the theory. Specifically, if two events stand in a causal relationship, their separation must be timelike. That is to say, that they are so related that a causal influence may be propagated from one to the other at a velocity not exceeding that of light. More formally expressed, if two events are causally related, the interval between them is timelike, and the four-vector formalism prohibits the transformation of a timelike into a spacelike vector.

In the space-time context, the fundamental role of proper-time comes to the forefront. Firstly, it should be noted that proper-time is definable in terms of the space-time interval as follows:

$$-c^2(d\tau)^2 = (dx^i)^2 - c^2(dt)^2, \tag{3.22}$$

from which we may immediately derive the already familiar result

$$d\tau = (1 - v^2/c^2)^{1/2}\, dt, \tag{3.23}$$

which is Lorentz invariant. In other words, from the standpoint of space-time geometry, τ is a scalar; its value is unaffected by a space-time or Lorentz transformation. Moreover, we note that the proper-time is the natural parameter for the interval between successive events in the life of a single material object. Such a succession of events is called the *world-line* of the object. In classical physics, it is natural to depict the trajectory or path of an object in terms of the distance $s(t)$ expressed as a function of time. In relativistic physics, the natural parameter for the trajectory of a particle is its proper-time τ. The utility of τ, of course, is that it is the same for all inertial observers.

The reader may probably be wondering why relativity physicists do not simply drop coordinate time in favour of the better behaved proper-time. The reason is that although proper-time may be utilized in many relativistic contexts, there are circumstances which are better adapted to the use of coordinate time. Think, for a moment, of a reference frame as a gigantic scaffolding made of rigid rods with a clock at each point of intersection. The coordinate clocks are all synchronous with respect to that frame. Suppose that an object travels from one part of the frame to another and one wishes to determine the duration of its journey. The elapsed time may be ascertained directly from the readings of the clocks at the endpoints of the journey, i.e. $t_2 - t_1$. By contrast, the proper-time of the journey corresponds to the time indications of a clock which travels with the moving object. Consequently, it depends on the actual path which is followed and not simply on the endpoints of the latter. Technically, this difference is expressed by saying that coordinate time is integrable whereas proper-time is not. Indeed, to say this is, in effect, to dispose of the twin paradox in a single sentence.

One final aspect of the character of four-vectors needs to be remarked. When dealing with classical physics we found it necessary to make a sharp distinction between a frame-transformation and a pure coordinate transformation. In particular, we noted that a three-vector is only a vector with respect to an inhomogeneous rotation. Thus, it is not a vector with respect to the fundamentally important Galilean transformation which is central to the principle of classical relativity. The great virtue of four-vectors in their application to physics is that this distinction no longer holds. That is to say, that from a formal standpoint, a four-vector *is* a vector with respect to the group of Lorentz transformations. In other words, when time plays the role of a fourth coordinate, a transformation to a different reference frame has the status of a coordinate transformation. In effect, such a frame transformation is represented by a rotation of a set of space-time coordinate axes. Thus, whereas the Lorentz transformation may be viewed as a geometric transformation, the Galilean transformation is not normally viewed in that way. That is not to say that a geometric reinterpretation of classical relativity is impossible. However, any such interpretation turns out to be somewhat artificial and unnatural. However, to clarify these matters, we should now pause to give an explicit geometric interpretation to spacetime and the Lorentz transformation.

It has already been noted that the space-time interval and the four-vector are analogues of the familiar spatial interval and three-vector, respectively, which are defined on a Euclidean manifold, E_3. In fact, we may now go one step further by introducing the notion of a *space-time manifold*, which, again, is an analogue of three-dimensional Euclidean space. Indeed, from the standpoint of strict mathematical development, the space-time interval and the four-vector are concepts which presuppose the existence of such a manifold. Moreover, although spacetime may be treated, in the context of special relativity, as a mere formal convenience, it turns out to be a conceptual necessity in the deeper context of general relativity. Consequently, it is highly desirable to accustom oneself to the space-time way of thought as soon as possible, if only to acquire the necessary conceptual preparation for the comprehension of the latter theory. Accordingly, we shall treat the four-vector not as a mere formal device but as the appropriate geometric object for representing physical processes in spacetime. Like Minkowski, we shall treat spacetime as the fundamental reality of which space and time are merely aspects. Accordingly, we must proceed to familiarize ourselves with certain of the basic ideas of space-time geometry.

While there are a few similarities between the two manifold geometries, there are also profound differences. In the first place, spacetime is not merely a generalization from three to four dimensions. Although space and spacetime are both flat and infinitely extended continua, space is a manifold of points whereas spacetime is manifold of "events", the latter being the commonly employed name for a point at a time. The most profound formal disanalogy between them is that the metric function of space is given by

$$(ds)^2 = (dx^1)^2 + (dx^2)^2 + (dx^3)^2,$$

whereas the metric of spacetime is given by

$$(ds)^2 = (dx^1)^2 + (dx^2)^2 + (dx^3)^2 - c^2(dt)^2.$$

The important distinction here is not the difference of dimensionality but the difference of *signature*. No matter what type of coordinate system happens to be chosen in which to express the metric of space, the metric coefficients will all have positive signs. Thus, one of the invariant or intrinsic properties of Euclidean space is its signature, which we represent by $(+ + +)$. A metric of this kind is said to be *positive definite*. In contrast, the invariant signature of Minkowski spacetime is $(+ + + -)$ or, as we shall see in later chapters, $(+ - - -)$. A signature of this kind is called *indefinite*. A pertinent difference between the two types of metric is that whereas the separation or interval between two non-coincident points is always positive in the former case, in the latter, the indefinite metric, it may be positive, negative or even zero. In fact, such a metric is sometimes called a *pseudo-metric*, since it violates certain of the basic properties which are normally imposed on metric functions. Thus, it would be a profound mistake to view Minkowski spacetime as a four-dimensional Euclidean manifold. However, in virtue of its flatness as well as its global topology, it is often described as pseudo-Euclidean. Minkowski deliberately down-played the non Euclidean character of spacetime by introducing the imaginary factor i. As we have seen, this enables one to represent the metric by means of a four-dimensional Kronecker symbol. However, there is no doubt that this is no more than an artifice which, although sometimes helpful, disguises the very real differences between the geometry of Euclid and that of Minkowski. For our part, we shall employ both the imaginary time axis ict and the real time axis ct on the basis of mere computational convenience. However, it is important to recognize that there is, in any case, nothing imaginary about the fourth or temporal dimension. Incidentally, it should also be noted that the time axis is represented by ct in preference to t merely for the sake of dimensional homogeneity, since the product of velocity and time is length. Thus, an interval of time Δt may be represented by $c\Delta t$ which is simply the distance traversed by a light-ray in the interval Δt. An appropriate unit along the time axis might, for example, be the *light-second*, which is the distance traversed by a light-ray during a one-second interval.

Wherever feasible, the use of diagrams to illustrate geometric ideas is invaluable as an aid to intuition. Unfortunately, no single type of space-time diagram is ideally suited to all one's needs. The most frequently employed one is the Minkowski diagram, which is a great intuitive aid but fairly burdensome when employed to express quantitative results. In addition, there is the Loedel diagram, so called after its Argentinian inventor, which yields the correct quantitative results in the easiest possible fashion, but which seems to be of inferior intuitive value for representing the qualitative features of space-time geometry. Consequently, in what follows I shall make use of each of them.

We begin with the Minkowski diagram of Fig. 3.3. It will serve to represent the fundamental trichotomy among timelike, spacelike and null vectors. The X-axis

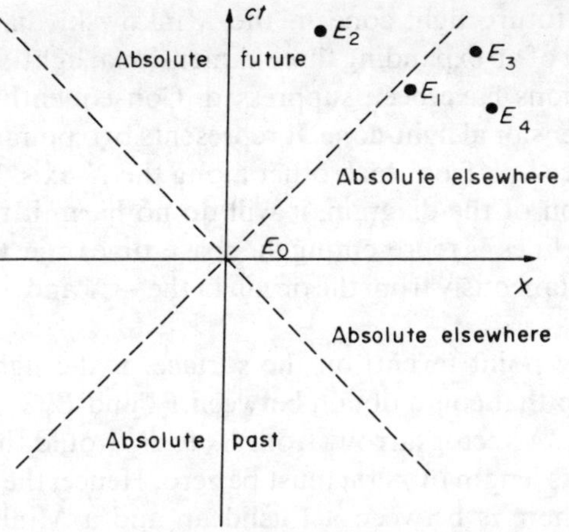

FIG. 3.3. The Minkowski light-cone

represents one of the three spatial directions while the *ct*-axis represents the temporal direction. For obvious reasons, two of the three spatial directions have been suppressed. These coordinates are assumed to be attached to an inertial or Lorentz frame of reference. Sometimes an inertial observer is located at the origin of the coordinate axes. This is, in effect, a stylistic decision. It is quite innocuous so long as it is remembered that no observer is required to be part of the physical situation which the diagram is intended to depict. In keeping with our general philosophic outlook—namely, *scientific realism*—we assume that those properties of the universe which are of interest to the physicist are independent, except trivially, of the existence of sentient observers. What is needed at the origin of the coordinates is an event E_0. In particular, we shall assume that E_0 is a light-pulse which immediately expands in all directions. The broken lines in the upper half of the diagram represent the space-time history of this light-pulse and constitute the *future light-cone* of E_0.

This novel terminology requires some explanation. Let us revert to the analogy of an expanding circular ripple on the surface of a pond, which we may idealize as a two-dimensional surface. One may treat it as a plane of simultaneous events relative to an observer who is at rest with respect to the pond. A circular ripple at time *t* is a one-dimensional manifold of simultaneous events. Such a manifold could be represented graphically in the form of a circle. If we add all of the interior points of the circle, we have a two-dimensional circular surface. To represent the history of the ripple as it expands from an idealized point location, one would need a separate diagram for each moment in this history. Suppose that we have drawn a very large number, in theory non-denumerably many, of such diagrams which have then been stacked on top of each other concentrically and in temporal sequence. One would then have what amounts to a three-dimensional diagram representing the history or world-tube of the ripple. It should be noted that this diagram or model is cone-shaped. A line passing through the axis of the cone plays the role of the time-axis. Hence, we see that an expanding circle in space may be represented by a cone in spacetime. We must now make the effort to think of an expanding three-sphere in space as represented by a cone in four-dimensional spacetime. This is precisely the significance of the future light-cone in the Minkowski diagram. It is understooa to represent the history of an expanding three-dimensional light-sphere. Of course, as we said earlier, two dimensions have been suppressed. Consequently, the diagram is, in fact, a model of a two-dimensional light-cone. It represents two points on the surface of the light-sphere which are receding from each other along the X-axis. While this is the essentially correct interpretation of the diagram, it will do no harm if the reader should prefer to interpret the broken lines as representing the space-time trajectories of two photons which have departed simultaneously from the origin in the $+X$ and $-X$-directions, respectively, at the time of E_0.

E_1 is an arbitrary point (event) on the surface of the light-cone. The only possible physical relationship that could obtain between E_0 and E_1 would be by means of a light signal from E_0 to E_1. A vector (arrow) from E_0 to E_1 would therefore be lightlike or null. That is to say, that its length or *norm* must be zero. Hence, the reader will appreciate how great a difference there is between a Euclidean and a Minkowski manifold! E_2 is an arbitrary event inside the light-cone of E_0. Any such event stands in a possible causal relationship to E_0. A vector from E_0 to E_2 is timelike. In particular, any line from E_0 to E_2 such that any vector tangent to it is timelike would be a possible trajectory or world-line of a material particle travelling at less than the speed of light. The totality of possible events in the future light-cone of E_0 constitutes E_0's *absolute future*. By this, we mean that since the light -cone is a Lorentz invariant, which is simply to say that the velocity of light is constant with respect to a Lorentz transformation, no physically allowable *boost* of the reference frame could possibly alter the future possibilities of E_0. A word of caution may be called for at this point. It is obvious that a change of reference frame will indeed affect the actual outcome of one's future. By taking an aircraft instead of a slower form of transportation, you will witness certain events which you would otherwise have missed. However, any event which you are able to witness in virtue of your air trip belongs absolutely to your possible future regardless of what choices are actually implemented.

E_3 and E_4 are arbitrary events which lie outside the light-cone of E_0. A causal connection between E_0 and E_3 or E_4 would have to be transmitted with a velocity in excess of c. That is to say, that no such connection is physically possible. A vector from E_0 to E_3 or E_4 is spacelike. Equivalently, E_3 and E_4 are in spacelike separation from E_0. It is sometimes said that such events belong to the *absolute elsewhere* of E_0 or to the *conditional present* of E_0. By this, it is meant that a Lorentz boost or rotation of the Minkowski coordinate system could result in E_3's lying on the X-axis along with E_0. E_0 and E_3 would then be in the relationship of simultaneity relative to the new frame of reference. Similar considerations apply to E_4. It should, of course, be noted that E_3 and E_4 are not distinguished by any peculiar characteristic. In particular, it is quite possible that their mutual separation be timelike. That is to say, that the difference of two spacelike vectors may be timelike. By parity of reasoning, the lower half of the Minkowski diagram represents the past light-cone of E_0. Any event within the lower lobe of the light-cone would be a possible cause of E_0. Hence, once again, the separation between E_0 and any such event would be timelike.

Clearly, Minkowski spacetime has a more complex structure than that of Euclidean space, although admittedly the comparison is not particularly apt since it would be more appropriate to compare it with Newtonian spacetime. Every punctal event in spacetime has a future and past light-cone. Moreover the structure is absolute in the sense of there being no Lorentz mapping which could modify it. The partitioning of pairs of events into timelike, spacelike and lightlike pairs is fixed. A spacelike vector cannot be transformed into a timelike one or conversely. Similarly, a timelike vector cannot be made lightlike or conversely. Thus, we see that the oft-repeated statement that every observer splits spacetime into space and time in his own way, although correct, could be seriously misinterpreted.

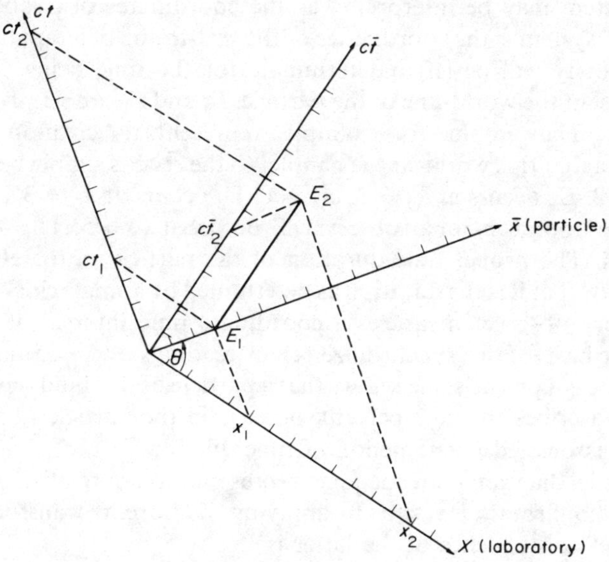

FIG. 3.4. Loedel diagram of a Lorentz transformation

As we have earlier remarked, the Minkowski diagram is well suited for representing qualitative properties in a reasonably perspicuous manner. However, for the purpose of representing the quantitative features of a Lorentz transformation, which would involve the rotation of a Minkowski coordinate system *plus* a change of scale, the Loedel diagram is to be preferred. Figure 3.4 is a Loedel representation of two Lorentz frames which are in motion relative to each other. It is quite important to understand how the diagram is constructed. We shall begin with the recipe for such a construction. Draw orthogonal coordinate axes similar to standard Cartesian coordinates but label one of them "X" and the other "$c\bar{t}$". Now rotate them through the acute angle θ and label the resulting axes "$\bar{X}$" and "ct". In Fig. 3.4, we have chosen $\theta = 53°$. The significance of this choice will emerge later. It follows by elementary geometry that the coordinates of an arbitrary point with respect to the two systems will satisfy the relation

$$\bar{X}^2 + c^2t^2 = X^2 + c^2\bar{t}^2. \tag{3.24}$$

But an immediate consequence of (3.24) is

$$\bar{X}^2 - c^2\bar{t}^2 = X^2 - c^2t^2. \tag{3.25}$$

(3.25) is simply the expression for the invariant space-time interval. Thus, it may be seen that the Loedel diagram is an ingenious device for preserving the length of space-time intervals under coordinate rotations. It is important to observe how the coordinates of a given event are determined. Instead of following the usual procedure of dropping perpendiculars from the point of interest to the coordinate axes, one draws a line from the point to a given axis, which is parallel to the latter's mating axis. For example, the reader will see that the X coordinate of E_2 has been obtained by drawing a line parallel to the ct-axis, which intersects the X-axis at x_2.

The (X, ct) system may be interpreted as the coordinates of a laboratory reference frame. The $(\overline{X}, \overline{ct})$ system is the coordinates of the rest-frame of a rapidly moving particle whose actual velocity will be left indeterminate for the time being. The heavy line is intended to represent the world-line of the particle. E_1 and E_2 are simply two events in the particle's history. They could, for example, represent its creation and consequent annihilation. Note that this world-line is parallel to the $\overline{ct}$-axis simply because the particle is at rest at $\bar{x}_1 = 3$. E_1 occurs at $\bar{x}_1 = 3$, $\overline{ct_1} = 0$. E_2 occurs at $\bar{x}_2 = 3$, $\overline{ct_2} = 6$. However, with respect to the laboratory (or an observer) E_1 occurs at $x_1 = 5, ct_1 = 4$ and E_2 occurs at $x_2 = 13$, $ct_2 = 14$. The proper-time duration of the particle is, therefore, $\overline{ct_2} - \overline{ct_1} = 6$ units. That is to say, that it is the duration as determined by a single clock which is at rest at $\bar{x}_1$. The laboratory observer measures a coordinate time interval of 10 units. This is determined on the basis of two synchronized clock readings at $x_1 = 5$ and $x_2 = 13$. On the assumption that he is a physicist, he knows that a particle of this kind typically survives for only 6 units but ascribes the 66.7 percent increase in the particle's lifespan to its high velocity and the associated phenomenon of time dilation.

Now if the Loedel diagram is an adequate representation of relativistic kinematics, one should be able to confirm these results by applying the Lorentz transformation. You will recall that a direct consequence of the latter is

$$\Delta\tau = (1 - v^2/c^2)^{1/2}\,\Delta t \tag{3.26}$$

Inserting the given values of $\Delta\tau$ and Δt into (3.26) and solving for $\mathbf{v}$ yields $|\mathbf{v}| = 0.8c$. Hence, the Loedel diagram is a correct representation, provided that the relative speed of the particle is indeed $0.8c$. Now the laboratory observer would compute the particle's velocity by dividing the distance by the time. $(13–5)/(14–4) = 0.8$. But $0.8 = \sin 53°$· It follows that the Loedel diagram consistently represents the Lorentz transformation on condition that the sine of the angle between the two sets of coordinate axes corresponds to the relative velocity. Accordingly, you may now add to the recipe for constructing Loedel diagrams the final instruction to choose $\theta = \sin^{-1} v$, where v is expressed as a fraction of c.

The reason for our having devoted so much attention to the Loedel diagram was twofold. Firstly, in the mathematical analysis of any physical situation, a visual representation is almost always an invaluable conceptual aid. The second reason is more profound. Despite the existence of rigorous mathematical derivations of the Lorentz transformation, the newcomer to special relativity may continue to suffer from the suspicion that relativistic kinematics is logically contradictory. In offering a graphical representation of the Lorentz transformation, which is drawn to scale, we have provided a physical *model* of the relevant relations. Were there an inconsistency in the theory, no such model would exist. It is impossible to construct a logically inconsistent drawing. The reader should, therefore, conclude that although the kinematics may be unexpected and even counterintuitive it is fully within the realm of the physically possible.

Relativistic Mechanics

When discussing classical mechanics, we noted that it was convenient to formulate physical laws in standard vectorial form. This ensured that the laws would be in conformity with the homogeneous and isotropic character of Newtonian or Euclidean space. However, it was also learned that the vectorial form of a basic law could not ensure that the law would be in conformity with the principle of classical relativity, since vectorial laws are not necessarily Galilei covariant. Thus, in constructing a possible candidate for a law of nature in classical physics there are two distinct formal conditions to impose on it. It should be added, of course, that these conditions are necessary but not sufficient. That is to say, that a Galilei covariant law is not necessarily a factually correct one. There are infinitely many possible laws of nature which have the correct covariance properties but the incorrect physical content.

Although relativistic mechanics is a more complicated theory than classical mechanics there is one respect, at least, in which the formulation of a relativistically correct mechanical theory is easier than that of its classical counterpart. With the aid of the Minkowski four-vector it is possible to formulate laws which are not only kinematically admissible, i.e. in conformity with the features of relativistic spacetime, but which are also in conformity with Einstein's principle of special relativity, i.e. they are automatically Lorentz covariant. In fact, the two requirements now turn out to be equivalent in the sense that any formula which satisfies the one must automatically satisfy the other. That is to say, that a Lorentz transformation may be regarded formally as a type of pure coordinate transformation. Hence, if we express a law in four-vector form we may be assured that if it holds in one frame of reference it will also hold in any other frame which is in uniform motion with respect to the first one. Another purely heuristic consideration which is of great aid in constructing relativistic theories is that classical laws have been found to be approximately correct at low velocities. Accordingly, the natural procedure to follow in constructing a relativistic theory of mechanics is to make a Lorentz covariant generalization of a classical law, i.e. a four-vector formulation of a classical law. However, again it should be stressed that the four-vector form is merely a necessary condition. We now proceed to sketch the principal ingredients of special relativity mechanics.

We should begin by establishing certain fundamental results from classical mechanics, which will provide a basis for subsequent generalization. The reader may recall that the *work* done by a force which acts on a particle is defined as the scalar product of the force and the displacement of the particle. Expressing this in differential form, we have

$$dW = \mathbf{F} \,.\, d\mathbf{s}. \tag{3.27}$$

This may be rewritten as

$$dW = m\frac{d\mathbf{v}}{dt} \,.\, d\mathbf{s}. \tag{3.28}$$

But $d\mathbf{s} = \mathbf{v}dt$, so that

$$W = m \int_{v_1}^{v_2} \mathbf{v} \,.\, d\mathbf{v}. \tag{3.29}$$

This integration yields the result

$$W = \tfrac{1}{2}\, m{v_2}^2 - \tfrac{1}{2}\, m{v_1}^2. \tag{3.30}$$

The quantity $\frac{1}{2}mv^2$ is called the *kinetic energy* of the particle. It is the energy which the particle possesses in virtue of its being in motion. (3.30) may be interpreted as asserting that the total work expended on a particle is equivalent to the energy gained by that particle.

The other classical result that we shall need is the principle of the conservation of linear momentum. Suppose that we have a mechanical system consisting of two particles of mass m_1 and m_2 moving with velocities $\mathbf{v}_1$ and $\mathbf{v}_2$, respectively. The two particles collide and then recede from each other with velocities $\mathbf{w}_1$ and $\mathbf{w}_2$. From Newton's third law, it follows that the total momentum of the two particles after the collision is equal to the total momentum before the collision. Therefore, we may write

$$m_1\mathbf{v}_1 + m_2\mathbf{v}_2 = m_1\mathbf{w}_1 + m_2\mathbf{w}_2 \qquad (3.31)$$

More generally, for any isolated or conservative system,

$$\sum_n m_n\mathbf{v}_n = \text{constant} \qquad (3.32)$$

We shall now provide a sketch of some of the more fundamental ingredients of relativistic mechanics. Our strategy will be to construct four-vector analogues of classical quantities and laws. We know that any four-vector equation is a valid candidate for inclusion in the laws of special-relativistic physics in virtue of its automatically satisfying the condition of Lorentz covariance. Secondly, one would expect that a relativistically correct law would bear a certain resemblance to a Newtonian law, since it is known that the latter are reasonably accurate representations of reality at very low or non-relativistic velocities. We begin by gathering the ingredients which are already at our disposal.

We have already defined the space-time interval:

$$(ds)^2 = \delta_{\mu\upsilon}dx^\mu dx^\upsilon = (dx^1)^2 + (dx^2)^2 + (dx^3)^2 - c^2(dx^4)^2. \qquad (3.33)$$

We also have the *four-scalar* proper-time, defined by

$$d\tau = (1 - v^2/c^2)^{\frac{1}{2}}\, dt. \qquad (3.34)$$

The familiar velocity is not a vector in spacetime. However, with the aid of what is already at our disposal, we may define the *four-velocity* in terms of the interval and the proper-time.

$$u^\mu = df\frac{dx^\mu}{d\tau} \qquad (3.35)$$

Applying the chain rule of elementary calculus, we may rewrite (3.35) as

$$\frac{dx^\mu}{d\tau} = \frac{dx^\mu}{dt}\frac{dt}{d\tau}. \qquad (3.36)$$

Then, with the aid of (3.34), we relate the four-velocity to the familiar three-velocity.

$$u^\mu = \frac{dx^\mu}{d\tau} = (1 - v^2/c^2)^{-\frac{1}{2}}\left(\frac{dx^i}{dt}, ic\right). \qquad (3.37)$$

It should be noted that the spatial components, i.e. the first three, of the four-velocity constitute the counterpart of the classical velocity. An appropriate name for this three-component object is *proper-velocity*. It is, in effect, the self-measured velocity of an object, which is obtained by dividing the distance which the object traverses by the proper-time. In contrast, the four-velocity has no direct classical analogue. This is clear from the fact that the norm or length of the four-velocity turns out to be a universal invariant. It is simply calculated as follows:

$$(U^\mu)^2 = \delta_{\mu\nu} U^\mu U^\nu = (1 - v^2/c^2)^{-1}(v^2 - c^2) = -c^2. \tag{3.38}$$

One would not expect the principle of momentum conservation to survive in its classical form. It has already been remarked that it is a direct consequence of the law of the equality of action and reaction. However, we learned, in Chapter 2, that the third law presupposes an instantaneous action at a distance or, at the very least, a unique standard of distant simultaneity. In fact, it may also be shown that the conservation of linear momentum is a consequence of the Galilean invariance of classical particle interactions. To find a Lorentz covariant counterpart of (3.31) or (3.32), we shall assume that m will be replaced by a four-scalar M and $\mathbf{v}$ by the four-velocity. The counterpart of (3.31) will then be

$$M_1 U_1^\mu + M_2 U_2^\mu = M_1 \bar{U}^\mu + M_2 \bar{U}_2^\mu \tag{3.39}$$

where the bars signify the four-velocities after the collision. More generally, we shall tentatively require

$$\sum_n M_n U_n^\mu = \text{constant}. \tag{3.40}$$

Although (3.40) is a four-vector equation and, thereby, relativistically admissible, one has only a vague idea, at present, of its actual significance. However, by inserting (3.37) into (3.40), we obtain

$$\sum_n M_n(1 - v^2/c^2)_n^{-\frac{1}{2}} \left(\frac{dx^i}{dt}, ic\right)_n = \text{constant}. \tag{3.41}$$

This suggests that we relabel M by m_0 and call it the *rest-mass* of a particle and relate it to the *relativistic mass* by the following definition

$$m \stackrel{\text{df.}}{=} m_0(1 - v^2/c^2)^{-\frac{1}{2}}. \tag{3.42}$$

The rest-mass of a body is an intrinsic property. It represents its inertia as measured in the body's own rest-frame. Relativistic mass, on the other hand, is a frame-dependent quantity. It is clearly a function of a body's velocity relative to a laboratory or rest-frame of an observer. Thus, the first striking result of relativistic mechanics is that the inertia or mass of an object increases with its velocity. An inspection of (3.42) reveals that if a particle with a non-zero rest-mass were, *per impossible*, accelerated to the velocity of light, its inertia would become infinite. It follows, of course, that no such particle can attain the velocity of light. Intuitively, if a given force were acting on a particle over an unlimited period of time, the velocity of the particle would constantly increase. However, since its mass would also increase the *rate* of acceleration would always be diminishing. It turns out that no matter how long a time the force is applied to the particle, the consequent increase in inertia will prevent the particle from attaining the velocity of light. Thus, c is not only a universal constant but also represents the upper limit to physically attainable velocities. It is worth remarking that the relativistic increase in mass is a highly confirmed phenomenon. It is particularly apparent when an electron is accelerated to within a fraction of 1 percent of the velocity of light in a linear accelerator. The reader may wonder how it is possible that a photon, a particle of light, is capable of travelling at the speed of light. The answer is that although a photon possesses relativistic inertia, it has a vanishing rest-mass.

We may now give formal recognition to the relativistic counterpart of classical momentum, which was implicit in (3.39) and (3.40). It is a four-vector which is defined by

$$P^\mu = df. \quad m_0 u^\mu. \tag{3.43}$$

Clearly, the first three of its components are

$$\frac{m_0 v^i}{\sqrt{1-v^2/c^2}}. \text{ Its fourth component will be } \frac{m_0 c}{\sqrt{1-v^2/c^2}}.$$

Consequently, the relativistic law of conservation telescopes the conservation of momentum and the conservation of mass. The reader should take note that P^μ as defined by (3.43) may be multiplied by the factor c. The fourth component then has the dimensions of energy. It then turns out that $(cP^i)^2 - E^2$ is an invariant. Hence, the possibility of constructing the four-vector P^μ leads to the combining of the principles of conservation of momentum and conservation of energy. The reason why the time component of P^μ may be variously treated as pertaining to both mass and energy will shortly emerge. Our final step is to construct a relativistic law of motion involving the four-force or Minkowski force F^μ. The relativistic law of motion is simply

$$F^\mu = \frac{dP^\mu}{d\tau} = m_0 \frac{du^\mu}{d\tau}. \tag{3.44}$$

It has already been remarked that the relativistic increase in mass or inertia is a highly confirmed experimental fact. We shall now establish another highly confirmed result for which special relativity is perhaps unjustly famous. Our derivation will be somewhat heuristic since the rigorous proof is a trifle complicated. We begin by making use of the purely formal result, which the more ambitious reader may derive for himself, that F^μ and P^μ are orthogonal vectors in spacetime. Formally expressed, this means that

$$\delta_{\mu\nu} F^\mu u^\nu = 0. \tag{3.45}$$

It immediately follows that

$$(1-v^2/c^2)^{-1}\left(\mathbf{f}\cdot\mathbf{v} - c^2\frac{dm}{dt}\right) = 0, \tag{3.46}$$

where **f** is the classical force. Since the first factor is not zero, we may conclude that

$$\mathbf{f}\cdot\mathbf{v} = c^2\frac{dm}{dt}. \tag{3.47}$$

But **f** . **v** is the classical expression for *power*, which is a measure of the *rate* at which work is being done. Bearing this in mind, we may write

$$\frac{dW}{dt} = c^2\frac{dm}{dt}. \tag{3.48}$$

The total work performed on a particle during a fixed interval may be obtained by integrating (3.48) with respect to time.

$$W = c^2 \int_{t_1}^{t_2} \frac{dm}{dt} dt = m_2 c^2 - m_1 c^2. \tag{3.49}$$

If we assume that at the initial time t_1, $\mathbf{v} = 0$, then (3.49) may be rewritten as

$$W = mc^2 - m_0 c^2. \tag{3.50}$$

Now W is just the kinetic energy that has been gained by the particle from t_1 to t_2. We have assumed that the initial velocity was zero. Consequently, the initial kinetic energy was zero. Hence, we must have

$$K = \frac{m_0 c^2}{\sqrt{1 - v^2/c^2}} - m_0 c^2. \tag{3.51}$$

The reader should verify that for low velocities the right-hand side of (3.51) reduces to approximately $\frac{1}{2} m_0 v^2$, which is just the classical expression for K which was derived in (3.30). We may conclude from (3.51) that when a particle of rest-mass m_0 is at rest, it must possess an *intrinsic* energy $m_0 c^2$. This represents many different kinds of energy including the binding energy of the atomic nucleus. It is precisely this energy which is released in the process of nuclear fission. The total energy of a particle is expressed by the best known but not the most interesting equation of special relativity: $E = mc^2$. This equation implies that mass may be converted into energy and that energy must possess inertia. However, it goes too far to say that mass and energy are one and the same. "$E = mc^2$" is a law of equivalence rather than a definitional identity. Nevertheless, it is certainly true to say that special relativity has revealed a very intimate relationship between two quantities which played an altogether distinct role in classical mechanics.

Having accomplished our principal purpose, which was to gain insight into the development and structure of relativistic mechanics, we may appropriately pause to reflect on the significance of our findings. If the reader reviews the path which we have followed, he will perceive it to have been a curious one. We began with the simple, albeit counter-intuitive, assumption that the wave-front equation is a relativistically valid law. This assumption was found to entail the Lorentz transformation and the consequent relativization of distance and time. It was then shown that the new conception of space and time could be modelled geometrically by Minkowski spacetime. The latter enabled us to construct various geometric objects on the space-time manifold. These space-time objects were then employed to construct a so-called relativistic analogue to classical mechanics. The ensuing development was found to lead to a series of truly remarkable physical consequences all of which have been experimentally confirmed. How, the reader may wonder, could so much possibly have been achieved with so little? In particular, how could the adoption of a new geometry of spacetime, which is surely just a conceptual construct, result in such dramatic consequences? Does geometry truly rule the world?

There are certainly many writers who by their words would have us think so. My own view is that when one examines the origins of special relativity the occult influence of geometry is found to recede. To begin with, one must be aware that there is virtually no limit to the number and variety of geometries that the human mind is capable of constructing. Minkowski geometry is rich in physical consequences only because it was specifically constructed for the representation of physical phenomena. To this one may retort that, nevertheless, the phenomena in question are purely kinematic. It is difficult to understand how the kinematic invariance of the wave-front or, equivalently, the constancy of the velocity of light could possibly lead, for example, to the mass-energy equivalence relation.

In fact, this argument makes two false assumptions. The first of these is the supposition that the entire edifice is based on the behaviour of light. As was indicated at the outset, the derivation of relativistic kinematics from the wave-front equation is no more than a simple paedagogical device. The invariance of c is just one of many consequences of Maxwell's electrodynamics. The truth of the matter is that long before the advent of special relativity, a relativistically correct, i.e. Lorentz covariant, theory was in existence. The Lorentz transformation is the offspring of electrodynamics. In fact, anyone who is familiar with electrodynamics, in its post-Einsteinian form, will know how harmonious its laws appear to be from the perspective of special relativity. Electrodynamics, in its classical form, involved two distinct but related field intensities, the electric intensity **E** and the magnetic intensity **B**. The two quantities are clearly related. For example, an electric charge at rest produces an electric field. When the charge is in motion, it is an electric current which gives rise to a magnetic field. The effect of the magnetic field is quite different from that of the electric field. This physical mystery is soon unravelled when electrodynamics is reformulated in a four-dimensional manner. Just as we found that space and time, energy and momentum, etc., may be combined to form four-vector quantities, so now it is found that the various ingredients of electrodynamics may be treated in the same way. In particular, the electric and magnetic field intensities may be combined to form the electromagnetic *four-tensor*, a many component object which is a generalization of the four-vector. In other words, electricity and magnetism are viewed as mere components of a single underlying electromagnetic field which appears in one phenomenal guise or another depending on one's frame of reference. One may conclude that the time was mathematically ripe for Minkowski space-time in the mid-nineteenth century. Accordingly, the new geometry is not merely a convenient model for light but also for a far-reaching theory which embraces one of the most pervasive aspects of physical reality. When one considers that mechanical objects have a microscopic structure which is largely electronic it seems less surprising that mechanics may be accommodated by the same geometric structure that houses electrodynamics.

The second false assumption of the argument is that nothing more than the Lorentz transformation or the geometry of spacetime is required for the construction of relativistic mechanics. This could hardly be less accurate. In fact, we modelled all of our relativistic laws on their classical counterparts. It is by no geometrical magic that such laws turn out to be highly confirmed. They are all, in effect, modifications of laws which are themselves known to be reasonably accurate at low velocities. Thus, the picture of relativistic physics as based on a scant physical hypothesis and a great deal of mathematics could hardly be more misleading. Special relativity is the product of a long history of scientific research and invention, which includes the whole of classical mechanics and electro-dynamics. Einstein's contribution consisted in grasping this tradition with scientific detachment and in drawing from it various inevitable but hitherto unthinkable consequences.

The Principle of Special Relativity

The most obvious consequence of the principle of classical relativity is that a uniform velocity has no effect on the laws of dynamics. That is to say, that the results of a dynamical experiment which has been carried out in an inertial reference frame are independent of the velocity of that frame. The significance of the principle of special relativity is to be distinguished from that of classical relativity on several grounds.

The most obvious difference is that the classical principle is limited in its scope to purely mechanical phenomena whereas the special principle is intended to apply to all laws of nature, mechanical or otherwise. In fact, if Einstein's view of time is correct, it follows that biological phenomena must be included within its scope.

The second obvious difference between the two principles is that the one recognizes the existence of an equivalence class of inertial frames which are related by the Galilean transformation, whereas the other recognizes an equivalence class related by the Lorentz transformation. While this distinction is well known, its full significance tends to be overlooked. Specifically, we have seen that Newton's conceptions of space and time do not specifically entail the Galilean transformation. In other words, the Galilean transformation cannot be interpreted as a consequence of the Newtonian metric: $\delta_{ij}dX^i dX^j$. It is perfectly possible to conceive of a universe having such a metric but in which the principle of relativity did not hold. In contrast, the Einsteinian or Minkowskian conception of space-time with the metric $\delta_{\mu\nu}dX^\mu dX^\nu$ actually entails the Lorentz transformation, so that the principle of special relativity is seen to be a necessary consequence of the structure of Minkowski spacetime.

We may render the distinction more precise by adding a little more to our mathematical equipment. In Chapter 1, we defined a vector as a geometric object with a linear transformation law given by

$$V^i \rightarrow \bar{V}^i = c_{ij}V^j, \tag{3.52}$$

where the C_{ij} are simply the cosines of the angles between the barred and unbarred coordinates. However, it was hinted at that time that this definition stood in need of further generalization. We shall now provide that generalization. Specifically, our discussion was limited to the restricted case of a transformation from one orthogonal Cartesian system to another. However, it often happens that the need arises to carry out transformations between systems of non Cartesian coordinates, such as the spherical coordinates which are employed to label points on the surface of the earth. A spherical coordinate system is an example of a *curvilinear* as distinct from a *rectilinear* system. Unlike the case of transformations involving Cartesian coordinates, those which involve curvilinear coordinates are *non-linear* transformations. Thus, the simple relationship expressed by (3.52) breaks down. The relationship between the Cartesian system (x, y, z) and the spherical system (r, θ, ϕ) is given by

$$\begin{aligned} r &\rightarrow x = r \sin\theta \cos\phi \\ \theta &\rightarrow y = r \sin\theta \sin\phi \\ \phi &\rightarrow z = r \cos\theta. \end{aligned} \tag{3.53}$$

The crucial point about (3.53) is that it involves the non-linear functions, *sine* and *cosine*. Consequently, the relationship between the Cartesian values of a point and the spherical values of a point will vary from point to point. However, all is not lost. Although the relationship between the two systems varies from point to point, the *rate* of this variation is constant. That is to say, that although the transformation is not linear in the coordinate variables, it is linear in the *derivatives* of those variables. Thus, instead of expressing a coordinate transformation as a linear relationship between coordinates, we shall express it as a linear relationship between coordinate differentials, as follows.

$$dX^i \rightarrow d\overline{X}^i = \frac{\partial \overline{x}^i}{\partial x^j} dX^j. \tag{3.54}$$

Thus, the matrix of the transformation is $C_{ij} = \partial \overline{x}^i / \partial x^j$. The partial derivatives of the new coordinates $\overline{X}^i$ with respect to the older coordinates X^i reduce to the cosines of the angles between the coordinate axes for the special case in which both sets of coordinates are Cartesian. Moreover, it turns out that the components of vectors also transform according to (3.54). Thus, the familiar (3.52) may be regarded as a special case of (3.54). Henceforth, therefore, the transformation law for the vector V^i, will be given by

$$V^i \rightarrow \overline{V}^i = \frac{\partial \overline{x}^i}{\partial x^j} V^j. \tag{3.55}$$

(3.55) tells us that a vector transforms in the same manner as the coordinate differentials. In other words, the infinitesimal displacement dX^i is to be regarded as the prototypical vector.

Let us suppose that the vector V^i is expressed in terms of its components in the spherical system by $V^1\hat{r} + V^2\hat{\theta} + V^3\hat{\phi}$. We now wish to determine the value of its components in the Cartesian system given by $V^1\hat{x} + V^2\hat{y} + V^3\hat{z}$. We first consult (3.53) for the general transformation law for coordinates between the two systems. We then refer to the general rule (3.55). The first component of $\overline{V}^i$, i.e. the $\hat{x}$ component, wiil be given by

$$\overline{V}^1 = \frac{\partial \overline{x}^1}{\partial x^j} V^j = \frac{\partial \overline{x}^1}{\partial x^1} V^1 + \frac{\partial \overline{x}^1}{\partial x^2} V^2 + \frac{\partial \overline{x}^1}{\partial x^3} V^3.$$

This is of the explicit form:

$$\begin{aligned} \overline{V}^1 &= \frac{\partial}{\partial r}(r \sin\theta \cos\phi) V^1 + \frac{\partial}{\partial \theta}(r \sin\theta \cos\phi) V^2 + \frac{\partial}{\partial \phi}(r \sin\theta \cos\phi) V^3 \\ &= \sin\theta \cos\phi \; V^1 + r\cos\theta\cos\phi \; V^2 - r\sin\theta\sin\phi \; V^3. \end{aligned}$$

In a similar manner we would compute the values of $\overline{V}^2$ and $\overline{V}^3$.

The foregoing mathematical treatment will be found to be essential in connection with general relativity. For the moment, however, we simply need it to gain insight into the non-geometric character of the Gaililean transformation. The essential point to recognize is that the Galilean transformation is one which involves not only the spatial coordinates but also the time t. That is to say, that it is a space-time transformation. The transformed coordinates depend not only on the old coordinates but also on the time. Thus, it may be represented in the following form:

$$X^i \rightarrow \overline{X}^i = X^i - vt = \overline{X}^i(X^j, t). \tag{3.56}$$

From (3.56) we have

$$d\overline{X}^i(X^j, t) = \frac{\partial \overline{x}^i}{\partial x^j} dx^j + \frac{\partial \overline{x}^i}{\partial t} dt. \tag{3.57}$$

(3.57) is simply obtained from the differential form of (3.56). We may now obtain the transformation law for the velocity vector dx^i/dt from (3.57).

$$\frac{d\bar{x}^i}{dt} = \frac{\partial \bar{x}^i}{\partial x^j}\frac{dx^j}{dt} + \frac{\partial x^i}{\partial t} \tag{3.58}$$

We note that (3.58) is *not* a homogeneous transformation like (3.55). It contains the additional (inhomogeneous) term $\partial x^i/\partial t$. It is in virtue of the inhomogeneous term that velocities may be transformed away under Galilean transformations. It should be noted that if the new coordinates are independent of time, (3.58) reduces to a pure coordinate transformation with respect to which velocity is, of course, a vector.

We may now proceed to formulate the principle of special relativity.

$Lx = x$ is a fundamental law of nature.
$Fx = x$ is a frame of reference.
$Lxy = x$ is related to y by a Lorentz transformation.
$Hxy = x$ holds with respect to y.

$$(\exists x)\{Fx \wedge (y)[(Ly \rightarrow Hyx) \wedge (z)(Fz \wedge Lzx) \rightarrow Hyz]\}$$

In words: There is at least one frame of reference with respect to which every fundamental law of nature holds and such that every other frame of reference which is related to it by a Lorentz transformation will be one with respect to which every such law holds. Certain of the salient points of this formulation may be brought out by first considering some of the numerous alternatives which have been proposed by physicists and philosophers.

(1) The laws of nature are Lorentz covariant

This version of the principle is clearly metatheoretical. It is to be construed not as a claim about the physical world but rather as one about the formal structure of certain natural-law *statements* as they occur in various physical theories. Doubtless, it is a correct claim about some, although clearly not all, laws of nature in the stated sense. The problem which this formulation raises is that it deprives the principle of special relativity of its factual status. Indeed, Bunge has made the point that there is no formal inconsistency between the two claims that Newtonian mechanics is Galilei covariant and relativistic mechanics is Lorentz covariant. Each is a true statement about the mathematical form of a particular theory. However, our account of the two relativity principles clearly presented them as rival physical hypotheses. Accordingly, we must conclude that (1) is defective as an interpretation of the principle of special relativity. The full force of this objection will only emerge in the course of the discussion of general relativity. There it will be shown that the theory of general relativity satisfies certain formal covariance requirements but that classical mechanics may be so reformulated that it will satisfy the same requirements. However, I hope that by now it is sufficiently clear that a principle of relativity has definite physical content. Such principles should have testable factual consequences. These remarks should not be construed to suggest that (1) is simply false but merely that it is incorrect to equate it with the principle of special relativity.

(2) The laws of nature, when properly formulated, are Lorentz covariant

Like (1), (2) is clearly metatheoretical. However, whereas (1) is a descriptive claim, (2) is a normative one. The addition of the clause "when properly formulated" signifies that all theories *ought* to be formulated in a Lorentz covariant fashion. Indeed, it was precisely in that spirit that our sketch of relativistic mechanics was constructed. We discovered that Lorentz covariance, interpreted as a type of prescription, has great heuristic value for the construction of putatively correct laws. Thus, it has the advantage over (1) of preserving the rivalry between the classical and the Einsteinian principles. Nevertheless, (2) generates certain philosophical objections which it fails to answer. In the first place, as it stands it runs the risk of being analytically true in the absence of any independent criterion of well-formedness other than Lorentz covariance itself. Secondly, it suggests that Newton, in constructing classical mechanics, was doing something that he ought not to have done, if only in retrospect. In effect, the principle states that the only good theories are ones which are Lorentz covariant. However, it raises the unanswered question that granted that there are some worthy theories that are Lorentz covariant, e.g. electrodynamics, why does it follow that all good theories should be Lorentz covariant? Thirdly, it happens that there are good theories which are not Lorentz covariant, e.g. the theory of acoustics. The solution to this third problem requires that one distinguish fundamental from non-fundamental theories. The demarcation between the two seems to be as follows. A fundamental theory must have global or universal significance. A necessary condition of this is that such a theory not be tied to a particular reference frame. Now acoustics is tied to such a frame—namely, that one in which the transmitting medium is at rest. In fact, acoustics is a theory which is, in principle, derivable from classical mechanics. One may compare the situation with that of electrodynamics in the nineteenth century. It was recognized that this theory lacked what was then deemed to be the desired covariance property, i.e. Galilei covariance. However, electrodynamics was not regarded as fundamental since it was supposed to be tied to a unique reference frame—namely, the rest-frame of the electromagnetic aether. Moreover, it was the belief of many nineteenth-century physicists that electrodynamics was in principle derivable from classical mechanics and that electromagnetic phenomena would ultimately be explained in terms of the mechanical stresses and strains in the aether. Accordingly, (2) is only applicable to fundamental theories.

(3) There can be no experimental determination of an absolute velocity

This is the version of the principle which is most frequently to be found in popular treatments of the theory of relativity. However, we shall see in the next chapter that at least one expert proposed precisely such an interpretation. (3) is clearly not a claim about the formal structure of physical theories. On the other hand, it is difficult to interpret it unambiguously as a claim about the physical world. Rather, it is to be construed as an *epistemic* claim to the effect that absolute velocity is beyond the range of one's experimental net. If this is what is intended, it is quite misleading. Its spirit is more in keeping with Newton and Locke than with Einstein. It implies, in effect, that absolute velocities are features of the real world but that it is either technically or in principle impossible to measure one. This, of course, is to miss the entire theoretical point which is not simply that absolute velocities are unobservable but that the very concept of absolute velocity is forbidden by the syntax and semantics of the theory. The empiricist interpretation of special relativity appears to trade on precisely this misunderstanding. It maintains that Einstein reinterpreted the concepts of space and time on the basis of his discovery that absolute velocity is unobservable. But how could a theory of the nature of reality be based on the human being's capacity to observe this or that feature of the world? These matters will be discussed, *in extenso*, in Chapter 4.

Our own formulation of the principle implies that it is a factual claim to the effect that there is a non-empty equivalence class of reference frames with respect to which the laws of nature may be expressed, or hold in their fundamental or global form. Moreover, it asserts that all such frames are kinematically related by the Lorentz transformation. Accordingly, the Lorentz transformation is physically interpreted as expressing the kinematic laws which hold among inertial frames of reference. Thus, we see that the Lorentz transformation need not be interpreted as a formal metamathematical principle in every context. In the present one, it is semantically interpreted as a factual claim concerning the law-like connections among the *physical* coordinates of the aforementioned reference frames. On our interpretation, the principle of special relativity is not equivalent to the principle of Lorentz covariance. However, the latter should be viewed as a necessary condition of the former. That is to say, that if a law of nature is in conformity with the principle of special relativity, then it will be a Lorentz covariant law. Thus, the physical principle is not to be construed as a consequence of the mathematical formalism. That would be a logical absurdity. On the other hand, the formalism is a consequence of the physical principle. The advantage of this interpretation is that it preserves the factuality of the principle without denigrating the heuristic role of covariance requirements and without committing the egregious fallacy of epistemological magic. By that I mean that it is not committed to the recurrent but untenable view that one may deduce matters of fact from matters of mathematics. At the same time, the heuristic success of Lorentz covariance is fully accounted for in so far as it is seen as a necessary condition for a factually correct theory, on the assumption, of course, that the principle of special relativity is factually sound.

Bibliography for Chapter 3

BORN, M., *Einstein's Theory of Relativity* (rev. edn.), Dover, New York, 1962. In my opinion, this remains the most successful popular account of relativistic physics. It begins with Newtonian mechanics and goes all the way to general relativity without making use of calculus. It includes a treatment of electrodynamics.

EINSTEIN, A., *On the Electrodynamics of Moving Bodies* (1905), in Einstein *et al.* (1923).

EINSTEIN, A., *Relativity: The Special and the General Theory* (15th edn.), Crown, 1952. A popularization by the creator of the theory. Remarkably clear and lucid.

EINSTEIN, A., H. A. LORENTZ, H. MINKOWSKI, and H. WEYL, *The Principle of Relativity*, Dover, New York, n.d. (original edn. 1923). A collection of several of the foundational papers, including Einstein's original contributions to both special and general relativity.

KATZ, R., *Introduction to the Special Theory of Relativity*, Van Nostrand, Princeton, 1964. A simple yet serious introduction to the subject.

MINKOWSKI, H., *Space and Time* (1908), in Einstein *et al.* (1923).

PROKHOVNIK, S. J., *The Logic of Special Relativity*, Cambridge University Press, Cambridge, 1967. A painstaking account of relativistic kinematics with particular attention to the twin paradox. Personally, I do not care for the philosophical aspects of the treatment.

SHADOWITZ, A., *Special Relativity*, Saunders, Philadelphia, 1968. A clear, sound introduction which pays considerable attention to space-time diagrams, including the Loedel diagram. A good introduction to four-vectors and tensors.

SMITH, J. H., *Introduction to Special Relativity*, Benjamin, New York, 1965. Elementary and readable. It includes a detailed analysis of the twin paradox.

TAYLOR, E. F., and J. A. WHEELER, *Spacetime Physics*, Freeman, San Francisco, 1966. An authoritative yet entertaining account. It is notable for its geometric approach, which includes a look at general relativity.

4

Empiricism, Rationalism and Special Relativity

EMPIRICISM and rationalism are two extremely disparate philosophic approaches to the interpretation of knowledge or of 'science' in the broadest sense of that term. My own view, which the reader will find running throughout these pages, is that both of these positions are too extreme and only contribute to the misunderstanding of the scientific enterprise. On my view, the correct approach to the interpretation of science, which avoids the excesses of these philosophies and is most consonant with scientific practice, is what may be called, for want of a better term, *scientific realism*. The point of the present chapter is to present characteristic examples of the two allegedly incorrect positions and to attempt to show how, in the view of the author, they go wrong in their interpretation of science in general and of the theory of relativity in particular.

As we are using the terms "empiricism" and "rationalism", they are rather loosely defined. Very broadly speaking, empiricism is the view that the knowing subject plays a relatively passive or minor role in the acquisition of knowledge. Scientific theories are formed neither from innate nor from rationally produced conceptual ingredients but are, at least in their substantive aspects, constructed from whatever ingredients happen to be provided by our collective experience. Our theories of the world, whatever the aspect of interest, are very closely tied to experience and may, in fact, be nothing but highly refined and economical descriptions of it.

Just as broadly interpreted, rationalism maintains that the knowing subject plays an active role in the formation of scientific theories. Reason provides us with an insight into those pervasive features of existence which could not possibly be different from what they actually are. Experience is fragmentary and corrigible and stands in need of revision by the method of pure ratiocination. That which *seems* to be the case very often turns out not to be, but rational insight, aided by the techniques of logical deduction, is able to reveal the world as it is in contrast to the way in which it merely appears. Moreover, the rationalist would argue that science, at least in its fundamental parts, has the status of *a priori* knowledge. That is to say, that it rests on principles which are universal and necessary. But experience yields only that which is partial and contingent. To claim that experience is the basic source of knowledge is to ignore the nature of scientific theories. In that sense, scientific knowledge, since it contains more than experience could possibly provide, must be logically independent of experience and, hence, prior to it. A given experience might always have been different from what it actually was whereas the world, as revealed in our theories, could not be.

It is both remarkable and instructive that proponents of two such antithetical epistemological doctrines should have perceived the theory of relativity as providing a vindication of their respective viewpoints. This should suffice to make one aware that the proper philosophical interpretation of science is a subtle undertaking which prohibits the drawing of conclusions with undue haste and incaution.

Neither empiricism not rationalism can be narrowly defined. Within each school of thought one finds great divergence of opinion and even bitter antagonism. Thus the critic of these schools is faced with the difficult decision as to precisely which particular versions he should select for analysis. He might concoct what he takes to be the viewpoints of the typical rationalist and typical empiricist and proceed to analyze and criticize them. But there is always a danger when criticizing a doctrine of one's own making that it turns out to be a man of straw. Accordingly, I have chosen to present for analysis, two representatives of each school, although one of them is not associated with a particular individual but rather with a movement. My criteria of selection were that they be post-Einsteinian, since it would be pointless to criticize an eighteenth-century philosopher for having failed to anticipate relativity theory, and that they be worthy representatives of the two movements, since little is to be gained from the criticism of blatently false or otherwise silly doctrines.

Relativity and the *A Priori* in the Philosophy of Eddington

Every student of the history of philosophy is aware of the fact that the term "idealism" is a label which is applicable to a multiplicity of divergent points of view. One need only recall that the term is associated with such diverse thinkers as Plato, Berkeley, Kant and Hegel. Dialectical materialists have even applied it to the position of Ernst Mach. Eddington's philosophy is an instance of that form of idealism which holds that one cannot make a sharp epistemological distinction between the subject of knowledge, the knower, and the object of knowledge. The world which we know is a construct of the human mind. Thus, it may be understood, broadly speaking, in contrast to the doctrine of realism according to which the object of knowledge has an independent existence in its own right, apart from the mental apparatus of either individual thinking beings or the human species. Thus, this form of idealism is to be construed as primarily epistemological, although it may well have profound metaphysical implications.

In the literature of twentieth-century philosophy, the theory of relativity has frequently been associated with epistemological idealism. Often, however, such associations have been made by journalistic writers on the basis of the scantiest knowledge of the details of Einstein's theory. Sir Arthur Eddington, Astronomer Royal, was a notable exception. During his lifetime he was widely recognized as a leading authority in the field of relativity theory. It is for this reason that we shall address ourselves to the Eddingtonian form of the idealistic interpretation of the theory of relativity. His most sustained account of this position is contained in *The Philosophy of Physical Science* (1939). We begin with an outline of Eddington's argument.

According to Eddington, as the science of physics has progressed, it has undergone a steady transformation in its character. Originally, it appeared in the form of a set of empirical generalizations which were vulnerable to experimental falsification. In the twentieth century, however, new physical theories appeared which were of such a character that no experiment could possibly confute them. In short, the fundamental laws of the "new" physics now enjoyed the status of *a priori* principles. According to Eddington, although such laws may have been suggested by the results of empirical investigations, they might just as well have been anticipated by means of an analysis of the manner in which those investigations were carried out. The *a priori* character of the laws of physics derives from the fact that such laws are not claims about an objective, 'ultimate' reality but about the nature of our knowledge. Accordingly, they may appropriately be described as "epistemic". The laws of relativity physics clearly manifest this epistemic character. For example, the denial of absolute distant simultaneity is not a claim about the nature of reality but an assertion to the effect that absolute distant simultaneity is "unobservable". But this, according to Eddington, is a statement which properly belongs to epistemology. It is a claim about the sensory and intellectual apparatus of the physicist.

Eddington is not arguing for the brand of subjective idealism propounded by Berkeley according to which existence can mean only either to perceive or to be perceived. He does not deny the existence of an independent "Reality". His position is somewhat closer to that of Kant. What he wishes to assert is that what we call "the physical world" is simply the world described by the physicist and that the laws of this constructed world have a subjective origin. Thus, one must distinguish the physical world from the real world. In consequence of the nature of his sensory and intellectual apparatus, the physicist *selects* those aspects of reality which are accessible to him and then formulates the laws which describe and correlate those scientific abstractions. Thus, Eddington labels his position *selective subjectivism.* He argues that the laws of relativity physics will never be experimentally confuted, since they reflect the manner in which the scientific selection has been carried out and it is impossible to select items which would fail to be in conformity with them.

Eddington explains the logic of his position by means of the following somewhat picturesque parable. Suppose an ichthyologist is investigating the nature of sea-creatures by casting a net into the sea and examining his catches. On the basis of what he catches, he proceeds to make the following generalizations:

1. No sea-creature is less than 2 inches long.
2. All sea-creatures have gills.

An onlooker is critical of the first generalization on the ground that it is not really a statement about the denizens of the ocean but is rather about the nature of the ichthyologist's net. Specifically, the size of the mesh is such that it is impossible to catch a fish which is less than two inches long. To this, the ichthyologist impatiently retorts that what his net cannot catch is not a fish and not in the purview of ichthyological knowledge. As a final note of contempt, he accuses the onlooker of being a metaphysician.

The significance of Eddington's parable is as follows. The ichthyologist represents the physicist. His net stands for the sensory and intellectual equipment that is at the physicist's disposal. The ocean with its fish of assorted sizes corresponds to the "objective reality" of the metaphysician. The generalization to the effect that all sea-creatures have gills is the counterpart of an ordinary empirical generalization which obviously runs a certain inductive risk. On the other hand, the claim about the length of sea-creatures is the counterpart of a fundamental law of physics, which, according to Eddington, is nothing but a disguised epistemic law. But perhaps the most important aspect of all, the contents of the ichthyologist's catch correspond to what he calls "observables".

On a superficial reading, Eddington's earlier contributions to the philosophy of physics appear to have a different thrust from that of his more mature work. In his earlier writing, he expounded a hard empiricist line to the effect that scientific theories were simply the coordination of "pointer-readings". It would appear that he drifted from positivism to idealism. In fact, the two positions are quite consistent. According to Eddington, a fundamental methodological principle of science is to reject all concepts which fail to correspond to observable entities. But clearly what constitutes the class of observables is dependent on the techniques that are at the disposition of the observer. This statement must be immediately qualified. An unobservable is not an entity which the current technology is incapable of capturing. An unobservable is unobservable *in principle*. To extend Eddington's simile, it is as though a net with a finer gauge of mesh would involve a logical impossibility. "Unobservability of a quantity arises from a logical contradiction in the definition which professes to specify the procedure for observing it" (1939, p. 41).

On Eddington's view, we have been misled into regarding the laws of relativity physics as corrigible physical hypotheses because they were suggested in the first place by experimental results. For example, he argues that the definition of absolute distant simultaneity which was implicit in classical physics was logically deficient. However, this deficiency might long have continued to escape notice had it not been for the experimental failure to detect the velocity of the electromagnetic aether. On the other hand, had one pursued the method of epistemological analysis advocated by Eddington, one could have anticipated the outcome of the Michelson–Morley experiment on *a priori* grounds.

It is obvious that Eddington's conception of the nature of that branch of human knowledge which goes under the general heading of physical theory deviates radically from the standard conception according to which a physical theory is a set of postulates from which testable theorems may be deduced. The postulates have the status of physical hypotheses which stand or fall on the degree of conformity of the theorems with experience. Eddington's position requires careful evaluation since to take it seriously is to revise one's entire conception of the scientific method.

In the specific case of the theory of special relativity, Eddington would appear to regard the concept of absolute distant simultaneity as crucial. Let us therefore examine his claim that this concept is logically defective. The argument is by now quite straightforward. In order to ascertain that two events are simultaneous it is necessary to locate synchronized clocks in different parts of the universe. But in the absence of a standard of distant simultaneity, we cannot know that our spatially separated clocks are, in fact, synchronous. Hence, the method of ascertaining distant simultaneity is circular. Alternatively, we may employ a single clock and make the assumption that the velocity of light is constant in all directions regardless of the velocity of its source. But, once again, we are caught in a vicious circle, since to assert that the time required for a light-ray to travel from A and B is equal to the time for it to travel from B to A is to suppose that there are clocks located at both A and B which are synchronized.

We have seen in some detail in the last chapter that the failure to attain a unique standard of distant simultaneity entails the inability to achieve a unique standard of spatial separation, which, in turn, requires a significant modification in the laws of physics. In particular, the Newtonian law of motion, $\mathbf{F} = m\mathbf{a}$, must be replaced by the four-vector law, $F^{\mu} = m_0 U^{\mu}$. It is Eddington's contention that the relativistic law is not a more accurate representation of objective reality than the classical law but simply a more consistent correlation of the subjectively selected phenomena. Although Eddington does not explicitly go this far, it would appear to follow from the train of his argument that in the

Real World, which is necessarily insulated from our conceptual probing, there may indeed be equivalence classes of absolutely simultaneous events. But such hypothetical equivalence classes would have no place in the world described by the physicist, since they are unobservables.

It must be stated at the outset that the theory of relativity does not, in fact, assert that distant simultaneity is a logically inadmissible concept. Indeed, it provides a well-defined method for determining when two events stand in the relationship of simultaneity. What it does assert is that distant simultaneity is a frame-dependent concept. That is to say, that two events which are simultaneous with respect to frame *A* are successive with respect to frame *B*. But does it follow that if the relativistic interpretation of simultaneity is consistent with experience, i.e. empirically confirmed, then the classical concept must be logically defective? There seems to be no adequate ground for any such conclusion. If the classical concept of absolute distant simultaneity were to be retained, it would follow that an event would have the same time value with respect to every frame of reference which is just to say that distant simultaneity would be a frame-independent concept. But how could such a relationship between events be ascertained to hold?

In a world in which signals could be transmitted with infinite velocities, the determination of absolute simultaneity would present no problem, at least on a theoretical level. So long, however, as there is a finite limit to the velocity of signal propagation, classical simultaneity would have the status of a purely theoretical construct which could not be empirically measured. But that would be no ground for rejecting it. In such a world the laws of nature might well be invariant with respect to the Galilean transformation. In short, physicists would be in a position to know that the theoretical construct of absolute simultaneity is consistent with the body of theoretical physics. The essence of Eddington's error is the supposition that every individual concept must be directly coordinated with an experimental procedure. In fact, the methodological requirement which we impose on all scientific theories is that they lead to testable consequences. It is essential, therefore, that every theory contain some concepts which are relatable to experience in a reasonably direct way. For example, statistical thermodynamics ascribes mechanical properties to the individual molecules of a gas in the absence of any experimental procedure for measuring such properties. The point is that the theory which incorporates the concept of the momentum of a single gas molecule leads to testable consequences involving such measurable properties as temperature and pressure. By the same token, absolute distant simultaneity is a perfectly admissible concept in a theory—namely, classical mechanics, which has testable consequences.

The essence of Einstein's discovery was not that distant simultaneity in the classical sense is a self-contradictory concept, like four-sided triangle, but that it is incompatible with the theory of electrodynamics. He therefore proposed an alternative concept of distant simultaneity which, as we have seen, is intimately connected with the Lorentz transformation. Now it happens to be the case that the relativistic concept of distant simultaneity can be coordinated with experimental procedures. However, it is clearly fallacious to infer that special relativity was the result of expunging an unobservable from the body of theoretical physics. Had it been the case that the laws of electromagnetism were invariant with respect to the group of Galilean transformations, we would have continued to retain the concept of absolute distant simultaneity despite our awareness of its being a purely theoretical construct. In such a world, the classical principle of relativity would be regarded as correct. Indeed, an idealist of Eddington's persuasion who inhabited such a world might be tempted to argue that absolute distant simultaneity has an epistemic status in virtue of the Galilean invariance of the laws of nature. He would argue that since the only invariant velocity of the Galilean group is the infinite velocity, one is obliged, *a priori*, to treat the contemporaneousness of distant events as absolute.

So much conceptual confusion has beclouded this issue that my remarks ought now to be amplified. Eddington is guilty of a widespread but, in my opinion, erroneous interpretation of the methodological foundations of the relativization of time. The essence of the error is the supposition that Einstein's critique of the Newtonian treatment of time and especially of distant simultaneity led fairly directly to the Lorentz time transformation. Let us, therefore, review the argument, step by step. Einstein's first point is simply that there is no method whereby distant simultaneity could be directly determined by a measurement procedure. Any such procedure would turn out to be circular, since it would presuppose either that two distant clocks simultaneously indicate the same time or that the velocity of a signal from A to B is identical with the velocity of a similar signal from B to A, which again pressupposes that such velocities have been measured with the aid of synchronous clocks. In effect, therefore, no *direct* method of observing distant simultaneity is at our disposal. The important point to note is that this argument is applicable to time in general, not specifically to Newtonian time. The only exception in the Newtonian universe would arise if, *per impossible*, the latter actually contained infinitely fast signals. Accordingly, the argument, to this point, has nothing whatsoever to do with electrodynamics or the Lorentz transformation.

The second step in the argument is to propose a method whereby the aforementioned circularity may be overcome. Essentially, it consists in having recourse to a theoretical construct of some kind. Einstein proposed an optical or electromagnetic signalling procedure for the synchronization of clocks. Given the theoretical constancy of light propagation, such a choice was natural but not essential. One might equally have proposed an acoustic procedure. The latter would, indeed, have been favoured by a Newtonian physicist on the ground that one knows when the air, the medium of transmission, is at rest. However, it is not the choice of signalling or synchronization procedure which divides classical and relativistic kinematics. This point is often overlooked or misunderstood simply because the velocity of light plays an essential role in the third stage of the argument. The essence of the second stage is, to repeat, that the epistemic circularity can only be overcome by recourse to theoretical constructs of one kind or another.

The third and final stage is to the effect that the frame independence of the laws of electrodynamic and optics can only be accounted for on the assumption that *whatever* theoretical procedure one may have selected for the determination of distant simultaneity, its results must be frame-dependent. More simply, the independence of the velocity of light from the reference frame in which it is measured leads to the Lorentz time transformation and the consequent relativization of distant simultaneity. It is only at this point that the bifurcation between classical and relativistic physics occurs. Thus, Eddington is quite wrong in arguing that the content of the laws of special relativity is epistemic. With the one fairly minor proviso to which we have alluded, the epistemic circularity, which leads some writers to characterize distant simultaneity as an "unobservable", would be as much an ingredient of a Newtonian universe as it is of a relativistic one. The relativity of distant simultaneity springs not from epistemology but from the factual circumstance of the nature of the laws of electrodynamics. The source of the confused interpretation of Einstein's original presentation is that he employed light-rays in a non-essential way in his epistemological critique of the Newtonian or common-sense view of time but subsequently employed them in an essential way, i.e. as a surrogate for Maxwell's theory in deriving the Lorentz transformation.

There is another aspect of Eddington's position which stands in need of criticism. Eddington maintains that all of the fundamental laws of nature are subjective. Now such a claim must be intended to mean that the content of the law refers not to objective reality but to the nature of the human mind. The source of this view seems to be that the distances and durations which we ascribe to pairs of events depend on the state of motion of the human subject who is carrying out the pertinent spatial and temporal measurements. However, the theory of relativity does not claim that spatial and temporal intervals are mind-dependent but simply that they are frame-dependent. If one chooses, one may replace the expression "frame of reference" by "frame of reference of an observer". However, nothing is added to the theory of relativity by any such linguistic transformation. Moreover, no reference is made to observers in any standard axiomatic presentation of the theory of electrodynamics. "Observer" may appear in the text but only for heuristic reasons. Some philosophers seem to make the mistake of supposing that the measurement of a frame-dependent or relativistic magnitude is somehow subjective; but this is the result of cloudy thinking. Of course, one may always argue that any measurement is a measurement performed by a perceiving subject, just as any theory is the invention of a thinking subject. If this is the ground for a subjectivist interpretation of physical knowledge, then it is logically unassailable. But like all unassailable positions it is utterly vacuous. Moreover, we shall learn more clearly from the later chapters that it is possible to formulate the laws of space-time physics in such a way that even the notion of a reference frame turns out to be unessential.

Our final observation concerning the relationship between the theory of relativity and epistemological idealism is the following. In the theory of relativity, many of the absolute quantities of classical mechanics were replaced by relativistic ones. On the other hand, the theory of relativity introduced its own set of absolutes or invariants. The most notable among then is the space-time interval. But more generally, as we have seen, any quantity which may be represented by a four-vector or four-scalar corresponds to an absolute feature of the world, which is independent of the frame of reference. Perhaps the most unfortunate feature of the theory of relativity is its name.

Maxwell's theory of electrodynamics affords an excellent example. In its classical form, the theory postulates the existence of two distinguishable fields of force, the electric field $\mathbf{E}$ and the magnetic field $\mathbf{B}$. Another basic quantity is the electric charge q. The force experienced by a charged particle in an electric field is given by the simple law: $\mathbf{F} = q\mathbf{E}$. However, if the source of the electric field is in motion, an electric current is produced which gives rise to the magnetic field $\mathbf{B}$. The force experienced by a charged particle under the influence of a magnetic field is rather more complicated. It is given by $\mathbf{F} = q/c(\mathbf{v} \times \mathbf{B})$, where $\mathbf{v}$ is the velocity of the particle. Thus, the total force acting on a charged particle, the so-called Lorentz force, is given by

$$\mathbf{F} = q\mathbf{E} + q/c(\mathbf{v} \times \mathbf{B}) \tag{4.1}$$

The peculiarity, if that is an appropriate word, of this law is that the second component on the right-hand side is clearly frame-dependent. In a frame of reference in which all the charges are at rest, the magnetic field vanishes. However, in the four-dimensional formalism of special relativity, the electric field and the magnetic field are united in a single quantity represented by the electromagnetic field, which has an intrinsic significance which is independent of the frame of reference. Thus, from the standpoint of relativity, Maxwell's theory exhibits a greater degree of objectivity than it appeared to possess in its classical three-dimensional version.

The Transcendentalism of J. R. Lucas

A more recent attempt at an *a prioristic* interpretation of special relativity is due to one of the more remarkable of contemporary philosophers of science, J. R. Lucas. Lucas combines qualities of technical acumen and erudition with a proclivity for bold, some would say wild, philosophical speculation in metaphysical and theological areas. The particular approach which he adopts towards special relativity is one which derives from Kant. It is known as the transcendental method. In brief, a transcendental epistemologist, unlike the traditional empiricists and rationalists, does not attempt to justify our knowledge by showing how it may be reconstructed on the foundation of certain basic, indubitable intuitions or perceptions. In short, it is not a reductionist epistemology like phenomenalism, which attempts to reduce all genuine factual knowledge to direct acquaintance with sense data. The transcendental approach takes the finished product—for example, a scientific theory—as the given starting-point. Science exists; it is otiose to investigate whether it is possible. The proper subject of epistemological investigation is not whether there is a genuine science but, given that there is, to uncover the conditions which make it possible, the necessary presuppositions of science. To do this is to provide a transcendental justification of the scientific principles of interest and, not incidentally, to show them to be *a priori* principles which manifest the conditions for the very possibility of experience in general.

Lucas claims in *A Treatise on Time and Space* (1973) to provide a transcendental deduction of the Lorentz transformation. We should stress the remarkable character of this claim, since it amounts to saying that the conditions of knowing are such that the perspectives of two knowing subjects are bound to be related by the Lorentz transformation. His starting-point is the concept of a person who is located in space and enjoys the possibility of communicating with others, from whom he is spatially separated, in a causally coherent fashion. He claims "that granted certain assumptions, in particular about parity of esteem and the featurelessness of space, the Lorentz transformations will emerge as the natural and rational way whereby widely separated observers—persons—can correlate their respective systems for labelling places and dating events" (1973, pp. 195 f.).

He argues that the Lorentz transformation may be best understood as a set of translation rules for going from one mathematical language to another. In particular, two such languages will differ in the way in which they assign places and dates. The gist of Lucas's argument is to show that given that two physicists are able to assign space-time coordinates to natural events, each in his own language or from his own perspective, in a rational and mutually agreeable fashion, so that intersubjective agreement may be secured, the two languages must be related by the Lorentz transformation, which turns out to be the only consistent rule of translation.

Let us suppose that there are two persons, Red and Blue, perhaps a Russian and a Greek, who are in mutual communication. Each will require a dictionary to make sense of the other's claims. But a single dictionary, Lucas argues, is not enough. It is necessary to discriminate not only from which language to which the translation is being carried out but also by whom. As he puts it: "We have not only Russian–Greek and Greek–Russian dictionaries, but dictionaries published in Russia and dictionaries published in Greece, and often there are significant differences between them" (1973, p. 207). Thus, we have, in effect, four sets of rules: Red's rules for translating red into blue, Red's rules for translating blue into red, Blue's rules for red into blue and Blue's rules for blue into red. Let us schematize the situation by using unbarred letters for expressions in Red's language and barred ones for those of Blue. Furthermore, "=" will represent interlinguistic synonymy according to Red's rules "≡" will represent interlinguistic synonymy according to those of Blue. Finally, let Red's system of translation from red to blue be the function R and for blue into red be $\mathscr{R}$. Let Blue's system for red into blue be the function B and for blue into red be $\mathscr{B}$. Schematically, we then have:

	Red's rules	*Blue's rules*
red into blue	$(x, y, z, t) = R(\overline{x}, \overline{y}, \overline{z}, \overline{t})$	$(x, y, z, t) \equiv B(\overline{x}, \overline{y}, \overline{z}, \overline{t})$
blue into red	$(\overline{x}, \overline{y}, \overline{z}, \overline{t}) = \mathscr{R}(x, y, z, t)$	$(\overline{x}, \overline{y}, \overline{z}, \overline{t}) \equiv \mathscr{B}(x, y, z, t)$

(The foregoing is simply a variant of Lucas's own notation and is not, unfortunately, the usual functional notation. That is to say, that R is not a function defined on $(\overline{x}, \overline{y}, \overline{z}, \overline{t})$ but on (x, y, z, t). It is a set of rules for translating expressions of red into expressions of blue in a *red* context.)

Consistency obviously requires that if Red translates a certain expression from his own language into the foreign one and then translates the resulting expression back into his own language, the final result must be essentially the same as the original expression. That is to say, that $\mathscr{R} = R^{-1}$. The rules are simply inverses of each other. Similarly, $\mathscr{B} \equiv B^{-1}$. However, Lucas points out that R and B are not inverses but converses. In the context of the Lorentz transformation, this is associated with the fact that $\mathbf{v} = -\overline{\mathbf{v}}$.

Lucas now proceeds to work out his transcendental derivation. Its thrust is to determine what rules each must adopt to translate the referring expressions of the other into his own language. He first argues that the speed of communication between them must be universal. This follows from a principle which he calls *parity of esteem*. In effect, so long as each participant is using the best possible means of communication, neither is entitled to assume that his own is better or worse than that of the other. Moreover, if communication is to be accessible to a causal account, it must take place with a finite velocity. For example, the participants would have to agree that the time at which a message is received is later than the time at which it was sent.

Let us suppose that Red and Blue exchange a series of communications. Clearly, Lucas cannot assume that they be of the electromagnetic variety, for that would be to commit a fatal *petitio principii*. Suppose, therefore, that they are in the form of letters sent via their respective postal systems. Red initiates the correspondence with the following letter.

t_1 Home (O, O, O)

Dear Blue,
Blah blah.
Regards, Red.

Blue then replies:

$\bar{t}_2$ Home $(\bar{O}, \bar{O}, \bar{O})$

Dear Red,

Thanks for your letter dated "t_1".

Regards, Blue.

Red's answer is:

t_3 Home (O, O, O)

Dear Blue,

Just received your letter dated "$\bar{t}_2$".

Regards, Red.

Then:

$\bar{t}_4$ Home $(\bar{O}, \bar{O}, \bar{O})$

Dear Red,

Thanks for the letter dated "t_3", which just arrived.

Regards, Blue.

On the basis of this exchange, Red knows the time at which he sent the first letter, the time at which he received a reply, and the time at which Blue, according to the latter's own reckoning, received Red's letter. Likewise, Blue knows the time of his first reply, the time he received Red's response and the time that Red claims to have received Blue's letter. Now neither is fully conversant with the postal service of the other, or with the method of assigning temporal labels in the other's language. What, however, Red may legitimately assume is that later times for Blue are later times for himself, i.e. that Blue's time reckoning is a strictly monotonic increasing function of his own. Moreover, he knows that when Blue answered his initial letter, it must have been at a time later than that at which he, Red, had sent it and earlier than that at which he received the reply. If t_2 is the time which Red ascribes to Blue's $\bar{t}_2$, it follows that

$$t_2 = \varepsilon t_1 + (1 - \varepsilon) t_3 \tag{4.2}$$

The reader will recall the general significance of ε from the previous chapter.However, in the present discussion it has nothing to do with light signals. It will also be recalled that systematic reasons were invoked for assigning the value of $\frac{1}{2}$ to that parameter. We should, however, try to forget such arguments for the moment and consider the matter strictly in the context of Lucas's transcendental argument.

Lucas points out that if one held one's own postal system in very high regard, one would set ε very close to 1, which is to say, that Red would suppose that the letters sent to Blue would arrive very shortly after he had sent them, whereas most of the delay in receiving a reply would be attributed to the relative tardiness of Blue's postal system.

In setting $\varepsilon = \frac{1}{2}$, Red invokes a principle of non-egocentricity, a special case of parity of esteem, in accordance with which he is not entitled to suppose that the postal system of his own country is better than that of Blue's. Moreover, it is possible to provide a mathematical demonstration that if the speed of transmission is the fastest possible speed, then the result of combining it with any other speed is always the fastest possible speed. This, in turn, implies that it is a universal speed. Lucas admits that the last detail of the argument depends on the empirical(?) assumption that the fastest possible speed in practice happens, also, to be the fastest in principle. However, he maintains that it is a reasonable assumption to make in the absence of any evidence to the contrary. Now since all of the foregoing considerations apply equally to the situation of Blue, it follows that they must agree on the magnitude of the universal speed of communication, which is thenceforth denoted by the constant c.

Red, is now in a position to assign not only a time, in his language, to the event of Blue's receiving his letter but also a distance from his own home (O, O, O) to that of Blue $(\bar{O}, \bar{O}, \bar{O})$. The former is given by

$$t_2 = \tfrac{1}{2}(t_1 + t_3) \tag{4.3}$$

and the latter by

$$l_2 = \tfrac{1}{2}c(t_3 - t_1) \tag{4.4}$$

By the same token, Blue will express in his language the time and location of the reply received by Red. The latter expressions are, of course,

$$\bar{t}_3 = \tfrac{1}{2}(\bar{t}_2 + \bar{t}_4) \tag{4.5}$$

$$\bar{l}_3 = \tfrac{1}{2}c(\bar{t}_4 - \bar{t}_2) \tag{4.6}$$

It is not necessary to trace the details of Lucas's argument beyond this point. Suffice it to say that in a spirit of mutual esteem according to which the claims of Red, (4.3) and (4.4), are no better and no worse than those of Blue, (4.5) and (4.6), they are presented with the task of correlating those claims so that their consistency may be assured. This, of course, leads with algebraic inevitability to the Lorentz transformation.

Lucas is quite well aware that he may be suspected of some form of epistemic magic in having contrived to pull this robust empirical rabbit from a rather evanescent transcendental hat. That is to say, that he has apparently reached the conclusion that the most central aspect of the theory of special relativity has the status of *a priori* knowledge. His response is that such qualms are not appropriate to the transcendental derivation of the Lorentz transformation. All that has been established is that for *purposes of communication* between persons, the formalism of the Lorentz transformation is needed for translating the space-time ascriptions of one person into those of another. That these transformations are applicable to so many aspects of the physical world is to be treated as no more than a fortunate empirical fact, which suggests that we are "living in, if not the best, at least the most rational, of all possible worlds" (1973, p. 227).

It must be allowed that the foregoing attempt at a transcendental derivation is a fairly ingenious exercise on the part of a talented philosopher. Moreover, when studied in detail, the communication argument does enhance our intuitive grasp of the Lorentz transformation. Nevertheless, we believe that like all attempts to arrive at physical results on the basis of *a priori* reasoning it must be ruled a failure, albeit an interesting one. Admittedly, Lucas's claims are more modest than some of their kind, in so far as they purport only to establish a rational presupposition of intersubjectivity. This does fall short of ascribing the property of necessity to our knowledge of the world. In this respect, Lucas's claims are not so far-reaching as those of Eddington. Nor is his position as flawed as some early attempts to deduce Newton's laws of motion from the principle of sufficient reason. However, one may still wonder whether his argument based on the nature of communication between persons would seem nearly so compelling in the absence of prior knowledge of the formal structure of special relativity. If his reasoning were cogent, it would have been possible for an epistemologist to have derived the Lorentz transformation without any knowledge of physics, and this we find implausible.

In fact it would seem that the principle of parity of esteem, when stripped of its unfortunate valuative connotation, is but another version of the principle of sufficient reason. In this instance, it is to the effect that in the absence of contrary evidence two communicators, Red and Blue, must be supposed to have access to the same means of communication. This supposition is buttressed by the claim that each has at his disposal the *best* possible, i.e. the fastest, method of communication. A crucial ingredient of the argument is that the fastest signal be finite. Lucas adheres to the view, and he is not alone in this, that the causal relationship involves succession in an essential way and that the ascription of instantaneous action-at-a-distance to a physical process is a type of occultism. He may well be correct in his belief. However, I fail to see why instantaneous causal propagation must be excluded from nature on *a priori* grounds. In other words, a universe similar to that conceived of by Newton, in which instantaneous propagation, whether or not actual, is at least permitted by the formalism of the system, is surely a conceptual possibility. I know of no demonstration to the effect that classical mechanics is inconsistent. We do have an abundance of empirical evidence that it is false, but that is irrelevant to Lucas's transcendental argument. In short, it is possible that our communicators might have at their disposal an instantaneous signalling device whereby they could share the identical system of labelling the times and places of events.

Moreover, Lucas's claim that the signalling or communication device which is in practice the best is also, in the absence of contrary evidence, theoretically the best is highly dubious. In the late nineteenth century it was not supposed that the Michelson–Morley experiment had revealed that optical signals represent the limiting velocity in the universe. That was only a post-Einsteinian interpretation of the result. Furthermore, it is not too far-fetched to imagine a universe in which there is a finite upper bound to the velocity of causal propagation but such that those propagations are not available for purposes of transmitting signals. For example, a universe which is devoid of electromagnetic phenomena but in which gravitational interactions take place with an associated velocity of c would provide one such model.

A more realistic picture is that prior to the inception of special relativity, 'the best postal system' would not be taken to connote either an instantaneous transmission or a universal but finite one. In this case, a natural assumption would be that the system of which to avail oneself is the one that yields the best results, *on average*. Under such circumstances, both Red and Blue would make use of that system, supposing it to be uniquely determined, but it could not serve as a standard for dating distant events. However, normal communication, as we know it, would still be possible without any causal anomalies resulting thereby.

Finally, a warning flag that should make us query the cogency of Lucas's argument is the latter's own admission that there are approximately true descriptions of the world which are characterized by Lorentz covariance. Lucas's argument does not pretend to account for this feature but treats it as a "fortunate fact". Fortunate, indeed! The coincidence is so great that it leads Lucas to shift from philosophy of science to theodicy—this is the most rational of possible worlds—in an attempt to account for it. Surely, it would be more reasonable to argue that the Lorentz transformation has a purely physical basis, which has been ascertained in the usual scientific manner, and that interlocutors as parts of the physical world with only physical means of communication at their disposal are quite naturally constrained by it. Like Einstein, I would recommend the methodological precept that one eschew the attribution of long-shot coincidences to the world.

One may only speculate as to why men who are well versed in physics, Eddington and Lucas are cases in point, should succumb to the temptation of raising physical laws to the status of *a priori* principles. I would suggest that in the case of the theory of relativity, it may have something to do with Einstein's characteristic style. In the case of both the special and the general theories, Einstein presented all of the essentials of the theory with noticeably little direct appeal to the results of actual experiments. Thus, one may be led to suppose that they are simply the fruit of rational insight. In fact, this may not be altogether false. However, one must remember that a set of hypotheses only acquires the status of theoretical knowledge when it has been subjected to and has withstood the scrutiny of experiment. Of this, Einstein was very well aware. Physics must avoid the fallacy of the astrologers which consists in recording successes while forgetting the failures.

The Radical Empiricism of Percy Bridgman

Virtually all recent empiricists have singled out special relativity as a scientific vindication of their doctrine. Einstein, himself, gave credence to this reaction by asserting his intellectual debt to Mach and Hume. But at other times, he stressed his conviction that scientific theories are free creations of the human intellect. One thing may be stated with assurance. With the passage of time, the views of Einstein became progressively divorced from the position of classical empiricism. But at the same time so did those of many contemporary empiricists. Thus, to discuss the question of the correctness of the empiricist interpretation of relativity theory is to get caught in the web of historical scholarship. One can hardly deal with such a question without first specifying to which brand of empiricism one is referring and at what stage of its development.

To render our discussion reasonably determinate, we shall concentrate on two versions of empiricism both of which are still quite influential in certain scientific circles; although both would be regarded as excessive by many contemporary philosophers of science. The more radical version was proposed by the American physicist P. W. Bridgman in *The Logic of Modern Physics* (1927). It is called *Operationalism.*

According to Bridgman, special relativity has an epistemological significance far beyond the specific revisions which it called for in our fundamental concepts of space and time. The most revolutionary aspect of Einstein's contribution lies in its requiring us to revise our ideas concerning the status of scientific concepts in general. According to Bridgman, in the light of special relativity, we must revise our views concerning the relationship which holds between scientific theories and the world. The singular psychological impact of special relativity in its early years resulted from the fact that the scientific community was totally unprepared to receive the profoundly new kinds of fact entailed by that theory. This was due to the failure of the pre-relativity scientist to pursue the path of "pure empiricism". Bridgman argues that if the scientist learns the appropriate epistemological lesson of Einstein's theory, no such conceptual shock could possibly result from any future scientific discovery. The attitude of pure empiricism is that there is no *a priori* principle which can limit the content of experience. Our thinking must be so adjusted that it is prepared for all future empirical contingencies. It will doubtless already have occurred to the reader that the principal philosophical lesson of relativity theory, as perceived by Bridgman, is the very opposite of that which was proposed by Eddington. According to the latter, relativity theory revealed that experience is determined by *a priori* features of the cognitive apparatus. According to Bridgman, relativity theory revealed that "Experience is determined only by experience".

Bridgman cites Newton's definition of absolute time as an example of an incorrect view of physical concepts. Absolute time was defined by Newton in terms of the *properties* which it was assumed to possess. There was no assurance, however, that any such concept could correspond to something in the physical world. It was left to Einstein to show that the concept was empirically empty. The appropriate way for science to define its concepts, Bridgman argues, is not in terms of properties but in terms of the experimental procedures with which it is associated. "In general, we mean by any concept nothing more than a set of operations; *the concept is synonymous with the corresponding set of operations*" (*The Logic of Modern Physics*, p. 5).

Consider, by way of illustration, the concept of length. The concept of length in physics is fully understood once the physicist has specified how to determine the length of any given object. The concept of length is, therefore, not to be defined in terms of its intrinsic properties or *essence* but in terms of the experimental operations which have been specified to determine a length. By the same token, the appropriate definition of absolute time would be the unambiguous specification of the set of operations by which absolute time is measured. However, when we examine any of the operations by means of which time is actually measured, we find them to be relativistic. Similarly, distant simultaneity, prior to Einstein, was regarded as an intrinsic property of pairs of events, so that any given pair would either have it or lack it. Einstein demonstrated that in fact such a property is not intrinsic but involves a relationship between two events and an observer. That is to say, that the ascription of simultaneity to a pair of events depends on the operation by means of which the time of those events is determined.

In fairness to Bridgman, we certainly admit that some of the key passages in Einstein's early writings seem to lend some support to Bridgman's interpretation. However, the doctrine of operationalism turns out to involve insuperable difficulties of a logical, methodological and semantic nature. Logically, to claim that a concept is to be defined operationally is to say that in any linguistic context in which that concept occurs, it may be replaced by a statement about a set of operations which preserves the meaning of the given context. This is what is meant by the word "definition". The concept to be defined, the *definiendum*, is simply an abbreviation for the defining expression, the *definiens*. On Bridgman's view, to attribute a certain theoretical property to X is equivalent, by definition, to saying that if a certain experimental operation is performed on X, then a specific test result will ensue.

Let us undertake a formal analysis of the relationship of simultaneity between two events, in the light of the foregoing considerations. We symbolize the events by E_1 and E_2. Let $I(C_1, C_2, E_1, E_2, O)$ be an instance of the relation of synchronized clock readings of E_1 and E_2 with respect to an observer O. $t(C_1, E_1)$ is the time of E_1 as given by the reading C_1. Finally, $S(E_1, E_2)$ is an instance of the binary relation of simultaneity between E_1 and E_2. Then the operational definition of simultaneity may be represented by the following schema.

$$S(E_1, E_2) = df. \quad I(C_1, C_2, E_1, E_2, O) \rightarrow t(C_1, E_1) = t(C_2, E_2)$$

In words, it is the assertion that E_1 and E_2 are simultaneous relative to O if and only if the result of the test procedure I performed on E_1 and E_2 is such that the two clock readings C_1 and C_2 are associated with a single number. But any student of logic should be able to perceive the flaw in this definition. Specifically, it is equivalent to the assertion that any pair of events on which the procedure I has *not* been carried out will be vacuously in the relationship of simultaneity. We see, therefore, that the very conception of an operational definition of simultaneity (or of any other theoretical concept) is logically untenable. The reader should note that we are not objecting to the nature of the test procedure, as such, but to the supposition that any test procedure could play the role of a definition.

Certain of the methodological difficulties of operationalism may be illustrated by a consideration of Bridgman's treatment of the concept of the electric field. An electric field is a field of force which is produced by a static configuration of electric charges. The presence of the electric field at a given point in space is determined by the effect which it has on a test charge q at that point. However, the introduction of the test charge will have an effect on the original static charge configuration which produced the field. That is to say, that the method for determining the intensity of the field at a given point will actually result in a change in the field. Clearly, to measure the field at a given point, we must make the test charge as small as possible. From the law $\mathbf{F} = q\mathbf{E}$, we define the electric intensity of the field by the limiting process

$$\mathbf{E} = \lim_{q \to 0} \mathbf{F}/q \tag{4.7}$$

But such a "definition" is, from a physical standpoint, a nonsensical formula for the simple reason that electric charge is a discrete quantity which occurs only in integral multiples of a basic elementary charge which is not further divisible. Thus, the notion of the test charge approaching zero in the limit is strictly incompatible with a fundamental physical fact.

It is probably for this reason that Bridgman classifies the electric field as a *construct*. He defines a construct as a conceptual invention which enables one to deal with physical situations which are not directly experienced but only inferred. For example, the inside of an opaque body is a concept which involves an element of invention. We cannot have direct experience of such an entity but find it to be a necessity of thought. Now Bridgman maintains that in some cases one is entitled to ascribe physical reality to theoretical constructs whereas in others one is not. In the latter case, our physical constructs are called *fictions*. Bridgman maintains that the electric field is one instance of such a fiction. It is a fiction in so far as we have no evidence for its existence beyond the operational procedure from which it is inferred. In other words, we infer the existence of the electric field from the accelerations imparted to test charges in the presence of other charges but have no basis for supposing it to exist which is independent of our operational definition of electric-field intensity. Thus, the electric field, for Bridgman, is simply a convenient way of describing the behaviour of charged bodies.

It appears to us that Bridgman is overlooking an important function of theories. It is not only the case that theories have predictive power. They are also designed to explain phenomena. In the absence of the electric field, the mutual interaction of electrically charged bodies would be incomprehensible. In addition to being instruments of prediction, physical theories are rational conjectures concerning the nature of reality. We see, therefore, that operationalism leads to a highly restrictive conception of the nature and role of theories. On this view, we could not say of a scientific theory that it is true or false but merely that it works or fails to work. Moreover, the foregoing treatment of the electric field is hardly viable from the standpoint of conceptual simplicity. If the electric field were nothing more than the reactions of test charges, it would follow that whenever a test charge is introduced the field comes into existence while in the absence of test charges there is no field. But this is akin to the view that has from time to time been entertained by philosophers to the effect that physical objects exist only so long as they are actually perceived. One of the many objections to such a view is that no explanation is provided of the spontaneous generation and annihilation of such objects. Moreover, it would seem likely that no explanation is possible which cleaves to the operationalist doctrine.

Sympathizers of Bridgman would no doubt maintain that my criticism is tendentious. Surely, they would argue, Bridgman's point is not that the electric field is an entity which appears and disappears in magical fashion but rather that there is no such entity in the first place. Now if the electric field does not exist, it is unclear why there should be any need to define it at all. Secondly, without the electric field it is impossible to account for the behaviour of the test charges when they are introduced, short of postulating an action-at-a-distance. But while action-at-a-distance cannot be ruled out on *a priori* grounds, it is surely more akin to the occult than is the electric field. Moreover, the mathematics of the theory of electromagnetism requires that the action of the source be propagated with a finite velocity—namely, the familiar *c*. But this leads only to further perplexity in so far as it would appear to involve the ascription of velocity, a physical magnitude, to that which is not. Bridgman's doctrine seems to generate an intricate philosophical web which derives from the inconsistent view that the electric field is a necessary conceptual ingredient of scientific theory but that the meaning of a scientific concept is exhausted by a set of test procedures. The reasonable solution to these difficulties is surely to allow that unseen fields may be real just as unseen material objects continue to exist, which is not, of course, to say that the latter is a precise analogy.

Finally, we shall briefly consider the thesis of operationalism from a semantic standpoint. If the meaning of a concept is equivalent to an associated test procedure, it follows that to be acquainted with the test procedure is to understand the meaning of the concept. However, such a view scarcely seems supportable. For example, there is an elementary standard test procedure for determining whether an electric current is flowing from *A* to *B*. It involves the simple operation of connecting a test instrument called an ammeter across the two points. If the needle of the ammeter is deflected, then a current is flowing, otherwise not. But it is obvious that young children and fools could be taught to carry out the test procedure without their having the least insight into the meaning of "electric current". Thus, to equate meaning with test procedure appears to be an egregious semantic confusion.

As physical theory becomes more advanced, as it increases in depth and scope, its basic theoretical concepts become increasingly remote from direct experience. Hence, we may expect to find a relatively greater number of "fictions" incorporated in the conceptual structure of the more advanced theories. It is noteworthy that Bridgman, who hailed Einstein's theory of special relativity as a conceptual revolution in scientific methodology, was led to reject the later theory of general relativity on the ground that it departed from its author's original operational viewpoint. It is instructive to cite Einstein's brief comment on Bridgman's criticism of the later theory.

"In order to be able to consider a logical system as physical theory it is not necessary to demand that all of its assertions can be independently interpreted and 'tested' 'operationally'; *de facto* this has never yet been achieved by any theory and can not at all be achieved. In order to be able to consider a theory as a *physical* theory it is only necessary that it implies empirically testable assertions in general" (1959, p. 679).

It is unfortunate that the scientific community at large has not been generally aware of Einstein's methodological views. Had it been so, it is doubtful whether the doctrine of operationalism would have enjoyed such wide acceptance. For example, social scientists, in an effort to be "scientific", supposed that they should endeavour to construct hypothetico-deductive systems analogous to the hypothetico-deductive theories of physics. However, they assumed that all of the concepts of a hypothetico-deductive system stood in need of operational definition. Thus, for example, we find that some psychologists identified "intelligence" with the score on an I.Q. test. Still others maintained that "the mind" is simply a fiction, since it is not accessible to direct observation. These are instances of the fact that scientific progress may actually be impeded by an incorrect philosophy of science.

In conclusion, we may say that Bridgman's principal error of interpretation was to focus on Einstein's treatment of spatio-temporal measurement to the virtual exclusion of all other aspects of the foundations of the theory of special relativity. As we have repeatedly emphasized, special relativity emerged from electrodynamics. In the absence of that theoretical underpinning, it is reasonably safe to speculate that Einstein's analysis of simultaneity would have been treated as an ingenious but scientifically insignificant footnote in empirical epistemology. Moreover, in the light of its intimate association with electrodynamics, it is hardly surprising that special relativity is essentially a field theory. We have already argued that Bridgman's treatment of the field concept is philosophically unsatisfactory. However, we shall subsequently learn that in any theory which is essentially field-theoretic, the field plays a physical role as a carrier of energy. Hence, it may hardly be set aside as a trivial methodological fiction.

The Liberalized Empiricism of the Logical Positivists

Like "idealism" and "rationalism", "empiricism" is a term which is associated with several diverse epistemological and semantic viewpoints. We have seen that operationalism is a version of empiricism to the effect that the meaning of individual terms is fixed by given experimental procedures. A rather less restrictive form of empiricism which is associated with the later phases of logical positivism is to the effect that although it may be either impossible or highly impracticable to specify the empirical significance of every descriptive term of a theory, it is necessary that every sentence of a theory have a literal or cognitive significance which admits of empirical interpretation. In other words, every individual statement of a theory must refer to an empirical state of affairs. Sentences rather than descriptive terms are now taken to be the fundamental bearers of semantic significance. This point of view has sometimes been encapsulated in the form of the semantic slogan: The meaning of a proposition is its method of verification. Let us now consider whether special relativity lends itself to this more liberal brand of empiricism.

We have seen that the theory is largely based on two fundamental postulates—namely, the principle of constancy and the principle of relativity. We shall begin by considering the empirical status of the former. There can be little doubt that the principle of constancy is accessible to experimental testing. If the velocity of light were dependent on the velocity of its source relative to an observer, it would be possible so to discern. However, that is not to say that the principle of constancy refers to an empirical state of affairs. What, one might ask, is the empirical state of affairs to which it refers? An empiricist might be tempted to reply that it refers to the outcome of numerous experiments such as the Michelson–Morley experiment. But it is surely of some significance that the principle was first enunciated by Einstein after a lapse of nearly twenty years from the time that the experiment was performed. Moreover, there is extremely solid evidence that the standard textbook presentation of special relativity, which maintains that the theory was formulated to account for the outcome of that experiment, is historically incorrect. This matter is well documented by Holton (1973) in *Thematic Origins of Scientific Thought*, chapter 9. However, our concern now, as always, is primarily methodological rather than historical.

The liberalized empiricism, to which we alluded above, would seem to uphold the view that a scientific hypothesis or postulate must be based on an inductive generalization from known phenomena. It is difficult to understand how the principle of constancy could be interpreted in this way. In the first place, the number of experiments which give *direct* support to the principle is quite small. Moreover, it should be noted that the Michelson–Morley experiment was designed not to establish the principle of constancy but to detect the effect of the earth's motion with respect to the aether. Michelson actually regarded his experiment as a failure rather than as a revelation of a new law of nature. Furthermore, the null-effect of the experiment might be explained in divers ways which do not involve recourse to a principle of constancy. In fact, several such attempts at explanation had been made by Einstein's older contemporaries including Lorentz. The inductivist view of science is, at best, naive when brought to bear on the interpretation of advanced physical theories. It may work, up to a point, when applied to the pre-theoretic, classificatory stage of science but is inadequate as an account of developed scientific theory. Rightly or wrongly, a case may be made for the claim that "all kangaroos are marsupials" is an inductive generalization. Both "kangaroo" and "marsupial" are observational concepts. Moreover, kangaroos have been observed in great abundance. Hence, one might plausibly argue that the foregoing generalization refers to the totality of observed instances of kangaroos which bear marsupial characteristics. However, the velocity of light is hardly an observational concept. On this point at least, Bridgman and I would be in agreement (cf. Bridgman's *The Logic of Modern Physics*, pp. 150 ff.). Moreover, the claim that this velocity is always independent of the velocity of its source cannot be plausibly interpreted as referring to a totality of observations.

Naturally, it is necessary that scientific hypotheses be in agreement with observation and experiment. However, that is a far cry from the claim that observation is the *referent* of those hypotheses. The role of observation in *advanced* science is principally a corrective one. It must be invoked for the modification or rejection of hypotheses. It is difficult to understand how this methodological truth could lead to the view that the *meaning* of a scientific hypothesis is to be found in its observational consequences. Nevertheless, it is not difficult to understand why Einstein's own presentation of his theory of 1905 has provided grist for the empiricist mill. Einstein did nothing less than to provide a revolutionary reinterpretation of the traditional and pervasive concepts of space and time. These concepts certainly appeared to be in perfect agreement with experience. Thus it was incumbent on him to provide a critique of the widely held world-view which he proposed to replace. Accordingly, it made excellent expository sense to dwell on the observational shortcomings of the traditional theory. In the same paper, however, Einstein makes it clear that the principle of constancy is an assumption. Indeed, how could so sweeping a generalization be anything else? The direct evidence, if any such existed, would have been so scant that it could not possibly have been taken as an inductive warrant for this postulate.

The principle of relativity was originally formulated by Einstein in the following form. "The laws by which the states of physical systems undergo change are not affected, whether these changes of state be referred to one or the other of two systems of coordinates in uniform translatory motion" (1905, p. 41).

One can hardly fail to be struck by the comprehensive character of this principle which Einstein describes as a "conjecture". It is neither logically self-evident nor observational in any obvious sense. As the reader knows, it amounts to the claim that there is no privileged frame of reference and hence no fundamental medium in or through which the fundamental processes of nature unfold. It is true that Einstein refers, albeit in a most cursory fashion, to some vaguely identified experimental results. However, the relevant data are not to be viewed as direct instances of the relativity postulate but merely as "unsuccessful attempts" to associate an absolute velocity with the earth's motion. From a logical standpoint, the mere failure to discern or to measure an absolute velocity cannot be taken as direct evidence for the relativity principle. The most that may be claimed is that the postulation of that principle is in accord with experience. However, as every student of methodology should know, indefinitely many mutually incompatible hypotheses may be formulated which are all in agreement with experience. Experience is, therefore, a guide to, or constraint on, hypothesis formation but not a source of new hypotheses.

In the second place, the principle of relativity is a claim about all fundamental processes whatsoever, both known and unknown, as referred to any non-accelerating frame of reference in the universe. The broadness of this claim could hardly find justification when interpreted in terms of an empiricist philosophy of science which would have us confine our assertions within the boundaries of experience. For example, we have no direct experience of the possible kinematic effects that could be generated by any of the myriad galaxies that are moving at velocities approaching that of light with respect to our own. The principle of relativity is a conjecture that physical processes will be the same when viewed from any such frame as they are when referred to our own galaxy. Some critics might argue that the mathematical form of the fundamental laws of nature together with that of the Lorentz transformation guarantee the principle of relativity. But that, of course, is to beg the question, since, if the principle of relativity were found to be false with respect to one or another galaxy then either the laws of nature or the Lorentz transformation would have to be abandoned. Mathematical formulae merely follow the facts.

Finally, it should be noted that when the two principles which we have been discussing are taken in combination, they lead to a multiplicity of consequences which appear to fly in the face of experience. We allude to the rejection of the classical view of space and time which seemed to represent the very fabric of all of our experience as well as to the rejection or, at least, to the profound modification of the principles of Newtonian mechanics, for which there is an abundance of supporting evidence. Had Einstein followed the methodologically conservative path of an empiricist it is surely doubtful whether he would have succeeded in his great theoretical innovation.

Bibliography for Chapter 4

AYER, A. J., *Language, Truth and Logic* (2nd edn.), Gollancz, London, 1946. This remains the clearest account of logical empiricism.

BRIDGMAN, P. W., *The Logic of Modern Physics*, Macmillan, New York, 1927.

BRIDGMAN, P.W., *The Nature of Physical Theory*, Dover, New York, n.d. Here, the author sets forth some of his objections to general relativity.

CARNAP, R., "The Methodological Character of Theoretical Concepts", in *The Foundations of Science and the Concepts of Psychology and Psycho-analysis* (eds. H. FEIGL and M. SCRIVEN), Minnesota Studies in the Philosophy of Science, vol. 1, University of Minnesota Press, 1956. The later views of the leading exponent of logical positivism, including a critique of operationalism.

EDDINGTON, A. S., *The Philosophy of Physical Science*, Cambridge University Press, Cambridge, 1939.

EINSTEIN, A., "Remarks to the Essays Appearing in this Collective Volume", in *Albert Einstein: Philosopher-Scientist* (ed. P. A. SCHILPP), vol. 2, Harper, New York, 1959.

FRANK, P., *Philosophy of Science*, Prentice-Hall, Englewood Cliffs, N.J., 1957. Concentrates on the philosophy of physics from an operationalist and empiricist standpoint.

HOLTON, G., *Thematic Origins of Scientific Thought*, Harvard University Press, Cambridge, Mass., 1973.

5

Special Relativity and Conventionalism

It has already been remarked that a key problem in the philosophy of science concerns the relationship between scientific theory and the world. The varieties of opinion on this issue range from the view of a scientific theory as a direct representation of reality, figuratively speaking as a diagram of the world, to that of its being a conceptual fiction whose status is the purely instrumental one of predicting phenomena. Most philosophers would naturally tend to favour some intermediate position. They would recognize that the scientist is concerned with the nature of reality. However, they would also hold that theoretical representations of the world do contain certain conceptual ingredients which are not actually determined by the world but are selected on such diverse grounds as aesthetic appeal, mathematical tractability, tradition, and so on. Accordingly, it is incumbent on philosophers to analyze the foundations of a physical theory in order to ascertain precisely which of its ingredients have a factual basis and which have been chosen on the basis of one or another convention. Those philosophers who claim that the conventional ingredients are more ubiquitous and fundamental than most scientists would commonly suppose are called *conventionalists*. However, there are obviously degrees of conventionalism and differences of opinion on important but quite specific details. In particular, there has been considerable debate concerning the status of the concept of simultaneity in the theory of special relativity. The principal aim of this chapter is to present the reader with the main issues surrounding this debate. However, we need first to set the stage by providing some historical and philosophical background.

Since the doctrine of conventionalism is to be discussed in detail in Chapter 11, the present introduction to the topic will be restricted to the briefest possible sketch. The three principal exponents of conventionalism are Henri Poincaré (1854–1912), Hans Reichenbach (1891–1953) and the contemporary philosopher of science, Adolf Grünbaum. Poincaré's theory of conventionalism was inspired by the development in the nineteenth century of various systems of non-Euclidean geometry. Prior to this revolutionary mathematical development, it had been naturally supposed by the overwhelming majority of scientists and philosophers that Euclidean geometry was the one and only possible formal theory of the structure of space. However, given the choice from among competing systems of geometry, the metric structure of space soon came to be perceived as an empirical question. Of the various competing systems of metric geometry, which one truly describes physical space? Poincaré responded by arguing that one cannot ascribe absolutely the features of any system of metric geometry to the world. The most that may be claimed is that one geometry provides a simpler and more convenient description of phenomena than any other. In fact, Poincare believed that Euclidean geometry occupied precisely that privileged place. At the same time, however, he maintained that whatever the facts of experience might be, the possibility of retaining Euclidean geometry, or, for that matter, any of the systems of non-Euclidean geometry, would always be a live option. Ultimately, the geometry which one adopts to describe the world is a matter of convention. Strictly speaking, the geometric features are not discovered but are decreed by human beings. However, it may prove to be the case that in order to retain a particular geometry for the description of the world, it will be necessary to make compensating adjustments in the fundamental laws of physics. Hence, the latter are also perceived to contain an ineradicable conventional ingredient.

Reichenbach's version of conventionalism is somewhat milder than the one which we have just described. Whereas Poincaré was apt to make sweeping and even extravagant claims concerning the conventional status of applied geometry and the principles of physics, Reichenbach was aware of the necessity of subjecting scientific theories to a detailed and painstaking analysis in order to determine precisely which of their ingredients are to be deemed conventional and which factual. His fundamental philosophic orientation was one of strict empiricism including a verificationist theory of meaning. Accordingly, he would maintain that when a particular aspect or concept of a theory is determined by experience, it should be regarded as factual. On the other hand, when the theory may be said to outrun experience, in the sense that a given ingredient is experimentally indeterminate, one may only conclude that such an ingredient has the status of a convention. The conventional aspects of a theory are, in effect, merely different modes of describing the numerically identical experimental facts. At the same time, he stressed that although there are alternative descriptions of the experimental data, the choice from among which is simply a matter of convention, it should not be supposed that all possible descriptions of the world are conventional equivalents. One set of conventions may differ in factual content from another set of conventional alternatives. Thus, once the choice of convention has been made, the task of theory construction is a factual problem which is to be pursued on the basis of experimental results.

Grünbaum, who is possibly the most influential and certainly the most prolific of the contemporary philosophical interpreters of the theory of relativity, has been deeply influenced by the work of Reichenbach. He tends generally to be in agreement with Reichenbach as to which ingredients of the theory of relativity are conventional and which are of the factual variety. However, he has always stoutly defended his contention that there is a profound philosophic difference between Reichenbach and himself. He rightly points out that Reichenbach's conventionalism has an epistemological foundation and is largely based on verificationism, a doctrine which Grünbaum explicitly rejects. He maintains that his own brand of conventionalism is rooted not in epistemology but in ontology. What he has in mind is that conventions ought only to be introduced into scientific theories when the nature of reality is such that no facts, whether directly knowable or otherwise, are available as a basis for a factual claim. For example, the geometry of space is a matter of convention, not because it defies experimental resolution but because, as a matter of ontological fact, space is not the kind of entity which could possess a metric in its own right. With this background in mind, we may now proceed to discuss Grünbaum's analysis of the status of simultaneity in the theory of special relativity.

The Conventionality of Simultaneity

Much of the recent discussion of conventionalism as it pertains to special relativity has been focused on the status of the concept of simultaneity, which, as the reader knows, is central to special relativistic kinematics. The bearing of the preceding discussion of geometric conventionalism on this issue is revealed by the following claim of Grünbaum. "The philosophical status of the Riemann–Poincaré conventionality of congruence is fully analogous to that of Einstein's conventionality of simultaneity" (1963, p. 28).

According to Grünbaum, the well-known procedure for the ascription of simultaneity to distant events, as prescribed by Einstein in his foundational paper of 1905, renders that relation conventional. At the outset of our discussion it is important to emphasize the obvious distinction between the thesis of the conventionality of distant simultaneity and the universally accepted doctrine of the relativity of distant simultaneity, since the two have been confounded on occasion. The latter is the thesis that once a physically acceptable standard of simultaneity has been adopted for a particular frame of reference, it turns out that a pair of events which are simultaneous in that frame and on the given standard will be non-simultaneous with respect to any other frame which is in motion relative to the first one. The doctrine which now confronts us, of which both Reichenbach and Grünbaum have been eloquent proponents, is to the effect that the aforementioned standard of simultaneity is actually a convention, one of many possible definitions of simultaneity which are compatible with the data and factually equivalent.

Reichenbach's account of the Einsteinian 'definition' of simultaneity runs as follows. In order to determine the simultaneity of two events which are located at the spatially distant points A and B, it is necessary to have a clock located at A and another at B. In addition, there must be a definite procedure whereby the two clocks are synchronized, so that if they are found to record the same number when one event occurs at A and another at B, the two events may be said to have occurred simultaneously. Obviously, if one of the two clocks were running 2 minutes ahead of the other, the concordance of their respective readings could hardly serve as an indication of simultaneity.

As the reader knows, Einstein proposed that the two clocks be synchronized in the following manner. A light-ray is emitted from A at t_1. It arrives at B at t_2, where it is immediately reflected so that it returns to A at t_3. The travel time of the light-ray from A to B is $t_2 - t_1$. Its travel time from B to A is $t_3 - t_2$. Now if we assume that the velocity of light is the same in each direction, we may infer that t_2 is the midway point in the total transit interval, $t_3 - t_1$. Hence, we may write

$$t_2 = t_1 + \tfrac{1}{2}(t_3 - t_1) \tag{5.1}$$

Both Reichenbach and Grünbaum have argued that (5.1) contains a significant conventional ingredient. Since it is based on the presupposition that the velocity of light is isotropic (independent of direction), it would have the status of a brute fact only if the identity of the velocity of light in various directions were also a brute fact. However, in order to measure a one-way velocity between two points, it is necessary to have clocks located at those points, which are known to be synchronized. *A fortiori*, synchronized clocks are presupposed for the determination of one-way light velocities. Hence, the attempt to base simultaneity on (5.1) turns out to be circular. One may only break out of the circle by *stipulating* the correctness of (5.1) as based on light-signals. In short, what we have at our disposal is not a method of measurement but a *definition* of simultaneity, a convention.

Reichenbach argues that there are infinitely many alternative definitions which, from a strictly epistemic standpoint, are as legitimate as that of Einstein. Specifically, any definition of the form

$$t_2 = t_1 + \varepsilon\,(t_3 - t_1),\ 0 < \varepsilon < 1 \tag{5.2}$$

where the restriction of the parameter ε to the open unit interval is to preclude such anomalies as an effect's preceding its cause. Reichenbach does freely admit that the choice of $\varepsilon = \frac{1}{2}$ results in great descriptive simplicity but holds that, apart from this aesthetic factor, any other choice of ε, within the specified range, would fit the "facts" just as well. In a nutshell, what is being claimed is that there is no fact of nature of any kind which requires or provides a unique standard for the setting of clocks and, consequently, for the determination of simultaneity. The apparent fact on which special relativity is based turns out not to be a fact at all but merely a convention.

The nature of the relationship of simultaneity may be exhibited in the diagram of Fig. 5.1 in which P and P' are two distinct spatial points. The solid arrows which point in the direction of time represent the world-lines of the two points.

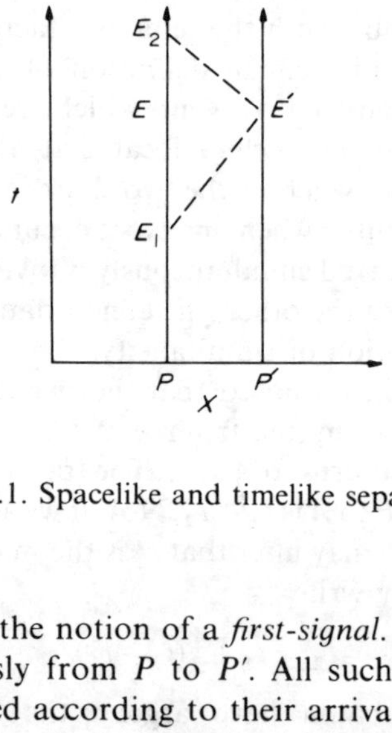

FIG. 5.1. Spacelike and timelike separations

Reichenbach now introduces the notion of a *first-signal*. Suppose that various kinds of signals are sent simultaneously from P to P'. All such signals, e.g. optical, acoustic, projectile, etc., may be ordered according to their arrival time at P'. The fastest signal between any two points will be called a first-signal. Then it is a basic law of the theory of relativity that light is a first-signal. Let E_1 in Fig. 5.1 be an event on the world-line of P corresponding to a first-signal's being emitted at P. E' is the event on the world-line of P' corresponding to the reception of this signal at P'. Simultaneously with the arrival of that first-signal at P', a second one is transmitted from P' to P, the event of its arrival on the world-line of P being E_2. In accordance with the rule for setting clocks as given by (5.2), it may be objectively asserted that E_1 is earlier than E' and the latter earlier than E_2. The objective basis for this is the existence of an asymmetric causal series which connects E_1 and E' and a second one which connects E' and E_2. These are indicated in the diagram by the broken lines. In the terminology of Reichenbach, E_1 and E' are connectible by a first-signal and likewise in the case of E' and E_2.

We now consider an arbitrary event E on the world-line of P in the open interval between E_1 and E_2. The question then arises as to the temporal ordering of E and E'. It is obvious from the diagram that the two events are so close to each other in time, relative to their spatial separation, that it would be impossible for them to be connected by a first-signal. That is simply to say that there is no physically realizable causal sequence between them. Thus, unlike the case of the other event pairs that we have considered, this one lacks the objective basis of a first signal whereby their temporal order could be established. One could arbitrarily declare E to be earlier than E' or conversely, or hold that the two events are simultaneous. Intuitively, E and E' could not be events belonging to the world-line of any clock. It is for this reason that physicists call intervals of the kind EE' *spacelike*. It follows, *mutatis mutandis*, that each and every event in the open interval E_1E_2 is in spacelike separation from E'. Reichenbach and Grünbaum prefer to describe the relation between E and E' as one of *topological simultaneity*. It should be noted that this is a very particular sense of simultaneity, since it fails to distinguish *unique* pairs of events. That is to say, that several distinct events at P bear the relationship of topological simultaneity to a unique event at P'. However, if one selects the rule (5.1), i.e. letting ε equal one half, a unique pair of events will be distinguished. Two such events are said to be *metrically simultaneous*. However, since the choice of ε is arbitrary, within limits, some other event pair might just as well have been singled out as metrically simultaneous. Thus, it turns out that whereas topological simultaneity is factual, metric simultaneity, what is normally understood by "simultaneity", is conventional.

It is at this point in the elaboration of the thesis of the conventionality of simultaneity that Grünbaum injects his characteristic viewpoint. Just as he argued that the conventionality of congruence is a consequence of the *fact* of the continuity and homogeneity of space, so now he would argue that the conventionality of simultaneity is also a consequence of an empirically established fact. It was noted in an earlier chapter that in Newtonian physics there is no upper limit, at least in principle, to the velocity of signal propagation. In the language of Reichenbach, this would mean either that there are no first-signals or that the velocity of propagation of a first-signal would be infinite or instantaneous. Intuitively, that is to say, that for any two events in the universe, it is possible in theory to transport a clock so rapidly that it will be in the space-time location of both of them. Equivalently, any two events will be on the world-line of a possible clock, so that there would be no spacelike separations in the spacetime of classical physics. The limiting case arises when two events are so related that the only clock on whose world-line both could occur would be one having an infinite velocity. Such event-pairs would uniquely constitute or determine a relationship of metric simultaneity. Grünbaum's point is that it is a matter of empirical fact that the velocity of light is an upper limit to causal propagation in general. Accordingly, it is *a matter of fact* that the basis for the determination of metric simultaneity rests on a convention.

It will probably have occurred to the reader that there is an inconsistency between the view that the choice of ε is conventional and the postulate of the constancy of the velocity of light. However, on the Reichenbach–Grünbaum interpretation of special relativity, it is argued that the principle of constancy refers only to round-trip light velocities. That is to say, for example, that if a light-ray is emitted from a certain point A, is then reflected back to A from B, the total time which elapses may be measured on a single clock located at A

and that $2AB/\Delta t = c$. On the other hand, it is argued that the theory is uncommitted to specific values for the incoming and outgoing velocities. Any set of values whose average is c is completely consistent with the empirical facts.

Since there is empirical confirmation that the round-trip velocity of light is constant, it follows that once a value has been conventionally assigned to ε for one direction, this value will then determine the one which is associated with the opposite direction. In the case of a simple to-and-fro trip from A to B and back to A, whatever the value of ε for the AB transit, its value for BA will be $1-\varepsilon$. Thus, only in the case of $\varepsilon = \frac{1}{2}$ is the value of ε the same in all directions. We see, therefore, that there is at least one respect in which the choice of $\frac{1}{2}$ is distinct from all other possibilities—namely that it is the only choice on the basis of which the method of synchronization is fixed once and for all.

Grünbaum considers the slightly more complex case of a light-ray which travels a closed path corresponding to the sides of an equilateral triangle as depicted in Fig. 5.2.

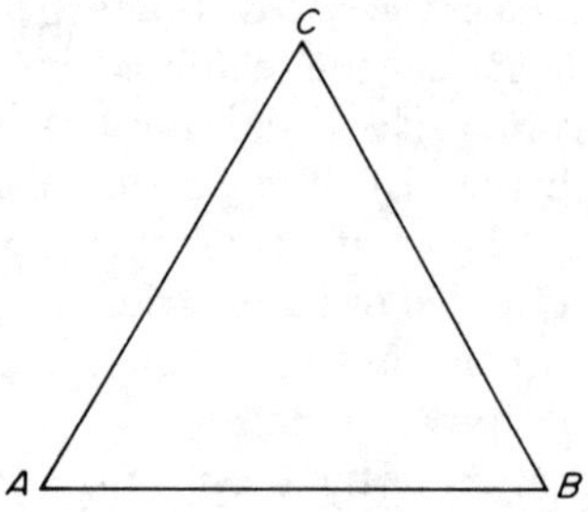

FIG. 5.2. Triangular light-path

If the round-trip time for ABA has been measured as $2T$, it may be presumed that the duration of the round trip $ABCA$ will be $3T$ as measured by a clock located at A. Now suppose a second clock to be situated at B and synchronized with the one at A according to the stipulation $\varepsilon = \frac{1}{2}$. If a light-ray leaves A at $t_1 = 0$, it will arrive at B at

$$t_2 = t_1 + \varepsilon(t_3 - t_1) = 0 + \tfrac{1}{2}(2T) = T \tag{5.3}$$

Similarly, if a third clock, located at C, is synchronized with the one at B on the basis of the same rule, the recorded arrival time at C of the ray will be $2T$. Such a choice of synchronization rule for B and C assures that the C-clock is automatically synchronized with the A-clock.

Grünbaum now considers the case where the rule for synchronizing B with A and C with B is based on the choice of $\varepsilon = \frac{2}{3}$. The arrival time at B of a light-ray which departs from A at $t_1 = 0$ will in this case be given by

$$t_2 = t_1 + \varepsilon(t_3 - t_1) = 0 + \tfrac{2}{3}(2T) = \tfrac{4}{3}(T) \tag{5.4}$$

Likewise, the one-way transit time from B to C will also be $\frac{4}{3}(T)$. Hence, the elapsed time for the path ABC will be $\frac{8}{3}(T)$. But since the clock at A must record an elapsed time of $3T$, it follows that the third and final leg of the trip must be completed in $\frac{1}{3}(T)$. Hence, in virtue of the fundamental synchronization rule, we have

$$t_2 = 0 + \varepsilon(2T) = \tfrac{1}{3}(T) \tag{5.5}$$

Whence, $\varepsilon = \frac{1}{6}$. It is obvious that the velocity of light is four times greater from C to A than for the other segments of the trip.

Grünbaum deals in a somewhat revealing fashion with the possible objection that the compensatory increase in the velocity of light is so inductively improbable that the antecedent assumption of which it is a consequence must be ruled out on factual grounds. His reply is that the very basis of such an objection constitutes a complete misconstrual of the situation. It is predicated, he argues, on the assumption that there is ultimately a factually "true" synchronization and, in consequence, a true or factual one-way light velocity. But the magnitude ascribed to a one-way light velocity is merely a convention ensuing from the conventional choice of ε. Hence, it can be no more inductively improbable than "the hundredfold increase in the numbers representing the extensions of all bodies when we shift from meters to centimeters as units of length!" (1963, p. 359).

I, at least, find this reply to be most perplexing. Grünbaum has repeatedly stressed that the only interesting claims concerning the conventionality of various ingredients of physical theories are of the non-trivial variety. Is it not strange, therefore, that he should put the alleged conventionality of one-way light velocities on all fours with the trivial type of convention involved in the choice of units? The latter is surely independent of any discoverable facts. We saw, for example, that Grünbaum claimed that the conventionality of congruence is non-trivial because it presupposes the "fact" that space is continuous. Now electric charge is definitely discrete, so that a strong case may be made for the claim that there is a natural unit of electric charge. However, this fact obviously has not inhibited the use of alternative units of electric charge.

Secondly, Grünbaum has frequently warned against an excessive empiricism of the kind which has been dubbed "experimenticism" (cf. 1973, p. 834). Nevertheless, he appears to be guilty of the empiricist error of supposing that the conceptual content of a theory is exhausted by its empirical consequences. Such a charge seems justified by his unequivocal claim that it is a mistake to suppose that there is a "factual one-way velocity behaviour of light along CA" (1963, p. 358). Surely, it is gross experimenticism to maintain that although there is a well-defined meaning attached to the round-trip velocity of light, no similar meaning may be attached to its one-way velocity. Can one seriously accept the claim that a light-ray which is reflected back to its source by a mirror along the path ABA has an objective velocity along the closed path although lacking an objective velocity from A to B and from B to A, on the flimsy ground that one is able to measure the elapsed time for the round trip without invoking a standard of synchrony, whereas one cannot do so in the case of either of the one-way trips? Grünbaum would appear to be relying on what is, today, the discredited equation of "experimentally ascertainable" and "objectively real". Even were such an equation defensible, it would surely refer to that which is experimentally ascertainable in principle, and Grünbaum falls short of producing a theorem to prove the impossibility in principle of measuring a one-way light velocity. Incidentally, there is reason to suspect that many philosophers, influenced by the expositions of relativity theory which have been provided by empirically motivated writers, have supposed that the impossibility of measuring one-way light velocities is in fact a theorem of special relativity. However, this view is incorrect. If it is indeed an ingredient of special relativity, it belongs not to the theory proper but to what may be called its "pragmatics".

A third area of difficulty faced by the Grünbaum–Reichenbach thesis is the matter of internal consistency. Is it possible to stipulate values of $\varepsilon \neq \frac{1}{2}$ in a consistent way? Consider the following straightforward extension of Grünbaum's example of the triangular trip, which may serve to illustrate the type of difficulty I now have in mind.

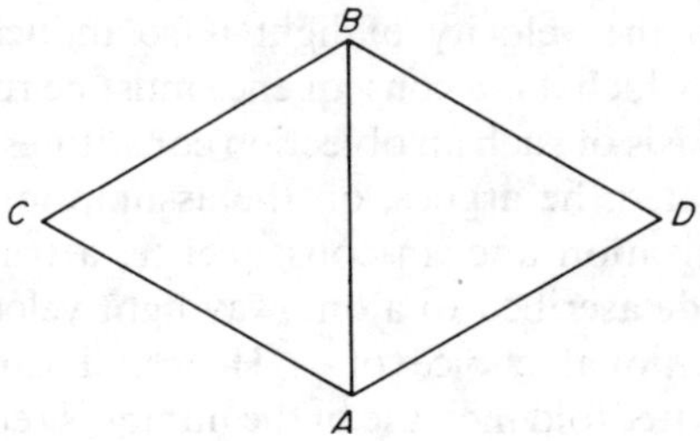

FIG. 5.3. Two overlapping triangular light-paths

Figure 5.3 schematizes a case similar to the previous one but with the addition of a second equilateral triangle with the common side AB. We now suppose that a clock is located at D and that two light-rays are transmitted simultaneously along AB and AD. There is abundant experimental evidence that the light-rays will return simultaneously to their source at A after having traversed the partially overlapping paths, $ABCA$ and $ADBA$. Let us suppose that we once again adopt the convention $\varepsilon = \frac{2}{3}, \frac{2}{3}, \frac{1}{6}$ for the path $ABCA$. We then adopt the same convention, *mutatis mutandis*, for the light path $ADBA$. Hence, the value of ε for the final leg of the latter path will be $\frac{1}{6}$. However, in accord with the round-trip light principle, since $\varepsilon = \frac{2}{3}$ for AB, it should equal $1 - \varepsilon = \frac{1}{3}$ for BA. That is equivalent to the claim that the "conventional time" of the second light-ray along the path BA is $\frac{2}{3}(T)$. But that is to say, that its "conventional velocity" along BA must be half that of another light-ray which makes the out-and-return trip ABA.

This is indeed a most serious and peculiar consequence. A change of unit or scale will, of course, alter the number which happens to be associated with a particular physical magnitude. That is hardly important from the standpoint of physical theory. However, when one adopts a standard of measurement whereby magnitudes which are known to be "factually" equal are declared to be unequal, something is seriously amiss. Independently of the concept of distant simultaneity and that of clock synchronization, two light-rays may be readily ascertained to have the same velocity along adjoining paths. Not only does this have an ample empirical basis but also reflects the universal conviction of competent physicists. Still, Grünbaum would presumably wish to maintain that it is possible to adopt synchrony standards whereby such an equality would be destroyed.

One may easily speculate on Grünbaum's probable response to my objection. I presume that he would argue that to contend that the choice of ε could destroy an empirically ascertainable equality is as absurd as to hold that the equality of two lengths could be destroyed by describing one of them as 2 feet and the other as 24 inches. All that is at stake is a descriptive convention, not a physical fact. But if that were the basis on which the conventionality of simultaneity hinged, then surely it would be trivial. While every physicist and engineer, not to mention the proverbial schoolboy, knows that the choice of units is a matter of convention, it is strictly inadmissible to change units in midstream, since so to do will lead to physically absurd consequences. However, we have seen that to adopt a non-standard synchrony compells one to do precisely that. Having adopted a value of $\varepsilon \neq \frac{1}{2}$, one is obliged constantly to readjust it to keep abreast of physical facts.

Adherents to Grünbaum's position have argued that non-standard synchrony may be shown to be consistent with the entire physical content of special relativity. J. A. Winnie (1970) has produced a notably elaborate defense of this view. But the true character of such arguments is now revealed for what it is. They amount to no more than the claim that should one be so obtuse as to change a scale of measurement in the course of expounding a physical theory, then whenever one is confronted by an apparent discrepancy between fact and theory, one may simply adopt yet another scale, *ad hoc*, to eliminate it. I am not now maintaining that such a procedure is logically impossible but merely contend that it is methododogically unjustified and philosophically unedifying. Far from being an instance of a philosophically interesting example of conventionalism, it is a banality, or so it seems to me, which is devoid of serious philosophic and scientific significance.

M. Friedman (1973), a recent critic of Grünbaum's views concerning geometry, has made the elementary but perceptive comment *à propos* of congruence that it is not enough when constructing a theory for a given concept to permit any interpretation of that concept which happens to satisfy a certain set of axioms. A theory is intended to explain certain *pretheoretic* notions. It is true that our common-sense ideas may be found to stand in need of revision in the light of the theory. However, the theory is also constrained by those ideas. If the gulf between the two is too great, the theory may cease to refer to the original pretheoretic domain. Friedman makes the pungent observation that "A property which is shared to an equal degree by the height of Mt. Everest and my mechanical pencil just is not *length*" (1973, p. 232). However, it might be that some method of measurement consistent with the axioms of congruence would declare the two to be equal.

It would seem to be relevant to bring such observations to bear on the concept of simultaneity. I would contend that even were it the case that a non-standard metric simultaneity were consistent with the axioms of special relativity, it would not thereby follow that such a concept is theoretically acceptable. In particular, even the refined common sense of the physicist would suppose this concept to refer to what logicians call an *equivalence relation*. I now wish to make a brief digression to consider the basic properties of such relations. (The reader should begin by reviewing, if necessary, the set-theoretic treatment of relations in Chapter 1).

A binary relation R is said to be an equivalence relation on a given domain of elements if and only if it is *reflexive*, *symmetric* and *transitive* in that domain. It will be helpful in explaining these notions to adopt a simple paradigm. Accordingly, we shall interpret R as "has the same weight as" in the domain of physical objects. That is to say, that xRy if and only if x has the same weight as y. R refers to the fact that x and y are equal in some specifiable sense. When we say that R is reflexive, we mean that every object in the intended domain bears the relation R to itself. Clearly, it is correct to say that every physical object has the same weight as itself. We say that a binary relation is symmetric if and only if whenever xRy, it is also the case that yRx. More formally, R is symmetric in the domain D if and only if

$$(x)(y)\{[(x, y \varepsilon D) \wedge \langle x, y \rangle \varepsilon R] \rightarrow \langle y, x \rangle \varepsilon R\}.$$

If x has the same weight as y then y has the same weight as x. Finally, R is transitive in D if and only if:

$$(x)(y)(z)\{[(x, y, z \varepsilon D) \wedge \langle x, y \rangle \varepsilon R \wedge \langle y, z \rangle \varepsilon R] \rightarrow \langle x, z \rangle \varepsilon R\}.$$

In our paradigm case, this simply means that if x has the same weight as y and y has the same weight as z, then x has the same weight as z. Every equivalence relation is said to *partition* the domain of elements on which it is defined. That is to say, that it divides the set of elements into a number of non-intersecting and exhaustive subsets. Thus, our paradigm relation partitions the set of physical objects into a set of subsets of physical objects having the same weight. For example, all physical objects weighing 5 pounds would be one such subset or *equivalence class*.

These considerations are germane to the present discussion since, on its usual interpretation, "metric simultaneity" is taken to be an equivalence relation which partitions the set of events into equivalence classes of simultaneous events. Now a fundamental difference between classical and Minkowski spacetime is that in the case of the former there exists a unique equivalence relation called absolute simultaneity and, correspondingly, a unique partitioning of events. Each equivalence class of events constitutes three-dimensional space at a given instant of time. In Minkowski spacetime, there are infinitely many equivalence relations, each of which is a relation of relative simultaneity. Each of them partitions spacetime into a set of equivalence classes, each of which constitutes a three-dimensional hypersurface of simultaneous events. As is sometimes said, each observer "splits" spacetime into space and time in his own way. However, the significant point, I believe, is that each method of splitting is a well-defined operation corresponding to an equivalence relation on spacetime. On the other hand, the counterpart of this relation, which is based on a non-standard synchrony is avowedly non-transitive and, therefore, not an equivalence relation. Abstractly considered, it is a perfectly good relation but not the one which refers to the notion of simultaneity.

Now one might raise the objection that my argument begs the question by simply assuming that metric simultaneity is an equivalence relation when, in fact, it is only the case that metric simultaneity based on standard synchrony is an equivalence relation. However, there are independent grounds for making such a claim. Specifically, although the notion of a spatial hypersurface of Minkowski spacetime is frame-dependent, it is a crucially important concept of special relativity. It is not the case that *any* cross-section of events constitutes such a hypersurface. The possible hypersurfaces are limited by the velocity of light. Now if the one-way velocity of light is treated as a convention, it follows that the distinction between space and time is likewise conventional. But a spacelike vector is one which is oriented in space, while a timelike vector is oriented orthogonally to a spatial hypersurface. Moreover, it is a theorem of special relativity that no admissible transformation can carry a spacelike vector into a time-like one, or conversely. One of the "absolutes" of special relativity is the timelike–spacelike distinction. However, a vector which is timelike could be made spacelike by changing the value of ε. Hence, to proclaim ε to be conventional is to reject special relativity!

Certain putative proofs of the factuality of $\varepsilon = \frac{1}{2}$ have been criticized on the ground that they have presupposed what is called the *reciprocity condition* according to which if object A has the velocity $\mathbf{v}$ with respect to B, then B must have the velocity $-\mathbf{v}$ with respect to A. This principle seems so obvious that it goes unmentioned and probably unnoticed in most textbooks. However, it is possible to show that the reciprocity condition entails $\varepsilon = \frac{1}{2}$.

Hence, if we abandon standard synchrony we must also abandon reciprocity. Far from detering certain philosophers from abandoning the conventionality of ε, it has merely led them to declare reciprocity to be a conventional ingredient of physical theory. However, following the suggestion of Friedman, we are led to pose the following possibly naive question. If a quantity s is attributed to a falling stone, while the quantity $0.9\,s$ is attributed to the earth, can s really designate what we call the relative speed of the two bodies?

Those who would insist on retaining the reciprocity principle may admittedly face a well-known philosophical dilemma. On the one hand, they would probably argue that reciprocity is factual rather than definitional, although, on the other hand, they would not countenance the possibility of its being subject to empirical confutation. Are they, then, to regard reciprocity as synthetic *a priori* or to reject the analytic–synthetic distinction as would be advocated by many contemporary pragmatists? An adequate discussion of such issues would require a separate chapter, if not a book, and I do not intend to wander so far afield. However, a couple of remarks on this matter may be in order.

In the first place, I shall state without defending my basic point of view on analytic statements. I believe, *pace* Carnap, that the analytic distinction is not one which ought to be made relative to a theory. A statement may be analytic or synthetic *in a language* but not in a theory, as such. Consequently, if the status of a theoretical statement is dubious, I believe that there is considerable antecedent probability that a careful analysis would reveal it to be a consequence of a factual assumption and hence, if we put aside the paradoxes of material implication, synthetic. The reader should also keep in mind that the connection between synchrony and reciprocity is almost self-evident. If the conventionality of synchrony as we have argued, amounts to no more than the possibility of arbitrarily changing the scale in the measurement of time and since velocity is a function of time, it follows that if the one is dependent on direction it is likely that the other will also be. But then the denial of reciprocity would be of no greater significance than the trivium that different numbers will be associated with a given velocity if we happen now to express it in metres/second and then in feet/second.

Grünbaum (1973, p. 682) has cited Bergmann (1942, p. 36), one of Einstein's collaborators, in a work which received Einstein's explicit approval. Bergmann indeed shows that reciprocity is implied by the standard Lorentz transformation, which, in turn, is implied by $\varepsilon = \frac{1}{2}$. However, Bergmann also remarks that its non-trivial nature is a consequence of the impossibility of making a direct comparison of time and distance units in two distinct frames of reference. Furthermore, just a few pages earlier in the same chapter, Bergmann asserts anent the transitivity of simultaneity on the basis of the light-signalling procedure that: "It must be understood that this assumption is a hypothesis concerning the behaviour of electro-magnetic signals" (1942, p. 30). This touches the core of the issue. There is a world of difference between hypotheses and conventions. Grünbaum seems curiously to have overlooked the obvious point that the lack of a strong empirical foundation for a factual conjecture does not turn the latter into a convention. Although Bergmann follows Einstein in talking about the "definition" of simultaneity, he does not suppose it to be incompatible with the factuality of $\varepsilon = \frac{1}{2}$. However, we shall have more to say of this in the concluding portion of this chapter.

Much recent discussion of this problem has taken a rather different approach from the one which we have been pursuing. It has been concerned with the possibility, both in principle and practice, of providing an objectively based method of synchronization which results in the securing of standard synchrony while avoiding the alleged circularity of the light-signalling method. The work which has come to be regarded as the *opus classicus* of this approach is a paper entitled "Conventionality in Distant Simultaneity" by Ellis and Bowman (1967). However, the basic idea which is ingredient to their thesis had earlier been suggested by Bridgman in his posthumously published *A Sophisticate's Primer of Relativity* (1962). Moreover, Grunbaum, himself, in the course of criticizing the methodological inferences which Ellis and Bowman drew from the existence of such a method, produced a version which overcomes some of the limitations of the other two.

The details are unimportant for the present discussion. It suffices to note that all parties to the dispute agree as to the existence of such a method. In broad outline, it rests on an empirical prediction of the theory of special relativity. As the reader knows, the theory predicts that a moving clock will be retarded relative to a "stationary" one. The amount of retardation will be a function of both the velocity and the length of path of the moving clock. Thus, if two or more clocks which are locally synchronous at A are transported along different paths with different velocities to a point B, they will be found no longer to be in synchronization. It is for this very reason that clock transport has generally been ruled out as an adequate basis for the measurement of distant simultaneity. However, attention is now drawn to the fact that the Lorentz time transformation has another aspect.

Consider the case of two stationary clocks, C_a and C_b, which are located at the distant points A and B. Let us suppose that C_a and C_b are in *standard signal synchrony*. A third clock C_t is locally synchronized with C_a and then transported to C_b with velocity $\mathbf{v}$. For simplicity, we shall assume that the departure time is zero. Let t be the arrival time as measured by C_b, while t' is the same time as measured by C_t. The reader will recall that t' is the proper-time of the interval. The basic prediction of special relativity is that

$$t' = t\sqrt{1 - v^2/c^2} \tag{5.6}$$

(5.6) is, of course, the basis for the prediction that C_t will be lagging behind C_b when the two coincide. However, it may be shown that regardless of the distance from A to B, the discrepancy $t - t'$ between the two clocks may be made as small as we choose and, in the limit, as $\mathbf{v}$ approaches zero so does $t - t'$. Accordingly, two clocks are said to be in *slow transport synchrony* if a third clock which is synchronous with the first is transported with infinitely small velocity to the second and is synchronous with the latter. Of course, it is not required that an infinitely small velocity be physically realized. On the basis of a series of readings at various velocities, the appropriate limit may be computed mathematically. Now the crucial point is made that since clocks which are in slow transport synchrony are also in standard signal synchrony, the former may be employed as an independent test of the one-way light principle.

Grünbaum (1969) offered a very elaborate critique of the Ellis and Bowman proposals. His point is not to show that slow transport synchrony is spurious but to demonstrate that it involves certain conventional ingredients. In particular, he argues that although the simultaneity relations as based on slow clock transport are unique, they are, nevertheless, conventional in so far as they presuppose a certain *time metric*, i.e. a rule for assigning the measure of an interval to pairs of time coordinates. According to Grünbaum, such a metric and standard of equality for time intervals is just as conventional, in the sense of Poincaré and Riemann, as spatial congruence.

Now it is certainly true that different congruence classes are yielded by different metrics. Moreover, the axiomatic definition of "metric" is very weak, in the sense of allowing great latitude in the choice of a particular metric. In particular, there is no formal requirement that a time metric be given by the absolute value of the difference between temporal coordinates, i.e. $\Delta t = |t_2 - t_1|$. However, it does not follow from such considerations that the metric is conventional in the context of physics. Mathematically, there are infinitely many possible metrics or infinitely many abstract spaces. It is trivially true that the act of selecting one such metric is based on an arbitrary choice. The problem for the physicist is to select a metric which "matches" reality. The sense in which such a matching is meaningful will be discussed in Chapter 11. However, it is by no means obvious that this matching is a conventional one. The physicist bases his geometry on the coincidence behaviour of physical objects, of which measuring rods are the paradigm. It is true that a certain amount of conjecture is involved in supposing that rods which are locally congruent will remain so, but I would hesitate to call such a conjecture a convention. Surely it is a factual conjecture. When faced with a choice among several conjectures all of which are equally satisfactory, we must adopt one of them on the basis of convention. But it has long been recognized that factual knowledge amounts to much more than can be given by the immediate deliverances of the senses. For example, there are sound empirical justifications for regarding some objects as better suited to be measuring rods than others. That is to say, that the choice of a standard of rigidity is not arbitrary but is based on such theoretical considerations as, for example, the molecular structure of the object of interest. Similarly, there are sound theoretical reasons for regardig some physical systems as better clocks than others. It is more than mere convention that leads us to treat the periods of certain systems as equal. Once it is admitted that there are objects which are particularly well-fitted to be basic measuring instruments, it follows that a metric which is selected on the basis of the behaviour of such instruments is not conventional.

At the risk of seeming repetitious, I repeat that there may be sound reasons for believing much of that for which there is no adequate basis in terms of immediate acquaintance. I do not think that Ellis and Bowman should be construed as arguing that slow transport synchrony is a fact of our immediate experience in the sense of precluding all other possibilities. Crudely put, if something looks like a lemon, then there are good physical reasons for classifying it as one, even though other possibilities are not precluded. In a nutshell, to posit something factual by the method of conjecture is not to introduce a convention. The basic error of the conventionalists is the supposition that what is not known with certainty on the basis of experience can only be asserted on the basis of convention.

Ellis (1971) wrote a reply to Grünbaum's critique in which he makes the point that there may be good physical reasons for adopting one convention in preference to others. He goes on to say that on Grünbaum's criterion, one would only be entitled to claim a good physical reason for a convention if a convention were logically superfluous. I agree with the spirit but not the word of Ellis's response. He too, it seems, has fallen into the trap of confusing convention and conjecture. However, he is correct in claiming that Grünbaum seems to classify as conventional any item which is not empirically determined. If this charge is correct, it is difficult to accept at face value Grünbaum's claim that the thesis of conventionalism is non-trivial.

Throughout this book it is our intention to avoid textual exegesis for its own sake. This is a task for scholars and historians. However, Grünbaum has made so much of the claim that the conventionality of simultaneity is contained in Einstein's original text (*On the Electrodynamics of Moving Bodies*, 1905) that we shall conclude this chapter with a brief examination of it. However, we suggest to the student that this is not the way to do the philosophy of science. The latter involves the analysis of theories not of personalities and the idiosyncracies of their writings. From our standpoint, the matter is at best of secondary importance, since even if it were shown that Einstein was a conventionalist in this regard, it would not follow that the conventionalist interpretation of special relativity is correct, even though it would doubtless contribute to its moral support. In fact, it is to do Einstein an injustice to treat all of his words with the degree of reverence that is usually reserved for infallible truths.

The basic textual evidence for the Grünbaum–Reichenbach interpretation consists in the following two passages.

"We have not defined a 'common time' for A and B, for the latter cannot be defined at all unless we establish *by definition* that the 'time' required by light to travel from A to B equals the 'time' it requires to travel from B to A" (1905, p. 40).

"In agreement with experience we further assume the quantity

$$\frac{2AB}{t'_A - t_A} = c$$

to be a universal constant—the velocity of light in empty space" (*ibid.*).

The first of these passages is taken to indicate that the equality of antiparallel light velocities is a matter of stipulation, while the second is supposed to suggest that, by contrast, the velocity of a round trip is determined on the basis of experiment. Admittedly, Grünbaum's interpretation has a reasonable prima facie justification. However, we shall endeavour to show that it is not the only possible interpretation. Firstly, it presupposes a rather sophisticated conception of the logical status of definitions on the part of Einstein. I refer to that conception which occurs in the context of logical syntax, in which a defined term is one which is introduced in a theory in terms of an expression containing only previously defined or primitive terms. That such definitions are purely stipulative is beyond dispute; but there is no evidence that Einstein intended to employ the term in that sense.

Viewed in their historical context, Einstein's statements lend themselves to a natural interpretation in terms of his desire to underline the fact that Newtonian dynamics rests on certain assumptions about time for which no adequate test criteria had been provided. Accordingly, when he alludes to a "definition of simultaneity", he may probably have had in mind something akin to what subsequent writers such as Bridgman were improperly to call "operational definitions", which were designed to provide a semantic interpretation of theoretical concepts. That this was how Einstein employed the notion of definition is quite clearly brought out in the following passage.

"If at the point A of space there is a clock, an observer at A can determine the time values of events in the immediate proximity of A by finding the positions of the hands which are simultaneous with these events. If there is at the point B of space another clock in all respects resembling the one at A, it is possible for an observer at B to determine the time values of events in the immediate neighbourhood of B. . . . We have so far defined only an 'A time' and a 'B time'" (1905, pp. 39 f.).

Here and in many other passages which it would be otiose to enumerate, there is ample evidence that Einstein sometimes equated defining with providing a method of measurement. With this in mind, we may interpret the first of the cited passages to mean not that the ingoing and outgoing velocities of light are stipulated, or decreed by convention, to be equal but that the *fact* of their equality may be exploited as a basis for measuring equal durations. Moreover, the passage literally means that a definition is offered for *time*, not for the character of a one-way light velocity.

I confess that Einstein's pronouncements are sometimes apt to be frustratingly ambiguous. Nevertheless, it is almost inconceivable that he wished to draw as sharp a distinction as Grünbaum suggests between the status of a one-way velocity and that of a two-way velocity, while neglecting to make an explicit verbal distinction between them. It is at least plausible that in the second of the cited passages, Einstein defined c as he did in virtue only of its methodological simplicity, in the sense of its being independent of the "definition" of simultaneity.

Many passages in Einstein's later work, *The Meaning of Relativity* (1953), provide additional support for our interpretation. In the course of covering the same ground dealt with in his 1905 paper, he reveals once more his peculiar view of the nature of definitions. For example: "It follows from what has gone before, that coordinates with respect to an inertial system are physically defined by means of measurements and constructions with the aid of rigid bodies" (*op. cit*., p. 27). He also makes the crucial observation that the principle of constancy is employed for "defining" time. "In order to complete the definition of time we may employ the principle of constancy of light in a vacuum" (*ibid*.). It is abundantly clear that the velocity of light is not defined to be constant but serves as a basis on which to "define" the equality of two durations. As to the principle of constancy, itself, Einstein formulates it in the following equation:

$$(\Delta X_v)^2 - c^2 \Delta t^2 = 0. \tag{5.7}$$

Clearly, (5.7) makes no reference to the round-trip velocity of light, nor is it tied to it in any way. In fact, without further ado, Einstein, in the later work, provided a method of synchronization based on the one-way velocity of light (cf. *op. cit*., p. 28). Could Grünbaum reasonably argue that the philosophical insight which Einstein allegedly possessed in 1905 had deserted him by 1921 when he was still at the peak of his creative talent and regarded by most as more sophisticated philosophically than he had been at the earlier date?

The most persuasive textual evidence in support of the Grünbaum–Reichenbach interpretation is to be found in Einstein's own popularization of his two theories (1952) where, in speaking of the *definition of simultaneity*, he writes: "That light requires the same time to traverse the path $A \rightarrow M$ as for the path $B \rightarrow M$ is in reality neither a *supposition nor a hypothesis* about the physical nature of light, but a *stipulation* which I can make of my own free will in order to arrive at a definition of simultaneity" (1952, p. 23, *Italics in the original*). He proceeds to add that on this *criterion for simultaneity*, the latter will be transitive and that the transitivity of simultaneity is indeed a physical hypothesis about light propagation.

In the quoted passage it seems fairly clear that Einstein would wish to distinguish a stipulation from a definition. In effect, one stipulates the equality of to and fro transit times and employs that stipulation as a criterion in the definition of simultaneity. In my opinion, it would be erroneous to assimilate Einstein's use of "stipulation" with the use to which it is put in the context of current discussions of conventionalism. I would suggest that Einstein intends the equality of to and fro light velocities to be stipulated in an analogous sense to that whereby it is stipulated that the length of the platinum-iridium metre bar or the wavelength of krypton are constants which may serve as physical standards for the determination of length and congruence. In this role, they are not to be construed as physical hypotheses about material entities. Nevertheless, if such standards of measurement were found to lead to factual claims which were inconsistent with well-established theories, they would be replaced by more satisfactory ones. Moreover, such physical standards are selected in preference to others on the basis of theoretical hypotheses. Thus, the equality of to and fro light velocities may first be derived from optical or electromagnetic hypotheses. It is then that the equality is employed in the "definition" of the simultaneity of distant events. Finally, we emphasize that Einstein proclaims that the aforementioned stipulation leads to the *factual* consequence that the relationship of simultaneity is transitive.

Two final remarks are here in order. Firstly, Einstein deals with this matter in a chapter which is entitled *On the Idea of Time in Physics*, which is largely devoted to the problem of finding an experimental criterion of simultaneity. As such, the discussion has nothing specifically to do with special relativity as distinct from classical physics. In fact, it occurs prior to the discussion of the foundations of special relativity. Secondly, it is noteworthy that whereas no convention could lead logically to factual consequences, Einstein explicitly proclaims that the stipulation of the equality of to and fro light velocities does so lead.

I conclude that the doctrine of the conventionality of simultaneity is either incorrect or trivial, and was never held by Einstein.

Bibliography for Chapter 5

BERGMANN, P. G., *Introduction to the Theory of Relativity*, Prentice-Hall, Englewood Cliffs, N. J., 1942. This is one of the standard introductions to the two theories of relativity. It contains an endorsement by Einstein.

BRIDGMAN, P., *A Sophisticate's Primer of Relativity*, Wesleyan University Press, Middletown, Conn., 1962.

EINSTEIN, A. (1952). See bibliographic reference in Chapter 3.

EINSTEIN, A., *The Meaning of Relativity* (5th edn.), Princeton University Press, Princeton, N.J., 1953. Einstein's own treatise on both special and general relativity. Despite its title, it is fairly advanced.

EINSTEIN, A., "Remarks to the Essays Appearing in this Collective Volume", in *Albert Einstein: Philosopher-Scientist* (ed. P. A. SCHILPP), vol. 2, Harper, New York, 1959.

ELLIS, B., "On Conventionality and Simultaneity—A Reply", *Aust. J. Phil.* **49**, 177 (1971).

ELLIS, B., and P. A. BOWMAN. "Conventionality in Distant Simultaneity", *Philosophy of Science*, **34**, 116 (1967).

FRIEDMAN, M., "Grünbaum on the Conventionality of Geometry", in *Space, Time, and Geometry*. (ed. P. SUPPES), Reidel, Dordrecht, 1973. This collection contains several excellent papers.

GRÜNBAUM, A., *Geometry and chronometry in Philosophical Perspective*. University of Minnesota Press, Minneapolis, 1968.

GRÜNBAUM, A., *Philosophical Problems of Space and Time*, Knopf, New York, 1963; and 2nd enlarged edn., Reidel, Dordrecht, 1973. The second edition contains extensive replies to criticisms and excellent bibliographies.

GRÜNBAUM, A., "Simultaneity by Slow Clock Transport in the Special Theory of Relativity", *Philosophy of Science*, **36**, 5 (1969).

MALAMENT, D., "Causal Theories of Time and the Conventionality of Simultaneity", *Noûs*, **11**, 293 (1977). This article is a significant contribution to the debate.

PROKHOVNICK, S. J. See reference in Chapter 3.

SALMON, W. F. "The Philosophical Significance of the One-Way Speed of Light," *Noûs* **11**, 253

(1977). A careful analysis by a disciple of Grünbaum, which takes account of the bearing of the issue on various physical theories.

WINNIE, J. A. "Special Relativity Without One Way Velocity Assumptions" (I and II), *Philosophy of Science*, **37**, 81 and 223 (1970).

6

The Commensurability of Classical and Relativistic Mechanics

THROUGHOUT this book, we have endeavoured to present a consistent view of physical theory which may be briefly characterized in the following way. A physical theory is an attempt to provide as accurate a representation as possible of the physical world. Such theories should not be characterized as empirical theories since they involve the use of theoretical concepts which are not directly connected to immediate observations. On the other hand, such theoretical systems must be accessible to experimental testing in order that their degree of verisimilitude to the physical world may be determined. If they are found wanting, they must be revised in the light of experience or else replaced by better theories, which are more accurate representations of the objective world. For want of a better term, I have characterized this view of the relationship between theories and the world as "scientific realism".

In recent years, this fairly widespread view of the development of scientific theory in terms of a series of successive approximations to the "truth" has been challenged by certain outstanding philosophers and historians of science. The historian of science who is especially noteworthy in this regard is Kuhn, who, in his *Structure of Scientific Revolutions* (1962), argued that such an evolutionary approach to scientific progress should be replaced by one which recognizes that revolutionary developments in scientific theory represent a much more radical departure than is suggested by the sort of account that I have been favouring. According to Kuhn, when one theory is replaced by another, it is not the case that a gradual development has occurred but, in effect, one *paradigm* of scientific explanation is replaced by another with which it has nothing in common. However, we shall focus our attention chiefly on the work of the philosopher Paul Feyerabend, who adopted virtually the same view of scientific change as Kuhn but developed it in greater systematic detail. Like Kuhn, Feyerabend holds that when an existing theory is replaced by a new one, the conceptual change that is involved is so profound that the two theories are literally *incommensurable*.

It appears that the subject of commensurability of scientific theories is among the most crucial topics to have emerged in the area of the semantics of science in the past several years. Indeed, it is a challenge to the very possibility of conducting rational research. The primary and manifest purpose of the present chapter is to level some criticism against the widely discussed Feyerabend–Kuhn thesis of the incommensurability of scientific theories. However, there is a second underlying purpose which is to show that many theses in the philosophy of science can only be adequately supported or refuted by means of a detailed analysis of actual theories as distinct from either the analysis of abstract theory *schemata* on the one hand or the casual reference to isolated parts of theories and to anecdotes from the history of science on the other. The former method was characteristic of the Vienna Circle, and the latter of Kuhn and Feyerabend.

Before dealing with the crux of the issue, as I perceive it, I shall make some critical remarks against Feyerabend's manner of attempting to establish his thesis. Thus, initially my treatment will be consciously *ad hominem*, but only by way of introduction. Feyerabend first formulated his position in the 1962 paper: "Explanation, Reduction and Empiricism." He considers the general case of a hitherto factually adequate theory T_1 which is supplanted by a new theory T_2, where T_2 is assumed to explain all of the phenomena explained by T_1. He now proceeds to attack the view, which he attributes to Hempel, Oppenheim and Nagel, among others, to the effect that since T_1 is adequate in a subset of the domain of T_2, it should be deducible from T_2 as a special case. This principle of deducibility in turn presupposes a second principle—namely, the principle of *meaning invariance*—which is simply the claim that within the context of a single derivation, as for example the derivation of T_1 from T_2, the meanings of all descriptive terms must be fixed. Now on Feyerabend's view, when T_1 is abandoned in favour of T_2 the change is much more radical than that which is suggested by the orthodox account. As he puts it, the ontology of the earlier theory is completely replaced by the ontology of the later one, which results in a change of meaning of both the theoretical and the observational terms. Thus, the two theories are, in fact, *incommensurable*. Accordingly, it is fallacious to claim that the one is derivable from or, equivalently, reducible to the other. By the same token, in fact, given two such competing theories, there could be no crucial test to decide between them, since they share no common observational vocabulary. Feyerabend's position seems to involve several assumptions, none of which is obviously true. (1) That the derivation of T_1 as a special case of T_2 is the central meaning of reduction. (2) That theories with radically different ontologies are not comparable and, in particular, cannot have comparable observational consequences. (3) That the theoretical terms of two such theories are never interdefinable. (4) That there are neither observational nor theoretical terms which are neutral with respect to T_1 and T_2.

While I believe that each of these assumptions is open to challenge, I shall refrain from adding to an already bulky literature on the subject. However, since Feyerabend's usual defense tends to be one of attack, I shall briefly comment on one of the counter-examples that he offers—namely, that of the alleged reduction of Galileo's law of falling bodies to Newtonian mechanics. Feyerabend points out that even though the two may be observationally indistinguishable in the domain of their intersection, the one cannot be a logical consequence of the other. It is hardly necessary to provide the details, since it is well known that Galileo's law requires a constant rate of acceleration whereas Newtonian mechanics entails that the gravitational attraction diminishes in proportion to the square of the distance. Thus, Feyerabend is of course correct in claiming that the one is not reducible to the other. But he has merely shown that a false T_1 cannot be reduced to a putatively true T_2, which is hardly surprising. Moreover, he does contend that the two are logically incompatible, which is very surprising indeed, since he is claiming that they are incommensurable. Thus, Feyerabend, himself, seems to find it difficult to take the incommensurability thesis too seriously.

Neither Feyerabend nor Kuhn, as philosophical writers, are models of logical clarity. However, in so far as any logic may be perceived in the argument against commensurability, it seems to be the following. If T_1 and T_2 are commensurable and in agreement in the domain of their intersection, then T_1 is reducible to T_2. But, in fact, T_1 is not reducible to T_2, hence T_1 and T_2 are incommensurable. Apart from the questionable logic of this form of argument, there is a more serious confusion—namely, between the notion of reducibility and that of commensurability. The relationship of reducibility between T_1 and T_2 is extremely strong, specifically that of entailment. On the other hand, T_1 and T_2 are commensurable just in case they are logically comparable, which is a very weak relationship—namely, that the intersection of their respective descriptive vocabularies be non-empty. Inversely, to say that T_1 is not reducible to T_2 is to make a rather weak claim, whereas to say that they are incommensurable is to make what I, at least, regard as an astonishingly strong one. If two theories are literally incommensurable, it makes no sense to regard them as agreeing in respect to any empirical data or to be such that even in principle any observation report which confirms the one simultaneously confutes the other. Thus, on this view, when a crucial opposition arises between two theories, the one will prevail on the basis of various non-rational criteria; for example, sociological and aesthetic considerations may be brought to bear, not to mention the appeal to ideological principles.

It would be unfair to omit from these opening remarks that Feyerabend has performed an invaluable service in directing the attention of philosophers of science to the dangers that are inherent in too hastily assuming that a particular reduction has been successfully carried out. Indeed, one sometimes has the impression that too frequently both the scientific and philosophical communities simply assume that one theory is reducible to another in the absence of any sustained effort to carry out such a reduction. No doubt, future attempts at this kind of reduction will be obliged to come fully to grips with the problem of meaning invariance. However, I suggest that Feryerabend has attempted to replace one dogma with his own dogmatic counsel of despair. I contend that instead of accepting the general claim that all significant scientific change involves essentially incommensurable theories, one should adopt the more cautious approach of attempting to resolve each individual case by systematic formal and semantic analysis of the theories of interest. If the sweeping claims of Feyerabend and Kuhn were passively to be accepted, all serious analysis of the relations between theories would automatically be precluded on the *a priori* ground that it would be bound to be fruitless. To me, it seems methodologically unsound to prejudge all such matters.

Both Feyerabend and Kuhn have cited the theories of Newtonian and special relativistic mechanics as an example of an incommensurable pair, although neither has provided a detailed analysis of them. My personal view is that the existing accounts of their interrelationship tend, at best, to be oversimplified. Moreover, while one is certainly better advised to treat them as distinct theories than to attempt the reduction of the one to the other, it is possible to make a rational comparison of them.

I would speculate that Feyerabend and Kuhn have reacted against the simplistic claims of both scientists and philosophers that in the limit as $c \to \infty$, the relativistic laws are transformed into the classical ones or, still worse, that the principle of special relativity is no more than a generalization of the principle of classical relativity from laws of mechanics to laws of physics. They may also have reacted against the heuristically valuable procedure which one finds in so many treatments of relativity physics, including our own, consisting in taking a classical law and modifying it in the appropriate manner to make it Lorentz covariant and, therefore, relativistically admissible, in seeming disregard for the fact that some or all of the symbols occurring in the two sets of formulas have indeed undergone a change of meaning. For example, Feyerabend correctly claims that the symbol "*m*" which occurs in classical mechanics should not be identified with its counterpart which refers to relativistic mass.

I make these remarks to underline the fact that I am not altogether unsympathetic to what I take to be the motives of Kuhn and Feyerabend in their rejection of the claim that these theories are commensurable. Their reaction is an understandable one. Nevertheless, I believe it to be an overreaction. One cannot argue from the breakdown of the arguments in support of commensurability, still less from those in support of reducibility, to the thesis of incommensurability. In what follows, we intend to show that with proper semantic caution the commensurability of the two theories may be demonstrated. Specifically, we shall show that theorems are deducible in the one that are semantically incompatible with theorems of the other. We shall conclude, therefore, that the theories are in the relationship of contrariety. The upshot will be that one may make a rational choice between them on the basis of experiment. Thus, while I would certainly support Feyerabend's contention that Newtonian mechanics is not reducible to relativistic mechanics, I also claim that their irreducibility does not imply their incommensurability.

Most contemporary philosophers would maintain that *meaning* is a multivocal or, at least, a multi-aspect, concept. The two most fundamental aspects of meaning are the denotative and connotative ones. The denotation of an expression is simply the set of entities, be they concrete or abstract, to which it refers. Roughly speaking, the connotative meaning of an expression is the set of all expressions which are entailed by it. Thus, the denotation of "flower" is the set of all flowers, whereas its connotation is the set of all expressions which denote properties that a thing possesses in virtue of its belonging to the set of flowers. Clearly, the connotative meaning is more complex than the denotative meaning of an expression. For example, the connotation of a scientific term such as "electron" will depend heavily on the theory in which it occurs. On the other hand, from the standpoint of scientific realism, which denies that the world is either a creation of the mind or of language, the denotation of such a term may be relatively constant across the boundaries of several theories. In what follows, our strategy will be to exploit the denotative aspect of certain scientific terms and, in particular, of the notions of spatial and temporal coordinates, which have the same denotation in different theories.

We shall begin by reviewing some very familiar and elementary aspects of relativistic kinematics and conclude with a rather more adventurous consideration of relativistic mechanics. In all cases, as was our previous practice, we shall restrict the treatment to point-particles and frames of reference which are in relative motion with respect to but one spatial coordinate, since the simplest cases are quite naturally the most clear-cut and perspicuous from a semantic standpoint. It should be added that virtually all of the results which we are about to employ have previously been introduced in Chapter 3. They are repeated here because it is vital to the following argument that we specify and keep track of each and every assumption.

Let K and $\overline{K}$ be two frames of reference which are in standard configuration. That is to say, that at an arbitrary zero-point of time the origins, O and $\overline{O}$, of the coordinate systems associated with the reference frames coincide, and $\overline{K}$ has the velocity $\mathbf{v}$ in the $+X$-direction with respect to K. It is also assumed that each reference frame is equipped with an adequate number of clocks located at various points along the X-axis. Finally, it is assumed that the clocks in a given frame have all been synchronized. Let us recall from previous chapters how the process of synchronization is carried out. Firstly, a master clock at the origin is set in motion. Then a time signal in the form of a light signal is sent from the master clock. When it arrives at the clock located at an arbitrary x, the latter is set in motion with the obvious correction for the velocity of the time signal. That is to say, if the signal is transmitted at t_0, upon its arrival at x, the clock at x will be set at $t_0 + x/c$.

Under the influence of the late N. R. Hanson, some philosophers may perhaps wish to voice the objection that our account is already "theory-laden", since it presupposes a method of synchronizing clocks which is specific to the theory of relativity as propounded by Einstein. However, such is not the case. One of the more frequent misinterpretations of relativistic kinematics is the belief that it is generated by a theory-specific method of synchronizing clocks. This has, of course, been bolstered by the fallacious operationalist belief that the theory of relativity is a theory of measurement. In fact, we are proposing a completely standard method of synchronizing clocks which is just an idealization of the method employed in everyday life.

Suppose that you are motoring from town A to town B and are not carrying a timepiece but wish to determine the duration of your trip. You might consult a clock on a public building in A which gives the reading t_1, then on arriving at B you look at a clock on another public building, which gives the reading t_2. You would then conclude that the duration of the trip was $t_2 - t_1$, on the assumption that the two clock had been synchronized by means of a signal from a master clock—for example, a time signal from a radio station, which was propagated with the velocity c. A sophisticated Newtonian might care to make the correction for the finite velocity of the signal, whereas a careless or merely down to earth Newtonian would hardly bother unless B happened to be located in a distant region of the solar system. The key point, for our present purpose, is that the method of synchronizing distant clocks is precisely the same whether one adopts Newtonian or Einsteinian physical theory. On the one hand, a relativity theorist could synchronize clocks with a sound signal instead of an optical one, and it is easily shown that, after suitable corrections, two clocks which are found to be optically synchronous will also be acoustically synchronous. On the other hand, a Newtonian might prefer an ideal method of synchronization involving, *per impossible*, a time signal with an infinite velocity, so that no correction would be required. Again, it may be demonstrated mathematically that if two clocks are optically synchronous, they will also be ideally synchronous in the foregoing sense. To return to our original theme, we see that just as one may speak of the spatial location of an event in a manner that is identical for both classical and relativistic physics, so one may also speak of its temporal location. In consequence of our method of synchronization, time is now effectively a fourth coordinate, and we shall call the referent of the coordinate variable t "coordinate time".

We shall now assume, as given, the Lorentz transformation for two frames which are in standard configuration.

$$x \to \bar{x} = \frac{x - vt}{\sqrt{1 - v^2/c^2}}; \quad \bar{x} \to x = \frac{\bar{x} + v\bar{t}}{\sqrt{1 - v^2/c^2}} \tag{6.1a}$$

$$t \to \bar{t} = \frac{t - vx/c^2}{\sqrt{1 - v^2/c^2}}; \quad \bar{t} \to t = \frac{\bar{t} + v\bar{x}/c^2}{\sqrt{1 - v^2/c^2}}. \tag{6.1b}$$

Although (1a) and (1b) belong to special relativity, it must be emphasized that v, t, and x

have the same denotative meaning in relativistic and classical physics. Moreover, the question of the origin of the transformation equations is irrelevant to our present interests. We are merely concerned to determine whether and in what way two distinct theories may be compared. Finally, it should be remarked that although it is not customary to treat time as a coordinate in the context of Newtonian physics, there is nothing to prevent one from so doing. It is, after all, the case that events in classical mechanics are labelled by the time of their occurrence as indicated by a clock which is a part of the observer's frame of reference. Furthermore, although time is an invariant in classical physics, it does not follow that it lacks transformation properties. Indeed, it is in virtue of the transformation law, $t = \bar{t}$, that it may be characterized as an invariant or scalar object.

We now consider a familiar ideal experiment. K and $\bar{K}$ are in standard configuration. A clock $\bar{C}$ is at rest in $\bar{K}$ at the point $\bar{x}_1$. At a given time, $\bar{C}$ coincides with a clock C_1 in K. C_1 records the time of coincidence as t_1, while $\bar{C}$ records $\bar{t}_1$. Subsequently, $\bar{C}$ coincides with a second clock in K, C_2, which records the time t_2 while $\bar{C}$ records the time of the coincidence as $\bar{t}_2$. The time which elapses between the two events in K is $t_2 - t_1$. Similarly, in $\bar{K}$ it is $\bar{t}_2 - \bar{t}_1$. Now, by means of (6.1b) we may deduce the following familiar result.

$$\Delta t = t_2 - t_1 = \frac{\bar{t}_2 + v\bar{x}_1/c^2}{\sqrt{1-v^2/c^2}} - \frac{\bar{t}_1 + v\bar{x}_1/c^2}{\sqrt{1-v^2/c^2}} = \frac{\bar{t}_2 - \bar{t}_1}{\sqrt{1-v^2/c^2}}. \tag{6.2}$$

As the reader knows, there is a significant difference between the quantities $t_2 - t_1$ and $\bar{t}_2 - \bar{t}_1$; namely, that the former is the result of the readings of two spatially separated clocks C_1 and C_2, whereas the latter is the result of two readings of a single clock $\bar{C}$ which was present at both events. It should hardly be necessary to remind the reader that the second quantity is the proper-time interval of the two events which is normally symbolized by τ. Hence, we may now rewrite (6.2) in the following form:

$$\Delta t = \frac{\Delta \tau}{\sqrt{1-v^2/c_2}}. \tag{6.3}$$

Physically, as the reader will recall, the significance of (6.3) is that a proper-time interval is always less than or equal to a coordinate time interval. It is also illustrative of the famous time-dilation effect. However, what is of immediate significance is that we have introduced the quantity τ, which plays a leading role in relativistic physics, in virtue of its having the same value in all inertial reference frames, by definition in terms of Newtonian concepts. Indeed, similar mathematical results were obtained by Einstein's older contemporary, Lorentz, who was able to interpret them in a manner which was consistent with the classical tradition.

The following case arises as an immediate corollary of (6.3). We have a frame of reference K_1 and clocks C_1 and C_2 which are synchronized at the origin at t_0. C_2 is set in motion along the X-axis with velocity $\mathbf{v}$. At the coordinate time t_1, C_2 is at the point $x_1 = vt_1$. In virtue of (6.3), C_2 will record the proper-time $t_1\sqrt{1-v^2/c^2}$. On arrival at x_1, it is returned to the origin with velocity $-\mathbf{v}$. It arrives at the origin at coordinate time $t_2 = 2t_1$, at which time it will record $2t_1\sqrt{1-v^2/c^2}$. Thus, when C_1 and C_2 coincide for the second time, C_2 is found to be lagging behind C_1 by a predictable amount. The discrepancy between the two readings is categorically not theory-laden. For example, the clocks could be so constructed that they would punch holes in a paper tape to record the number of elapsed time units; then any observer would be in a position to compare the number of holes in the two tapes. A layman wholly innocent of both Newtonian and relativistic physics would observe the discrepancy if one existed. Consequently, there is a genuine logical incompatibility between the two theories from which the possibility of a crucial experiment emerges. It follows that the two theories are commensurable.

Incidentally, it is worth remarking that time dilation is part of the everyday experience of the nuclear physicist who encounters elementary particles travelling at close to the speed of light. For example, the half-life of the muon, the most abundant constituent of cosmic radiation, is increased by a factor of about nine in virtue of its great velocity. However, perhaps it could be justly argued that the expression "half-life of a muon" is indeed theory-laden. However, quite recently, more perspicuous results have been obtained involving the round-the-world jet transportation of atomic clocks. These results have apparently confirmed the relativistic predictions to a very high degree of experimental accuracy.

Let us now consider a second kinematic experiment. At the time at which K and $\bar{K}$, in standard configuration, have coincident origins a particle begins to move along the $\bar{X}$-axis with the uniform velocity $\bar{\mathbf{u}} = \bar{x}/\bar{t}$. We now pose the question: What is the velocity of the particle with respect to K? As the reader knows, Newtonian kinematics gives the answer $\bar{\mathbf{u}} + \mathbf{v}$, whereas the result according to relativistic kinematics may be obtained by dividing the second of (6.1a) by the second of (6.1b). After some simple algebraic manipulation the result obtained is

$$u = x/t = \frac{\bar{u} + v}{1 + \dfrac{\bar{u}v}{c^2}}. \tag{6.4}$$

(6.4) is the well-known relativistic law for the composition of velocities, according to which, velocity is a subadditive quantity. Once again, the mathematical discrepancy between the two theories is suggestive of a fairly direct experimental resolution. Specifically, we introduce a third reference frame $\bar{\bar{K}}$, which is just the rest-frame of the particle. Although the concept of reference frame is somewhat loosely defined, any frame of reference which is suitable for special relativity—for example, a radar station—will be equally suitable for Newtonian mechanics. Accordingly, it may be categorically asserted that the expression "particle at rest with respect to frame K" is an observational one that has the same meaning in the two theories. Hence, the Newtonian and the relativity theorist would agree that the denotation of $\bar{\bar{K}}$ is unambiguous. Then, in principle, in order to achieve an experimental result which is in agreement either with the one theory or the other but not both, one has merely to measure the velocity of K with respect to $\bar{\bar{K}}$.

Let us now proceed from the rather straightforward topic of relativistic kinematics to the semantically hazardous area of relativistic mechanics. We shall now find that four-dimensional Minkowski coordinates are particularly useful. However, we remind the reader that they are nothing more than a formal geometric device without any inherent physical significance.

From the differential form of (6.3), we may introduce the classically definable expression:

$$\frac{dt}{d\tau} = (1 - v^2/c^2)^{-\frac{1}{2}}. \tag{6.5}$$

We also introduce the concept of the space-time interval by the expression:

$$dX^\mu = (dX^i, icdt). \tag{6.6}$$

From (6.5) and (6.6), we construct the concept of the four-velocity, defined by

$$U^\mu = \frac{dX^\mu}{d\tau} = \frac{dX^\mu}{dt}\frac{dt}{d\tau} = (1 - v^2/c^2)^{-\frac{1}{2}}\left(\frac{dx^i}{dt}, ic\right). \tag{6.7}$$

Thus, we see that the concept of the four-velocity, while quite extraneous to Newtonian mechanics, is, nevertheless, wholly definable in classical terms. We are obliged at this point to introduce a new primitive concept—namely, the proper-mass of a particle, m_0, which is specific to the theory of relativity. We, of course, do not pretend that m_0 is definable in classical terms; but it is not our intention to argue that the denotations of the two theories are coextensive. Now we make use of the spacelike four-vector F^μ to write the relativistic law of motion:

$$F^\mu = m_0 \frac{dU^\mu}{d\tau}. \tag{6.8}$$

It can hardly be overemphasized that (6.8), like its Newtonian counterpart, is not to be viewed as a definition of force but as a testable physical hypothesis. We now come to the most crucial point in the development of our argument. We introduce the *physical postulate* that F^μ is related to the Newtonian force vector f^i by

$$F^\mu = (1 - v^2/c^2)^{-\frac{1}{2}}\,(f^i, (i/c)v^i \cdot f^i). \tag{6.9}$$

Needless to say, (6.9) has not been miraculously plucked from the air. In many elementary presentations of relativistic mechanics, it is customary to deduce (6.9) from Newton's second law through the erroneous identification of Newtonian and relativistic mass. Clearly, no defender of Feyerabend would permit so facile a step. Moreover, anyone who makes the mistake of confusing physical laws with definitions is likely to object to (6.9) on the ground that it involves the mixing of concepts from semantically distinct theories. In particular, it may be argued that the left-hand side of (6.9) contains a relativistic concept while the right-hand side contains the classical force which is defined in terms of Newtonian mass. In fact, Newtonian force is not defined in terms of Newtonian or classical mass but is a distinct physical quantity which constitutes part of the external environment of physical objects which possess mass. To repeat, the two concepts are related by a testable law and not by definition. The experimental justification of (6.9) may be found in the covariant formulation of electrodynamics in which we have the force-law:

$$F^\mu = (q/c)F^{\mu\nu}U^\nu \tag{6.10}$$

in which F^i is the well-known Lorentz force of classical electrodynamics. $F^{\mu\nu}$ is a new type of mathematical quantity called a *tensor*, the nature of which will be fully explained in the next chapter. For the moment, it suffices to note that a tensor is a generalization of the notion of a vector. In this instance, its various components represent the components of the electromagnetic field.

Having established a relationship between the Newtonian f^i and the relativistic F^μ in equation (6.9), we now derive an important physical result. The invariant norm of F^μ is given by

$$(F^{\mu})^2 = \left(|f^i|^2 - \frac{|v^i|^2\,|f^i|^2}{c^2} \cos^2\theta \right)(1 - v^2/c^2)^{-1}. \tag{6.11}$$

Since F^{μ} is a spacelike vector, it is permissible to transform to the rest-frame of a particle on which it acts, with respect to which the time component of F^{μ} vanishes. Physically, this merely amounts to the force's acting on a particle whose instantaneous velocity is zero. The norm of $\bar{F}^{\mu}$ is given by

$$(\bar{F}^{\mu})^2 = (\bar{f}^i)^2 \tag{6.12}$$

Equating (6.11) and (6.12), we have

$$(\bar{f}^i)^2 = (f^i)^2 \left(\frac{c^2 - v^2 \cos^2\theta}{(c^2 - v^2)} \right). \tag{6.13}$$

Whence

$$\bar{f}^i > f^i. \tag{6.14}$$

Accordingly, we conclude that the magnitude of the Newtonian force which acts on a particle depends on both the magnitude and direction of the velocity of that particle. The Newtonian case would only hold rigorously for a moving particle when the force and velocity vectors are parallel.

The philosophical significance of (6.14) is that it has the status of a *theorem of relativistic mechanics expressed in the language of Newtonian mechanics.* Thus, it is explicitly incompatible with Newtonian mechanics, not only formally but also factually. I believe that a proponent of the thesis of incommensurability would be hard-pressed to argue that the derivation of (6.14) depended on the exploitation of a radical meaning change. Any such argument would have to rely on the premise that Newtonian force is *defined* as the product of mass and acceleration, where mass is a term whose meaning is determined within the context of classical physics. However, such a premise represents a methodologically unsound position.

In conclusion, we claim to have shown that it is possible to derive theorems of relativistic physics *in the language of classical physics.* Hence, while it remains true that one must be on guard against the semantic mixing of two distinct theories, it is, nevertheless, possible to establish contacts between them. With a modicum of semantic care, one may demonstrate that one theory is incompatible with another. Thus, the possibility of rational choice between rival theories is preserved; something, I suspect, that every scientist knows.

Bibliography for Chapter 6

FEYERABEND, P. K., "Explanation, Reduction and Empiricism", in *Scientific Explanation, Space and Time* (eds. H. FEIGL and G. MAXWELL), Minnesota Studies in the Philosophy of Science, vol. 3, University of Minnesota Press, Minneapolis, 1962.

KUHN, T. A., *The Structure of Scientific Revolutions*, University of Chicago Press, Chicago, 1962 (2nd edn. 1970).

7

More Mathematics

Motivation

THE mathematical foundations of both Newtonian and special relativistic physics may be kept at a fairly elementary level. If you found them taxing, it was because you were expected to grasp in a single chapter a host of concepts that would normally be associated with several undergraduate courses. However, given a more leisurely pace, you would almost certainly find all of the mathematics to this point both easy and routine. One very important reason for the relative simplicity of the mathematics is that it has been possible to presuppose that the space of physics has a flat or Euclidean character. When space is Euclidean, it is always possible to represent it by means of a Cartesian coordinate system in which the various mathematical operations take on a particularly simple form. Now it is true that we took a brief glance, for purposes of generality, at more complicated systems of coordinates such as the spherical ones. However, it is never obligatory to make use of them. It merely happens that they are conveniences to be exploited when the physical system of interest has a spherical symmetry. So long as the geometry of the space may be assumed to be Euclidean, i.e. the distance is given by $(ds)^2 = \delta_{ik} dX^i dX^k$, one may always resort to Cartesian coordinates. Now when we approached special relativity from the four-dimensional or space-time standpoint, we found that the geometry was not quite Euclidean. Accordingly, it became more convenient to have recourse to Minkowski coordinates to represent the inertial frames (Lorentz frames) of special relativity. Nevertheless, the Minkowski coordinates, although unfamiliar, were essentially hardly more difficult to handle than the customary Cartesian ones.

Now, however, as we approach the theory of general relativity, the situation changes. In the first place, general relativity is essentially a space-time theory. Its laws must be formulated in four-dimensional terms. That, in itself, poses no great problem. However, whereas the four-dimensional manifold of special relativity was flat (pseudo-Euclidean), that of general relativity is curved. To one who is unfamiliar with it, the notion of the curvature of space may seem elusive, not to speak of the curvature of spacetime. However, the reader should be assured that this concept is mathematically well-defined. In due course it will be adequately explained. For the moment, it will suffice to reflect on the nature of a surface of a sphere. This is an example of a two-dimensional curved manifold. The geometry of such a manifold is markedly different from that of the Euclidean plane. For example, the angle-sum of a triangle on such a surface is always greater than two right angles and will vary with the area. Now the description of the geometry of such a surface is fairly easy to visualize since the surface is embedded in a three-dimensional Euclidean space. One cannot visualize the curvature of a higher dimensional manifold, since it is impossible to imagine its being embedded in a Euclidean manifold of greater dimensionality. Nevertheless, it is certainly possible to detect the curvature or deviation from Euclidicity of any manifold by reference to its *intrinsic* properties. For example, one could determine that a given surface is spherical simply by ascertaining the properties of triangles inscribed thereon. By the same token, certain measurements may be performed to determine whether a three-dimensional manifold possesses intrinsic curvature.

Let us consider how the notion of the curvature of a manifold may be handled in a quantitative manner. One of the simplest curved manifolds is the ordinary one-dimensional circle. Let us compare two such manifolds—namely, a circle of radius 1 inch and another of radius 2 inches. If you reflect on these, you will perceive that the smaller circle curves more rapidly or sharply than the larger one. Indeed, as you know, the curvature of a great circle on the surface of the earth is hardly discernible, which is why the earth appears to be flat. Accordingly, we may express the curvature of a circle as the reciprocal of its radius: $1/r$. This clearly satisfies our original intuition that the larger the circle the smaller is its curvature. How could one handle the curvature of an arbitrary curve drawn on a plane surface? Although one can no longer speak of its radius, it is the case that so long as the curve is sufficiently smooth, then one may consider at any point on the curve that circle whose circumference most closely approximates the curve in the infinitesimal neighbourhood of the point of interest. If the curve happens to be turning sharply at the given point then the circle of curvature at that point will be relatively small. On the other hand, if the curvature of the variable curve is very slight, the circle of curvature will be relatively large. Accordingly, the measure of curvature of a variable or arbitrary plane curve may be specified by the reciprocal of the radius of the circle of curvature at a given point.

So far our treatment has been entirely intuitive. However, it may be directly related to some fundamental analytic notions that were established in Chapter 1. You will recall that the derivative of a function measures the rate of change of that function. Furthermore, it was found that the derivative could be given a geometric interpretation as the tangent to the graph of the given function. In fact, one may interpret the first derivative at a point as a vector which is tangent to the given point of a curve. Hence, the derivative along the curve constitutes a field of tangent vectors. Intuitively, the tangent vector at any point on the curve is a vector which is orthogonal to the radius of the circle of curvature at that point. Let us call this tangent vector field $\mathbf{t}$. As the curvature of the curve changes, the direction of $\mathbf{t}$ will also change. The rate of change of $\mathbf{t}$, i.e. the second derivative of the original function, will then represent the curvature of the curve. The magnitude of the second derivative will be equal to the reciprocal of the radius of curvature. So far, we have spoken of the derivative of a curve but have failed to specify the independent variable. Although any suitable one may be chosen, it is convenient to express the curve parametrically in terms of length or, more specifically, what is called *arc-length* s. When the curve is represented parametrically in terms of s, $\mathbf{t}$ is a *unit* tangent vector. More specifically, let $\mathbf{r}(s)$ be the position or radius vector of a given curve. Then $d\mathbf{r}/ds = \hat{t}$ is the unit tangent vector and $d\hat{t}/ds$ is the curvature $\kappa = 1/\rho$, where ρ is the radius of curvature. These are some of the basic ideas of *differential geometry*, which is simply the application of calculus to geometric matters.

Having seen how the curvature of a curve or one-dimensional manifold may be handled, we may proceed to consider how one may treat the curvature of a surface. Take an arbitrary point P on a curved surface. Consider a vector $\mathbf{n}$ which is normal to the surface at that point. Imagine a plane in which $\mathbf{n}$ is lying. Clearly, this plane will slice through the surface and thereby define a certain curve. This plane may be rotated about the axis $\mathbf{n}$. In any position a particular curve will be defined. One such curve will have a maximum curvature κ_1 at P while another will have a minimum curvature κ_2 at P. Then $\kappa_1 \kappa_2$ is a measure of the Gaussian or *intrinsic* curvature of the surface at the point P. Thus, a scalar curvature field may be defined on any arbitrary, smooth, two-dimensional manifold. However, it turns out that when one generalizes from two to n dimensions, it is no longer possible to represent the curvature by a scalar field. What is now required is a tensor field, called the curvature tensor. This final generalization is due to Riemann (1826–66).

In the case of general relativity, the situation is still more complicated. In the first place, as already remarked, the space or manifold of interest is a four-dimensional space-time continuum. Secondly, its curvature varies from region to region according to the distribution of matter. The basic task of experimental relativity, in effect, is to determine the nature and degree of curvature in the various parts of the universe.

Since, in the aforesaid sense, there is no *a priori* geometric structure to the universe but all is determined *a posteriori*, it follows that there can be no privileged coordinate system, either Cartesian or Minkowskian, for the description of the universe. Consequently, the laws of physics should be formulated in such a way that they reflect no particular property of one coordinate system or another. In other words, if one coordinate system is as good as any other, the laws of physics should be expressible in a form which is completely independent of the chosen system of coordinates—which is to say, that they will have the same form in all such systems. The requirement that the laws of physics have a form which is the same in every coordinate system is known as the *principle of general covariance*. In what follows we shall frequently have occasion to refer to it and to reflect on its philosophical significance.

The mathematical formalism which we shall employ to achieve complete independence of any given system of coordinates is called tensor analysis. While it is not the only formalism capable of doing the job, it was the only one available to Einstein in 1916 when he published his fundamental paper on general relativity. We make use of it since most of the treatises on the subject employ it and because it is probably the most intuitive formalism for the beginner in the subject. Accordingly, the following sections of this chapter will attempt to present the reader with the basic ingredients of tensor analysis. Now you are really about to embark on "higher" mathematics! If the going becomes difficult, screw up your courage and restudy the difficult section. Above all, don't become discouraged. If you find the subject difficult, remember that Einstein did too!

Curvilinear Coordinates

Suppose that the coordinates of a point in a Cartesian system are given by X, Y, Z or X^i. Suppose that we now select more or less arbitrarily a second *curvilinear* system U^j so that each of the X^i is a function of the U^j.

$$X^i = X^i(U^j) \; (i, j = 1, 2, 3) \tag{7.1}$$

or in less condensed form:

$$\begin{aligned} X^1 &= X^1(U^1, U^2, U^3) \\ X^2 &= X^2(U^1, U^2, U^3) \\ X^3 &= X^3(U^1, U^2, U^3). \end{aligned} \tag{7.2}$$

In general, with respect to the curvilinear coordinates U^i, it will not be the case that the metric will have the simple form $\delta_{ij}dU^i dU^j$. We now wish to consider the problem of expressing the metric in arbitrary curvilinear coordinates. To make the discussion as concrete as possible, we shall first tackle it in terms of the reasonably familiar spherical coordinate system (r, θ, ϕ). Hence, as a special case of (7.1), we have:

$$x = r \sin\theta \cos\phi, \ y = r \sin\theta \sin\phi, \ z = r \cos\theta. \tag{7.3}$$

Now we already know the expression for the metric in Cartesian coordinates—namely,

$$(ds)^2 = (dx)^2 + (dy)^2 + (dz)^2. \tag{7.4}$$

Our first step, therefore, will be to substitute the expressions of (7.3) into (7.4). This yields

$$(ds)^2 = (dr \sin\theta \cos\phi)^2 + (dr \sin\theta \sin\phi)^2 + (dr \cos\theta)^2. \tag{7.5}$$

We know, of course, how to compute the differentials on the right-hand side of (7.5) by elementary calculus. If you have forgotten, review the salient section on differentiation, paying particular attention to the differentiation of the trigonometric functions. The first one goes as follows.

$$\begin{aligned} d(r \sin\theta \cos\phi) &= \frac{\partial}{\partial r}(r \sin\theta \cos\phi)dr + \frac{\partial}{\partial \theta}(r \sin\theta \cos\phi)d\theta + \frac{\partial}{\partial \phi}(r \sin\theta \cos\phi)d\phi \\ &= \sin\theta\cos\phi\, dr + r\cos\theta\cos\phi\, d\theta - r\sin\theta\sin\phi\, d\phi. \end{aligned} \tag{7.6}$$

We now do the same thing for the second and third partial differentials as follows.

$$\begin{aligned} d(r \sin\theta \sin\phi) &= \frac{\partial}{\partial r}(r \sin\theta \sin\phi)dr + \frac{\partial}{\partial \theta}(r \sin\theta \sin\phi)d\theta + \frac{\partial}{\partial \phi}(r \sin\theta \sin\phi)d\phi \\ &= \sin\theta \sin\phi\, dr + r\cos\theta\sin\phi\, d\theta + r\sin\theta\cos\phi\, d\phi. \end{aligned} \tag{7.7}$$

And,

$$\begin{aligned} d(r\cos\theta) &= \frac{\partial}{\partial r}(r\cos\theta)dr + \frac{\partial}{\partial \theta}(r\cos\theta)d\theta + \frac{\partial}{\partial \phi}(r\cos\theta)\, d\phi \\ &= \cos\theta\, dr - r\sin\theta\, d\theta. \end{aligned} \tag{7.8}$$

We now have the tedious but routine task of squaring each of (7.6), (7.7) and (7.8). The formula for the square of an expression containing three terms is simply

$$(a+b+c)^2 = a^2 + b^2 + c^2 + 2ab + 2ac + 2bc. \tag{7.9}$$

Thus, the square of (7.6) is

$$\begin{aligned} &\sin^2\theta\cos^2\phi(dr)^2 + r^2\cos^2\theta\cos^2\phi(d\theta)^2 + r^2\sin^2\theta\sin^2\phi(d\phi)^2 \\ &+ 2r\sin\theta\cos\theta\cos^2\phi\, dr d\theta - 2r\sin^2\theta\sin\phi\cos\phi\, dr d\phi \\ &- 2r^2\sin\theta\cos\theta\sin\phi\cos\phi\, d\theta d\phi. \end{aligned} \tag{7.10}$$

The square of (7.7) is

$$\sin^2\theta \sin^2\phi (dr)^2 + r^2 \cos^2\theta \sin^2\phi (d\theta)^2 + r^2 \sin^2\theta \cos^2\phi (d\phi)^2 + 2r \sin\theta \cos\theta \sin^2\phi \, dr d\theta + 2r \sin^2\theta \sin\phi \cos\phi \, dr d\phi + 2r^2 \sin\theta \cos\theta \sin\phi \cos\phi \, d\theta d\phi. \tag{7.11}$$

Finally, the square of (7.8) is simply

$$\cos^2\theta (dr)^2 + r^2 \sin^2\theta (d\theta)^2 - 2r \sin\theta \cos\theta \, dr d\theta. \tag{7.12}$$

We must now perform the seemingly monumental task of adding (7.10), (7.11) and (7.12). But do not despair, some pleasant surprises are in store. Firstly, two of the terms in (7.10) simply cancel two of those in (7.11), so that we are left with

$$\begin{aligned} &\sin^2\theta \cos^2\phi (dr)^2 + \sin^2\theta \sin^2\phi (dr)^2 + \cos^2\theta (dr)^2 \\ &+ r^2 \cos^2\theta \cos^2\phi (d\theta)^2 + r^2 \cos^2\theta \sin^2\phi (d\theta)^2 + r^2 \sin^2\theta (d\theta)^2 \\ &+ r^2 \sin^2\theta \sin^2\phi (d\phi)^2 + r^2 \sin^2\theta \cos^2\phi (d\phi)^2 \\ &+ 2r \sin\theta \cos\theta \cos^2\phi \, dr d\theta + 2r \sin\theta \cos\theta \sin^2\phi \, dr d\theta \\ &- 2r \sin\theta \cos\theta \, dr d\theta. \end{aligned} \tag{7.13}$$

We shall now make a substantial simplification of (7.13) by exploiting the important trigonometric identity which we learned in Chapter 1—namely,

$$\sin^2\theta + \cos^2\theta = 1. \tag{7.14}$$

Firstly, we rewrite (7.13) as follows:

$$\begin{aligned} &\sin^2\theta (dr)^2 (\sin^2\phi + \cos^2\phi) + \cos^2\theta (dr)^2 \\ &+ r^2 \cos^2\theta (d\theta)^2 (\sin^2\phi + \cos^2\phi) + r^2 \sin^2\theta (d\theta)^2 \\ &+ r^2 \sin^2\theta (d\phi)^2 (\sin^2\phi + \cos^2\phi) \\ &+ 2r \sin\theta \cos\theta \, dr d\theta (\sin^2\phi + \cos^2\phi) - 2r \sin\theta \cos\theta \, dr d\theta. \end{aligned} \tag{7.15}$$

Employing (7.14) to simplify (7.15), we get

$$\begin{aligned} &\sin^2\theta (dr)^2 + \cos^2\theta (dr)^2 \\ &+ r^2 \cos^2\theta (d\theta)^2 + r^2 \sin^2\theta (d\theta)^2 \\ &+ r^2 \sin^2\theta (d\phi)^2 \\ &+ 2r \sin\theta \cos\theta \, dr d\theta - 2r \sin\theta \cos\theta \, dr d\theta. \end{aligned} \tag{7.16}$$

Employing (7.14) now on (7.16), we end up with

$$(dr)^2 + r^2 (d\theta)^2 + r^2 \sin^2\theta (d\phi)^2. \tag{7.17}$$

Here is an excellent example of a phenomenon which is a constant source of delight to all who pursue mathematical subjects. We have discovered that underlying the apparent chaos of (7.10), (7.11) and (7.12) was the startling and beautiful simplicity of (7.17). The latter, of course, is precisely the expression for the metric, $(ds)^2$, in spherical coordinates.

However, our analysis does reveal that the expression for the metric is, in fact, somewhat more complex than its counterpart in Cartesian coordinates. In general, in place of the familiar expression which involves the Kronecker delta, we shall expect the coefficients of the metric to be not constants but functions of the coordinates. The general expression for the metric in any arbitrary coordinate system will be

$$(ds)^2 = g_{ik} dX^i dX^k. \tag{7.18}$$

If the space or manifold is three-dimensional, (7.18) is actually a sum of nine terms. However, in the particular case that we have just examined in detail, there were only three terms. This happens to be a property of all those coordinate systems which are said to be *orthogonal*. Briefly, an orthogonal system of coordinates may be characterized as follows. If one of the coordinate variables is held constant at some fixed value, the other two variables define a coordinate surface. Obviously, by holding each variable constant in turn we arrive at three intersecting coordinate surfaces. Now if the three surfaces intersect at right angles, the coordinate system is orthogonal and the metric is substantially simpler than in the more general case of oblique coordinates.

We may treat the g_{ik} as the matrix display

$$g_{ik} = \begin{pmatrix} g_{11} & g_{12} & g_{13} \\ g_{21} & g_{22} & g_{23} \\ g_{31} & g_{32} & g_{33} \end{pmatrix}. \tag{7.19}$$

Thus, if the coordinate system is orthogonal, the off-diagonal elements of g_{ik} vanish. That is

$$g_{ik} = 0 \text{ for } i \neq k. \tag{7.20}$$

Again, for spherical coordinates, we have $g_{11} = 1$, $g_{22} = r^2$, $g_{33} = r^2 \sin^2\theta$. Finally, it is certainly to be remarked that g_{ik} is an example of a tensor. It is a rather simple but extremely important one. It is known both as the *fundamental* tensor and the *metric* tensor, according to the particular role which it is made to perform.

Not only is g_{ik} the most important geometric object in the study of tensor analysis and the differential geometry of manifolds but it also turns out to play a very fundamental role in the theory of general relativity. Accordingly, it is appropriate to take a second look at it from a somewhat more general perspective than that of the coordinate system (r, θ, ϕ). Let (x, y, z) be a standard three-dimensional coordinate system and $\overline{X}^i$ be an arbitrarily chosen system of general curvilinear coordinates such that $x = x(\bar{x}^1, \bar{x}^2, \bar{x}^3)$, $y = y(\bar{x}^1, \bar{x}^2, \bar{x}^3)$, $z = z(\bar{x}^1, \bar{x}^2, \bar{x}^3)$ are all one-one differentiable functions or *diffeomorphisms*. In terms of (x, y, z), the metric is, of course, given by the familiar expression

$$(ds)^2 = (dx)^2 + (dy)^2 + (dz)^2. \tag{7.21}$$

We obtain expressions for the coordinate differentials in terms of the general coordinates $\overline{X}^i$ as follows.

$$\begin{aligned} dx &= \frac{\partial x}{\partial \bar{x}^1} d\bar{x}^1 + \frac{\partial x}{\partial \bar{x}^2} d\bar{x}^2 + \frac{\partial x}{\partial \bar{x}^3} d\bar{x}^3 = \frac{\partial x}{\partial \bar{x}^i} d\bar{x}^i \\ dy &= \frac{\partial y}{\partial \bar{x}^1} d\bar{x}^1 + \frac{\partial y}{\partial \bar{x}^2} d\bar{x}^2 + \frac{\partial y}{\partial \bar{x}^3} d\bar{x}^3 = \frac{\partial y}{\partial \bar{x}^i} d\bar{x}^i \\ dz &= \frac{\partial z}{\partial \bar{x}^1} d\bar{x}^1 + \frac{\partial z}{\partial \bar{x}^2} d\bar{x}^2 + \frac{\partial z}{\partial \bar{x}^3} d\bar{x}^3 = \frac{\partial z}{\partial \bar{x}^i} d\bar{x}^i. \end{aligned} \tag{7.22}$$

We now insert (7.22) into (7.21) and make use of the familiar summation convention on repeated indices, which, as you will now see, makes life much easier.

$$
\begin{aligned}
(ds)^2 &= \frac{\partial x}{\partial \bar{x}^i} d\bar{x}^i \frac{\partial x}{\partial \bar{x}^j} d\bar{x}^j + \frac{\partial y}{\partial \bar{x}^i} d\bar{x}^i \frac{\partial y}{\partial \bar{x}^j} d\bar{x}^j + \frac{\partial z}{\partial \bar{x}^i} d\bar{x}^i \frac{\partial z}{\partial \bar{x}^j} d\bar{x}^j \\
&= \frac{\partial x}{\partial \bar{x}^i} \frac{\partial x}{\partial \bar{x}^j} d\bar{x}^i d\bar{x}^j + \frac{\partial y}{\partial \bar{x}^i} \frac{\partial y}{\partial \bar{x}^j} d\bar{x}^i d\bar{x}^j + \frac{\partial z}{\partial \bar{x}^i} \frac{\partial z}{\partial \bar{x}^j} d\bar{x}^i d\bar{x}^j \qquad (7.23) \\
&= \left(\frac{\partial x}{\partial \bar{x}^i} \frac{\partial x}{\partial \bar{x}^j} + \frac{\partial y}{\partial \bar{x}^i} \frac{\partial y}{\partial \bar{x}^j} + \frac{\partial z}{\partial \bar{x}^i} \frac{\partial z}{\partial \bar{x}^j} \right) d\bar{x}^i d\bar{x}^j \\
&= \bar{g}_{ij} d\bar{x}^i d\bar{x}^j.
\end{aligned}
$$

In effect, we have just now repeated in a more abstract or general fashion what we previously did in the concrete case of (r, θ, ϕ). Clearly, the effect of the metric coefficients g_{ij} is to preserve the *invariance* of ds under any arbitrary change of coordinate system. To know g_{ij} is to know which line segments are to be treated as congruent. Clearly, the arc-length s will be determined by the explicit form of the g_{ij}. Specifically, it is given by

$$
s = \int_{s_1}^{s_2} \left(g_{ij} \frac{dx^i}{ds} \frac{dx^j}{ds} \right)^{1/2} ds. \qquad (7.24)
$$

But it has already been shown that once s has been specified along a given curve, the intrinsic geometry of the curve may be determined. More generally, if g_{ij} is given for a particular manifold, then the entire intrinsic geometry of that manifold may be deduced. Having made our first formal contact with a tensor, we may now seize the opportunity to get to know them with greater intimacy.

Tensors and their Algebra

Like vectors, tensors are geometric objects having a linear, homogeneous law of transformation. In fact we have already encountered tensors in the early chapters. Specifically, scalars and vectors are just two kinds of tensors. A scalar is said to be a tensor of rank zero and a vector is said to be a tensor of rank one. The metric tensor is an example of a tensor of rank two. Any tensor of rank two may be represented by the matrix of its components. The latter helps us to visualize the sort of mathematical entity that a tensor is, but it cannot be extended to tensors of higher rank than two.

We shall begin by restricting our attention to vectors and then proceed to generalize our findings. So long as our attention was principally restricted to Cartesian coordinate systems, it was possible to refer quite simply to the *components* of a vector. However, in curvilinear coordinate systems, the vectors which form the basis (recall $\hat{x}^1, \hat{x}^2, \hat{x}^3$) may be constructed in two quite different ways. Let $\mathbf{r} = x\hat{x} + y\hat{y} + z\hat{z}$ be the position vector (arrow) of an arbitrary point P in Cartesian coordinates. Now $\mathbf{r} = \mathbf{r}(U^1, U^2, U^3)$, where U^i are the variables of some arbitrary curvilinear system. In the new system, it is possible to resolve the vector $\mathbf{r}$ in three directions which are *tangent* to the coordinate curves at P. These three tangents to the curves are given by $\partial\mathbf{r}/\partial u^1, \partial\mathbf{r}/\partial u^2, \partial\mathbf{r}/\partial u^3$ respectively. Hence, the vector $\mathbf{r}$ may be represented as

$$
r_1 \frac{\partial \mathbf{r}}{\partial u^1} + r_2 \frac{\partial \mathbf{r}}{\partial u^2} + r_3 \frac{\partial \mathbf{r}}{\partial u^3}.
$$

The three components r_1, r_2 and r_3 are said to be the *contravariant* components of **r**. They are always represented by a superscript r^i. Incidentally, it should be remarked that the vectors (arrows) $\partial \mathbf{r}/\partial u^i$ unlike the $\hat{x}^1, \hat{x}^2, \hat{x}^3$ system are not generally unit vectors. We could construct unit base vectors simply by dividing each tangent vector by its magnitude but it happens to be unnecessary.

Before dealing with the alternative basis associated with a general set of coordinates, it will be convenient to introduce an important concept of classical vector analysis. Suppose that we have an arbitrary differentiable scalar field Φ in a certain region of space—for example, a temperature distribution—then it is possible to associate with it a certain vector field. That is to say, at each point of space at which Φ is defined we associate a vector. The vector field may be denoted by $\nabla\Phi$ (read "del phi"). Such a field is called the *gradient* of Φ. Each vector represents the direction and magnitude of the greatest rate of change of the scalar field at any given point. ∇ is an example of a *vector operator*, which behaves in the course of symbolic manipulations much as a vector behaves. The definition of the gradient operator is given by

$$\nabla = \frac{\partial}{\partial x}\hat{x}^1 + \frac{\partial}{\partial y}\hat{x}^2 + \frac{\partial}{\partial z}\hat{x}^3. \tag{7.25}$$

Whence,

$$\nabla\Phi = \frac{\partial\Phi}{\partial x}\hat{x}^1 + \frac{\partial\Phi}{\partial y}\hat{x}^2 + \frac{\partial\Phi}{\partial z}\hat{x}^3. \tag{7.26}$$

Now it happens to be the case that for any constant surface, the gradient calculated at any point on the surface is always perpendicular to it. The proof of this is very elementary although we shall omit it. Intuitively, if you were climbing a hill and wished to follow a path which would increase your elevation more rapidly than any other, it would be one which was perpendicular to the so-called contour line or lines of equal elevation around the hill. This property of the gradient provides a route to another coordinate basis. Thus, given the coordinates U^1, U^2, U^3, for each coordinate surface, we may define a vector which is perpendicular to it. Thus, ∇U^1 is perpendicular to the surface U^2U^3, etc. Hence, we may construct the vectors $\nabla U^1, \nabla U^2, \nabla U^3$. Thus, in the system U^i, we may represent the vector (arrow) **r** by $R_1\nabla U^1 + R_2\nabla U^2 + R_3\nabla U^3$. The components R_i of **r** are said to be the *covariant* components of the vector. It should be noted that this has nothing to do with the other usage of the term "covariant" as in "Lorentz covariant", etc. Furthermore, you should note that in much of the literature of physics, reference is made to covariant and contravariant vectors. Indeed, in what follows we shall adhere to this convenient and natural practice. But it is little more than an idiom of physics usage. One and the same vector may be represented either by its contravariant or by its covariant components. To put it somewhat poetically, the vector exists as an objective geometric object in its own right; it is we who break it up into its components, be they covariant or contravariant, when we introduce our man-made coordinate systems. Thus, there is no great methodological or philosophical implication to be drawn from the distinction. For that reason, we have not made the discussion of it too laborious. On the other hand, it is essential to be aware of it, since it plays an important role in the mechanics of classical tensor analysis.

An excellent example of a covariant vector is precisely the gradient of a scalar. Thus, $\nabla\Phi$

$= A_i$ is a covariant vector. Such a vector is always represented in component form with a subscript, which refers to the fact that we have chosen the covariant components. Now what is of immediate interest to us is the transformation law for a covariant vector. That is, given the vector A_j in one coordinate system, what will be its components $\overline{A}_p$ in some other system? The answer is given by the law

$$\overline{A}_p = \frac{\partial x^j}{\partial \overline{x}^p} A_j. \tag{7.27}$$

Any quantity which transforms in this way is said to be a covariant vector. There are several routine but important observations to be made about (7.27). In the first place, of course, the summation convention applies to repeated indices in a single term. However, the repeated index in a tensor equation will always occur once as a subscript and once as a superscript. The X^j are simply the original coordinates, while the $\overline{X}^p$ are the new or "barred" coordinates. The repeated index j is called a *dummy* index since it really signifies a summation. The index which occurs only once in each term is called a *free* index. It is a true index which may take any one of a fixed set of values depending on the dimensions of the manifold. An easy way of remembering (7.27) is simply to note that p is the *subscript* of the barred vector and occurs in the *denominator* of the coefficient. The dummy index j occurs once as a superscript and once as a subscript. Thus, the use of subscripts to denote covariant components is most appropriate. Finally, since you are still quite new to the game, it would not be inappropriate to rewrite (7.27) in its expanded form. Assuming three dimensions, we have

$$\begin{aligned}
\overline{A}_1 &= \frac{\partial x^1}{\partial \overline{x}^1} A_1 + \frac{\partial x^2}{\partial \overline{x}^1} A_2 + \frac{\partial x^3}{\partial \overline{x}^1} A_3 \\
\overline{A}_2 &= \frac{\partial x^1}{\partial \overline{x}^2} A_1 + \frac{\partial x^2}{\partial \overline{x}^2} A_2 + \frac{\partial x^3}{\partial \overline{x}^2} A_3 \\
\overline{A}_3 &= \frac{\partial x^1}{\partial \overline{x}^3} A_1 + \frac{\partial x^2}{\partial \overline{x}^3} A_2 + \frac{\partial x^3}{\partial \overline{x}^3} A_3.
\end{aligned} \tag{7.28}$$

How fortunate that we shall be able to operate with the compact notation of (7.27)!

Without further ado, we now point out that the contravariant components of a vector are denoted by a superscript as, for example, A^j, and its transformation law is given by

$$\overline{A}^p = \frac{\partial \overline{x}^p}{\partial x^j} A^j \tag{7.29}$$

Note that the barred coordinates now occur in the *numerator* of the coefficient. So, once again, we find the notational convention to be helpful. It is interesting to note that the coordinate differentials transform in the manner of a contravariant vector. This is known from ordinary calculus.

$$d\overline{X}^p = \frac{\partial \overline{x}^p}{\partial x^j} dX^j. \tag{7.30}$$

Thus, it is useful, although not obligatory, to denote coordinates by superscripts; X^i rather than X_i.

We may now form a tensor of rank two by multiplying, component by component, two tensors of rank one, i.e. two vectors. Thus,

$$\begin{aligned} A^j B^k &= C^{jk} \\ A^j B_k &= C^j{}_k \\ A_j B_k &= C_{jk}. \end{aligned} \tag{7.31}$$

The three tensors (strictly, tensor *representatives* by analogy with vector representatives) are of the same rank but are said to be of different type. The first is twice contravariant, the second once contravariant once covariant, and the third twice covariant. This form of multiplication results in what is called the outer product or tensor product of two tensors. However, it is not the case that every tensor of higher rank is the product of tensors of lower rank. A useful typographical convention to indicate the type of tensor is $\binom{2}{0}$ for the first, $\binom{1}{1}$ for the second, and so forth.

The transformation law for a tensor of a given rank and type follows directly from (7.27) and (7.29). For example,

$$\overline{B}^{qr}_{stu} = \frac{\partial \overline{x}^q}{\partial x^j}\frac{\partial \overline{x}^r}{\partial x^k}\frac{\partial x^l}{\partial \overline{x}^s}\frac{\partial x^m}{\partial \overline{x}^t}\frac{\partial x^n}{\partial \overline{x}^u} B^{jk}_{lmn}. \tag{7.32}$$

The profound significance of (7.32) is that no matter what the tensor or what the nature of the coordinate transformation, so long as the latter is reversible, the transformation law is of the linear, homogeneous variety. That is to say, that if each component of a tensor has a particular value at a given point of the manifold as evaluated with respect to a given coordinate system, the value of each component of the same tensor at the same point as evaluated with respect to a second arbitrary coordinate system is obtained by multiplying the component in the first system by a number, or more accurately by a set of numbers, the results of which are then summed. The all important consequence of this is that if a tensor vanishes in one coordinate system, i.e. all of its components are zero, then it must vanish in every system. This, in turn, implies that a tensor equation which holds in one coordinate system must hold in every coordinate system. Thus, tensors are precisely the sort of mathematical entity needed for general relativity.

Given the transformation law of a tensor, it is easy to show that the sums and differences of two tensors of the same type are also tensors of that type. Moreover, the product of any two tensors results in a tensor whose rank is the sum of the ranks of the multiplicands. We shall prove the last-mentioned property of tensors to give the reader some flavour of the nature of such proofs. Suppose that A^j_k and B^l_{mn} are tensors. We wish to show that their exterior product $A^j_k B^l_{mn} = C^{jl}_{kmn}$ is also a tensor. In effect, we have only to show that the exterior product transforms like a tensor of the appropriate kind.

$$\overline{A}^p_q = \frac{\partial \overline{x}^p}{\partial x^j}\frac{\partial x^k}{\partial \overline{x}^q} A^j_k \tag{7.33}$$

$$\overline{B}^r_{st} = \frac{\partial \overline{x}^r}{\partial x^l}\frac{\partial x^m}{\partial \overline{x}^s}\frac{\partial x^n}{\partial \overline{x}^t} B^l_{mn}. \tag{7.34}$$

We now equate the products of each side of (7.33) and (7.34).

$$\overline{A}^p_q \overline{B}^r_{st} = \frac{\partial \overline{x}^p}{\partial x^j}\frac{\partial x^k}{\partial \overline{x}^q} A^j_k \frac{\partial \overline{x}^r}{\partial x^l}\frac{\partial x^m}{\partial \overline{x}^s}\frac{\partial x^n}{\partial \overline{x}^t} B^l_{mn}. \tag{7.35}$$

We now simply rearrange the terms in (7.35) to get

$$\overline{A}^p_q \overline{B}^r_{st} = \frac{\partial \overline{x}^p}{\partial x^j} \frac{\partial x^k}{\partial \overline{x}^q} \frac{\partial \overline{x}^r}{\partial x^l} \frac{\partial x^m}{\partial \overline{x}^s} \frac{\partial x^n}{\partial \overline{x}^t} A^j_k B^l_{mn} = \overline{C}^{pr}_{qst} \tag{7.36}$$

which is the required result.

Another extremely useful result is that δ^p_q which behaves just like the familiar Kronecker delta except that it must have one contravariant and one covariant index is a tensor. We now proceed to justify this claim.

If δ^p_q is a tensor of type $\begin{pmatrix}1\\1\end{pmatrix}$, then its transformation law must be

$$\overline{\delta}^r_s = \frac{\partial \overline{x}^r}{\partial x^p} \frac{\partial x^q}{\partial \overline{x}^s} \delta^p_q. \tag{7.37}$$

Firstly, any expression of the form $\partial x^p / \partial x^q$, where the coordinate variables of numerator and denominator belong to the same system, must be equivalent to δ^p_q, since if $p = q$ the expression is identically equal to unity, and if $p \neq q$, the expression is equal to zero, as distinct coordinate variables must be functionally independent. Now the coordinates X^p are functions of $\overline{X}^r$ which, in turn, are functions of X^q. Hence, by the chain rule of elementary calculus,

$$\frac{\partial x^p}{\partial x^q} = \frac{\partial x^p}{\partial \overline{x}^r} \frac{\partial \overline{x}^r}{\partial x^q} = \delta^p_q. \tag{7.38}$$

We now apply the result of (7.38) to (7.37). The right-hand side of (7.37) reduces to

$$\overline{\delta}^r_s = \frac{\partial \overline{x}^r}{\partial x^p} \frac{\partial x^p}{\partial \overline{x}^s} \tag{7.39}$$

in virtue of the property of the Kronecker delta. But by the chain rule, the right-hand side of (7.39) is simply $\partial \overline{x}^r / \partial \overline{x}^s$. So we are left with the identity $\overline{\delta}^r_s = \overline{\delta}^r_s$. Hence, the assumption of (7.37) to the effect that δ^p_q is a tensor of type $\begin{pmatrix}1\\1\end{pmatrix}$ is valid.

From a tensor of rank n, we may form a related tensor of rank $n-2$ by a process known as contraction. This is accomplished by equating any contravariant index with any covariant index and then performing the summation which is implied by the repetition of indices. For example, from the tensor B^{pq}_{rst}, we form the related tensor

$$B^{pq}_{rpt} = C^q_{rt}. \tag{7.40}$$

We now prove that the result is, in fact, a tensor.

$$\overline{B}^{jk}_{lmn} = \frac{\partial \overline{x}^j}{\partial x^p} \frac{\partial \overline{x}^k}{\partial x^q} \frac{\partial x^r}{\partial \overline{x}^l} \frac{\partial x^s}{\partial \overline{x}^m} \frac{\partial x^t}{\partial \overline{x}^n} B^{pq}_{rst}. \tag{7.41}$$

We now equate j and m in (7.41) to get

$$\overline{B}^{jk}_{ljn} = \frac{\partial \overline{x}^j}{\partial x^p} \frac{\partial \overline{x}^k}{\partial x^q} \frac{\partial x^r}{\partial \overline{x}^l} \frac{\partial x^s}{\partial \overline{x}^j} \frac{\partial x^t}{\partial \overline{x}^n} B^{pq}_{rst}. \tag{7.42}$$

We now rearrange the terms on the left-hand side of (7.42) to get

$$\overline{B}^{jk}_{ljn} = \frac{\partial \overline{x}^j}{\partial x^p}\frac{\partial x^s}{\partial \overline{x}^j}\frac{\partial \overline{x}^k}{\partial x^q}\frac{\partial x^r}{\partial \overline{x}^l}\frac{\partial x^t}{\partial \overline{x}^n} B^{pq}_{rst}. \tag{7.43}$$

But by the previously established result (7.38), (7.43) reduces to

$$\overline{B}^{jk}_{ljn} = \delta^s_p \frac{\partial \overline{x}^k}{\partial x^q}\frac{\partial x^r}{\partial \overline{x}^l}\frac{\partial x^t}{\partial \overline{x}^n} B^{pq}_{rst}. \tag{7.44}$$

In virtue of the property of the Kronecker delta, (7.44) is further reduced to

$$\overline{B}^{jk}_{ljn} = \frac{\partial \overline{x}^k}{\partial x^q}\frac{\partial x^r}{\partial \overline{x}^l}\frac{\partial x^t}{\partial \overline{x}^n} B^{pq}_{rpt}. \tag{7.45}$$

(7.45) establishes that B^{pq}_{rpt} transforms like a tensor of type $\binom{1}{2}$, so that the symbolization C^q_{rt} is fully justified.

A closely related operation to that of contraction consists in forming the outer product of two tensors and then contracting with respect to an index from each of the original tensors. This process is called *inner multiplication*. For example, an inner product of A^p_q and B^r_{st} would be $A^p_q B^r_{pt} = C^r_{qt}$. We have, in fact, already encountered an instance of this in the case of the ordinary inner or scalar product of two vectors. You will recall that the scalar product of two vectors $\mathbf{A} \cdot \mathbf{B}$ was defined by $\delta_{ij} A^i B^j$. This operation may now be viewed as the contraction of the outer product $A^i B_j$. We now suggest that the reader establish, as an exercise, that C^r_{qt} is, in fact, a tensor of type $\binom{1}{2}$.

The final item which we shall treat in connection with the algebra of tensors is the *raising* and *lowering* of indices. The tensor which results from the raising or lowering of one or more indices is said to be an *associated* tensor. In the simplest case, A^i and A_i are instances of associated tensors. In fact, in the case of a manifold endowed with a metric, which is the only kind which is of practical interest to us, A^i and A_i are simply two different representations, the one by contravariant components and the other by covariant ones, of a single geometric object. The two sets of components are related by the metric tensor as follows.

$$g_{ik} A^k = A_i. \tag{7.46}$$

That is to say, that the index of a tensor may be lowered by taking the inner product of that tensor with the metric tensor. In addition to the familiar metric tensor g_{ik}, there is a *conjugate* metric tensor which is defined by the relation

$$g^{ik} g_{jk} = \delta^i_j. \tag{7.47}$$

It may be employed to *raise* an index by the rule

$$g^{ik} A_k = A^i. \tag{7.48}$$

Tensor Calculus

Having at our disposal the basic ingredients of tensor algebra, we now proceed to the considerably more difficult subject of tensor calculus or tensor analysis. However, we implore our industrious reader, yet again, not to be deterred by the conceptual difficulties which are inherent to this subject. Although we are now in a mathematically sophisticated area, nothing will be presupposed that has not already been touched on in this book.

Our basic objective is to arrive at an operation which is analogous to that of ordinary differentiation but which possesses a *coordinate-independent* significance. That is to say, that we wish to be able to compute the derivative of a tensor, so that the result is also a tensor. That ordinary differentiation is not adequate to the task at hand may be demonstrated in a purely mechanical fashion. Suppose that we try to form the derivative of a covariant vector A_j.

$$\overline{A}_p = \frac{\partial X^j}{\partial \overline{X}^p} A_j. \tag{7.49}$$

We differentiate $\overline{A}_p$ with respect to $\overline{X}^q$ to obtain by the rule for the differentiation of a product

$$\frac{\partial \overline{A}_p}{\partial \overline{x}^q} = \frac{\partial x^j}{\partial \overline{x}^p} \frac{\partial A_j}{\partial \overline{x}^q} + \frac{\partial^2 x^j}{\partial \overline{x}^q \partial \overline{x}^p} A_j. \tag{7.50}$$

Applying the chain rule to the first term on the right-hand side of (7.50) we have

$$\frac{\partial \overline{A}_p}{\partial \overline{x}^q} = \frac{\partial x^j}{\partial \overline{x}^p} \frac{\partial x^r}{\partial \overline{x}^q} \frac{\partial A_j}{\partial x^r} + \frac{\partial^2 x^j}{\partial \overline{x}^q \partial \overline{x}^p} A_j. \tag{7.51}$$

(7.51) reveals that the quantity $\partial A_j/\partial x^r$ does not transform like a tensor. Specifically, its transformation properties differ from those of a tensor in virtue of the presence of the inhomogeneous term involving second partial derivatives. Were it not for the inhomogeneous term, $\partial A_j/\partial x^r$ would be a tensor of type $\binom{0}{2}$. Now the preceding discussion was fairly straightforward. What is less clear, however, is why the ordinary process of differentiation should break down, in the sense of failing to yield a new tensor as the derivative of a given tensor. We shall now spend some time on the discussion of this fundamental question.

With this end in view, let us again consider the derivative of a vector. Suppose that the vector $\mathbf{R}$ represents the position of a particle moving through space. Let us further suppose that each coordinate is a function of time. We are therefore able to write $\mathbf{R} = \mathbf{R}(t)$. In this context, time is said to be a parameter rather than a coordinate as it is usually regarded in this book. Now consider how one would arrive at the time derivative or velocity of the particle which, of course, is symbolized by $d\mathbf{R}/dt$. At a given moment the location of the particle is at $\mathbf{R}(t)$; a short time later the location of the particle is at the nearby point $\mathbf{R}(t+\Delta t)$. The change in location is, therefore, $\mathbf{R}(t+\Delta t) - \mathbf{R}(t)$. Furthermore, the average velocity of the particle during this finite displacement was

$$\frac{\mathbf{R}(t+\Delta t) - \mathbf{R}(t)}{\Delta t}.$$

Then, by an argument which should be familiar from our discussion of the derivative of a function in Chapter 1, the instantaneous rate of change of position or the velocity of the particle is given by

$$\lim_{\Delta t \to 0} \frac{\mathbf{R}(t+\Delta t) - \mathbf{R}(t)}{\Delta t} = \frac{d\mathbf{R}}{dt}. \tag{7.52}$$

To make this situation absolutely clear, we employ a diagram in which the third dimension is suppressed without serious loss of generality.

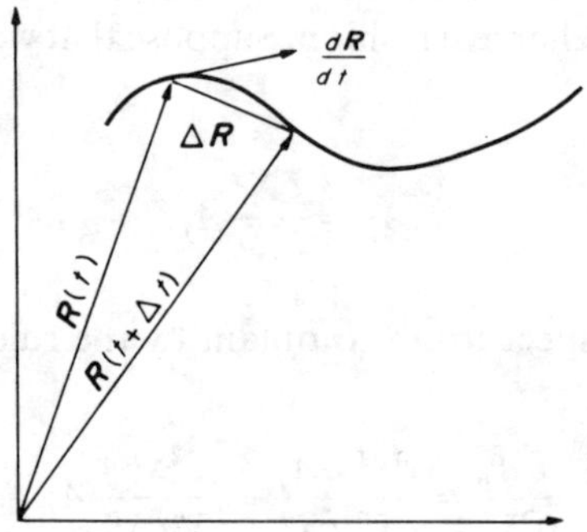

FIG. 7.1. Comparison of two vectors at neighbouring points

It follows that to make sense of the derivative of a vector, it is necessary to be able to give an unambiguous significance to the process whereby two vectors at neighbouring but nevertheless distinct points are compared and, in particular, subtracted. Now so long as we are working in a flat Euclidean manifold with a Cartesian basis ($\hat{x}^1, \hat{x}^2, \hat{x}^3$) there is no problem. We compare the vector at the point P with a second vector at a neighbouring point P' by sliding the first vector along the path from P to P' while holding its components fixed. In effect, the vector is transported parallel to itself. Then we have two vectors defined at the same point P', and we may substract the one from the other component by component. Equivalently, if we "parallel-transport" a vector around a small closed path, on returning to the starting-point the transported vector will coincide in direction and magnitude with the original vector. That vectors may be transported in this way in a Euclidean manifold without undergoing change is expressed by saying that Cartesian vectors are *free* vectors. That is to say, that they may be freely moved from place to place while retaining their original identity.

However, this straightforward process is not available to us when we are working in a curved manifold to which a Cartesian coordinate system is not applicable. Once we employ curvilinear coordinates, we are faced with the difficulty that the basis vectors change from point to point. Accordingly, no unambiguous meaning attachable to the notion of a self-identical vector at two distinct points is immediately apparent. No matter how hard we may try, we cannot prevent a vector from twisting and turning as we move it from point to point. The reader is no doubt sceptical of this impossibility or, at least, will find it difficult to visualize. To make the matter more intuitive we shall present the situation "writ large".

Figure 7.2 represents the surface of the earth, idealized as a sphere, with a triangle whose base is a segment AB of the equator which subtends an angle of 90° at the North Pole, B.

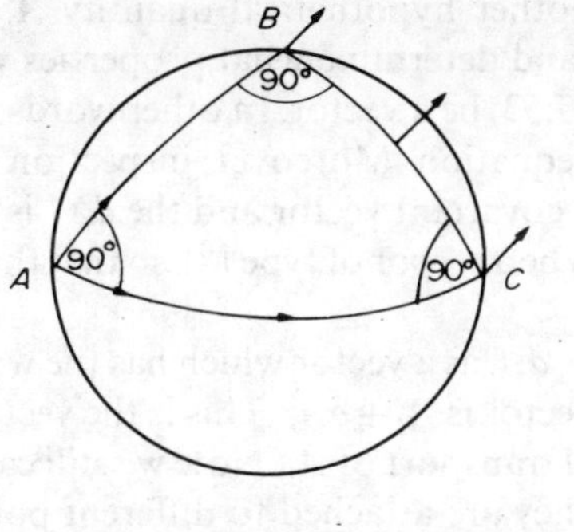

FIG. 7.2. The ambiguity of parallel transport on a curved manifold

We start with a vector at the point A, which we shall endeavour to parallel transport around the spherical triangle. (This example is taken from Bergmann (1968), p. 72.) At all times we transport the vector so that it is held as nearly parallel to itself as possible. Yet when it is returned to the starting-point it is clearly orthogonal to the original vector!

From such considerations as these, it emerges that it will be necessary to define or construct a standard of parallel transport before we undertake the process of differentiation in a non-Euclidean manifold. Such a problem is, strictly speaking, in the province of affine geometry, a subject which we choose largely to ignore in this work. To define parallel transport is to endow the manifold with an *affine connection* or *affinity*. Moreover, it turns out that the choice of affinity is virtually arbitrary. Thus, there are many possible standards of parallel transport, although once one is selected it must be rigorously adhered to thereafter.

However, affine geometry, which contains all of the apparatus necessary for tensor analysis, is not of direct concern to us since it lacks a full concept of distance. We are interested in a metric space, a space in which it is possible to define the inner product of a vector with itself, hence the magnitude of a vector. Thus, our aim is to discover a particular affine connection called a *metric connection*. Intuitively, the metric affinity is a rule of parallel transport such that a vector will remain not only autoparallel but also self-congruent. The latter, as one might expect, will be intimately tied to the metric tensor g_{ik}. It turns out that when the affine connection is defined in terms of the metric it is uniquely determined. In what follows we shall make use of a new notational convention. Instead of symbolizing the partial derivatives of vectors and tensors by $\partial A^i/\partial x^k$, we shall often employ $A^i{}_{,k}$. Second partial derivatives would appear, for example, as $A^i{}_{,jk}$.

Let us recall the current problem. Suppose that a certain vector field is defined on a finite region of a given manifold. That is to say, that the manifold has a vector defined at each point. At two neighbouring points the respective vectors will be A_i and $A_i + dA_i$. However, since the coordinate basis varies from point to point, the difference between the neighbouring vectors, namely dA_i, will not be a vector. But dA_i is given by

$$dA_i = \frac{\partial A_i}{\partial x^r} dX^r. \tag{7.53}$$

Now dX^r is a combination of coordinate differentials, which is the prototype of the contravariant vector. Hence, the partial derivative $A_{i,r}$ must be the wrong sort of quantity.

Let us replace it with another hypothetical quantity $A_{i;r}$ (where the comma has been replaced by a semicolon) and determine what properties would be required of it in order that the left-hand side of (7.53) be a vector. In other words, we would wish the counterpart of (7.53) to be a tensorial equation. Moreover, inspection of (7.53) suggests that since the left-hand side should be a covariant vector and the dX^r is a contravariant vector, that the counterpart of $A_{i,r}$ should be a tensor of type $\binom{0}{2}$, so that the indices on the two sides will be properly balanced.

Let us now think of $A_i + dA_i$ as a vector which has the wrong sort of components. Let us suppose that the correct vector is $A_i + \delta A_i$. This is the vector which would be defined at P' as the result of the parallel transport of A_i. Now we still cannot make a direct comparison of A_i and $A_i + \delta A_i$ since they are attached to different points. However, we can certainly compare $A_i + dA_i$ with $A_i + \delta A_i$ which are both vectors at P'. Moreover, we shall define A_i in such a manner that $dA_i - \delta A_i$ will be a vector. Therefore, employing (7.53) as our model, we may write

$$dA_i - \delta A_i = A_{i;r} dX^r. \tag{7.54}$$

The new derivative which emerges from our definition of parallel transport should be of the form

$$\frac{dA_i - \delta A_i}{dX^r} = A_{i;r}. \tag{7.55}$$

Hence,

$$A_{i;r} = A_{i,r} - \frac{\delta A_i}{dX^r}. \tag{7.56}$$

Of course, the significance of the second term on the right-hand side is, still to be determined.

We may continue in our search for a satisfactory redefinition of differentiation by recalling that (7.56) should be consistent with the situation that prevails in a flat Euclidean manifold. As you know, in this special case, parallel transport presents no problem since a vector may be transported parallel to itself without undergoing a change in its components. In more wordaday terms, one may add and subtract vectors which have been defined at different points of the Euclidean manifold since their components have the same significance everywhere. Let us denote the vector A_i in Cartesian coordinates by $\overline{A}_j$. The important thing to remember is that $\delta \overline{A}_j = 0$. That is equivalent to saying that the components of $\overline{A}_j$ are unchanged when the latter is transported to P'. We must have

$$\overline{A}_j = \frac{\partial x^i}{\partial \overline{x}^j} A_i$$
$$A_i = \frac{\partial \overline{x}^j}{\partial x^i} \overline{A}_j. \tag{7.57}$$

Hence, from the second of (7.57), we have

$$\delta A_i = \delta\left(\frac{\partial \overline{x}^j}{\partial x^i} \overline{A}_j\right). \tag{7.58}$$

However, since one of the factors must vanish in virtue of the vanishing of δA_j, (7.58) reduces to

$$\delta A_i = \delta\left(\frac{\partial \bar{x}^j}{\partial x^i}\right)\bar{A}_j. \qquad (7.59)$$

By the calculus, it follows from (7.59) that

$$\delta A_i = \frac{\partial^2 \bar{x}^j}{\partial x^i \partial x^r}\, dX^r \bar{A}_j. \qquad (7.60)$$

We now substitute the first of (7.57) into (7.60) and rearrange terms to get

$$\delta A_i = \frac{\partial^2 \bar{x}^j}{\partial x^i \partial x^r}\frac{\partial x^j}{\partial \bar{x}^i} A_j dX^r. \qquad (7.61)$$

Accordingly, the unknown quantity $\delta A_i/dX^r$ of (7.56) must, on the basis of (7.57), be

$$\frac{\delta A_i}{dX^r} = \frac{\partial^2 \bar{x}^j}{\partial x^i \partial x^r}\frac{\partial x^j}{\partial \bar{x}^i} A_j. \qquad (7.62)$$

As an abbreviation for the right-hand side of (7.62), we write Γ^j_{ir}. Putting this into (7.56) we may finally write

$$A_{i;\,r} = A_{i,\,r} - \Gamma^j_{ir} \qquad (7.63)$$

(7.63) is the basic rule for the *covariant derivative* of a covariant vector. The Γ^j_{ir} is a new kind of geometric object which differs in its transformation properties from a tensor. It should be noted that it is so constructed as to be symmetric with respect to its subscripts which may, therefore, be freely interchanged. This geometric object is precisely the *affinity* to which we earlier alluded.

(7.63) is not a general rule for covariant differentiation but the special case for the covariant derivative of a covariant vector. We shall return to this subject shortly to formulate the general rule which applies to any type of tensor, but before so doing, we wish to lay some more groundwork. It has already been remarked that there is a very great freedom in the choice of the affinity or the standard of parallel transport. However, we are interested in a particular definition which will be fully consistent with the requirements of metric geometry. A sufficient condition for the satisfaction of this requirement is that the invariant length or "norm" of a vector be preserved under parallel transport. It is obvious that this is a perfectly natural requirement to impose on our geometry. In other words, when a vector A^i is moved parallel to itself, we require that $g_{ik}A^iA^k$ be invariant. Another way of satisfying the demands of metric geometry is to treat the metric tensor as a constant from the standpoint of covariant differentiation. That is to say, that we must choose the affinity in such a way that $g_{ik;r}$ vanishes identically throughout the manifold. We shall omit the mathematical details and simply state the general result that the *metric affinity* is given by

$$\Gamma^l_{ik} = \tfrac{1}{2}g^{ls}\left(\frac{\partial g_{is}}{\partial X^k} + \frac{\partial g_{ks}}{\partial X^i} - \frac{\partial g_{ik}}{\partial X^s}\right) = df\left\{\begin{matrix} l \\ ik \end{matrix}\right\}. \qquad (7.64)$$

The symbol on the right-hand side of (7.64) is called the *Christoffel symbol* of the second kind. The Christoffel symbol of the first kind is then defined by

$$[ik, s] = \frac{1}{2}\left(\frac{\partial g_{is}}{\partial X^k} + \frac{\partial g_{ks}}{\partial X^i} - \frac{\partial g_{ik}}{\partial X^s}\right). \tag{7.65}$$

From (7.64) and (7.65), it follows that

$$\begin{Bmatrix} l \\ ik \end{Bmatrix} = g^{ls}[ik, s]. \tag{7.66}$$

Neither of the Christoffel symbols is a tensor, but their indices do behave like those of a tensor.

We have at last reached the point where we are able to formulate a general rule for the covariant differentiation of tensors. In the first place, we have the case of a covariant vector, which is immediately obtainable from (7.63) upon replacing the affinity symbol by the metric affinity or Christoffel symbol.

$$A_{p;q} = A_{p,q} - \begin{Bmatrix} s \\ p\,q \end{Bmatrix} A_s. \tag{7.67}$$

The corresponding rule for a contravariant vector is

$$A^p_{;q} = A^p_{,q} + \begin{Bmatrix} p \\ q\,s \end{Bmatrix} A^s. \tag{7.68}$$

These rules may be easily remembered from the following observations. In the case of a covariant vector, having a lower index, the Christoffel term is subtracted. In the case of a contravariant vector, having an upper index, the Christoffel term is added. This corresponds to the symbol $\pm$. Secondly, the index of the vector appears in the same position in the Christoffel symbol, upper or lower as the case may be. Thirdly, the index of differentiation is always in the lower position. Finally, a third index is chosen for the remaining place in the Christoffel symbol and forms an inner product with the original vector.

The generalization to tensors of any type is now quite straightforward. We shall consider the case of B^{jk}_{lmn} as a concrete example.

$$B^{jk}_{lmn;q} = B^{jk}_{lmn,q} + \begin{Bmatrix} j \\ q\,s \end{Bmatrix} B^{sk}_{lmn} + \begin{Bmatrix} k \\ q\,s \end{Bmatrix} B^{js}_{lmn} - \begin{Bmatrix} s \\ l\,q \end{Bmatrix} B^{jk}_{smn} - \begin{Bmatrix} s \\ n\,q \end{Bmatrix} B^{jk}_{lms} - \begin{Bmatrix} s \\ m\,q \end{Bmatrix} B^{jk}_{lsn}. \tag{7.69}$$

A scrutiny of (7.69) indicates that the first term is always the ordinary derivative and that there are as many Christoffel symbols as indices in the original tensor. Each term is derived by taking each tensor index and treating it in the manner appropriate to a vector having an index of the same kind, while leaving the remaining indices untouched. Finally, it should be noted that covariant differentiation of a product of tensors goes just like ordinary differentiation of products. Letting T_1 and T_2 represent arbitrary tensors, we have

$$(T_1\,T_2)_{;q} = T_1(T_{2;q}) + (T_{1;q})T_2. \tag{7.70}$$

At this point, the reader may wish to heave a deep sigh and go for a stroll in the park. You have now come a very long way indeed, having at your disposal the mathematical groundwork for one of the deepest physical theories yet devised. Let us now assume that you have returned from your stroll and are eager to develop your skill in calculating covariant derivatives of tensors.

As our first exercise, we shall attempt to derive explicit expressions for Christoffel symbols in a particular coordinate system. Since we already have at our disposal some important information about spherical coordinates, we shall once more take them as our model. So let us, then, attempt to work out the various Christoffel symbols for spherical coordinates. Now, in general, a three-index symbol in three dimensions will have twenty-seven components. So it appears that we shall have to do a great deal of work. However, one never proceeds blindly in that fashion. One begins by analyzing the symmetries and other special properties of the particular case so that one may take advantage of them to reduce the workload.

In the first place, spherical coordinates are orthogonal. Thus, g_{pq} always vanishes when $p \neq q$. Correspondingly, the Christoffel symbol vanishes when g_{pq} vanishes. Hence, we have only to consider the cases where $p = q$. Secondly, since the Christoffel symbols are functions of the first derivatives of the metric tensor, it follows that when the latter is constant, the former will vanish. Finally, we take advantage of the fact that $g_{pq} = g_{qp}$. Letting the Christoffel symbol of the first kind be denoted by $[pq, r]$, we have three general cases.

I. $$p = q = r$$

In this case, the Christoffel symbol is of the form

$$[pp, p] = \frac{1}{2}\left(\frac{\partial g_{pp}}{\partial X^p} + \frac{\partial g_{pp}}{\partial X^p} - \frac{\partial g_{pp}}{\partial X^p}\right) = \frac{1}{2}\frac{\partial g_{pp}}{\partial X^p}.$$

II. $$p = q \neq r$$

Now we have

$$[pp, r] = \frac{1}{2}\left(\frac{\partial g_{pr}}{\partial X^p} + \frac{\partial g_{pr}}{\partial X^p} - \frac{\partial g_{pp}}{\partial X^r}\right) = -\frac{1}{2}\frac{\partial g_{pp}}{\partial X^r}.$$

Recall that g_{pr} must vanish since $p \neq r$.

III. $$p = r \neq q.$$

This gives

$$[pq, p] = \frac{1}{2}\left(\frac{\partial g_{pp}}{\partial X^q} + \frac{\partial g_{qp}}{\partial X^p} - \frac{\partial g_{pq}}{\partial X^p}\right) = \frac{1}{2}\frac{\partial g_{pp}}{\partial X^q}.$$

Now you will recall that in spherical coordinates, $g_{11} = 1$, $g_{22} = r^2$, $g_{33} = r^2 \sin^2 \theta$. Hence, the derivatives of g_{pq} vanish for $p = q = 1$. Accordingly, we have only to consider the cases in which $p = 2$ or $p = 3$. Finally, before we begin, remember that $x^1 = r$, $x^2 = \theta$, and $x^3 = \phi$.

I. $$[pp, p]$$

(1) $$p = 2$$

$$[22, 2] = \frac{1}{2}\frac{\partial g_{22}}{\partial X^2} = \frac{1}{2}\frac{\partial r^2}{\partial \theta} = 0$$

(2) $p = 3$

$$[33, 3] = \frac{1}{2}\frac{\partial g_{33}}{\partial X^3} = \frac{1}{2}\frac{\partial r^2 \sin^2\theta}{\partial\phi} = 0.$$

II. $[pp, r]$

(1a) $p = 2,\ r = 1$

$$[22, 1] = -\frac{1}{2}\frac{\partial g_{22}}{\partial X^1} = -\frac{1}{2}\frac{\partial r^2}{\partial r} = -r$$

(b) $p = 2,\ r = 3$

$$[22, 3] = -\frac{1}{2}\frac{\partial g_{22}}{\partial X^3} = -\frac{1}{2}\frac{\partial r^2}{\partial\phi} = 0$$

(2a) $p = 3,\ r = 1$

$$[33, 1] = -\frac{1}{2}\frac{\partial g_{33}}{\partial X^1} = -\frac{1}{2}\frac{\partial r^2 \sin^2\theta}{\partial r} = -r\sin^2\theta$$

(b) $p = 3,\ r = 2$

$$[33, 2] = -\frac{1}{2}\frac{\partial g_{33}}{\partial X^2} = -\frac{1}{2}\frac{\partial r^2 \sin^2\theta}{\partial\theta} = -r^2\sin\theta\cos\theta.$$

III. $[pq, p]$

(1a) $p = 2,\ q = 1$

$$[21, 2] = \frac{1}{2}\frac{\partial g_{22}}{\partial X^1} = \frac{1}{2}\frac{\partial r^2}{\partial r} = r$$

(b) $p = 2,\ q = 3$

$$[23, 2] = \frac{1}{2}\frac{\partial g_{22}}{\partial x^3} = \frac{1}{2}\frac{\partial r^2}{\partial\phi} = 0$$

(2a) $p = 3,\ q = 1$

$$[31, 3] = \frac{1}{2}\frac{\partial g_{33}}{\partial X^1} = \frac{1}{2}\frac{\partial r^2 \sin^2\theta}{\partial r} = r\sin^2\theta$$

(b) $p = 3,\ q = 2$

$$[32, 3] = \frac{1}{2}\frac{\partial g_{33}}{\partial X^2} = \frac{1}{2}\frac{\partial r^2 \sin^2\theta}{\partial\theta} = r^2\sin\theta\cos\theta.$$

We are now able to deduce the values of the Christoffel symbols of the second kind by the defining relationship (7.66). It will be recalled that in any orthogonal coordinate system $g^{pq} = 1/g_{pq}$. Hence, we have

$$g^{11}[22, 1] = -r = \begin{Bmatrix} 1 \\ 2\,2 \end{Bmatrix}$$

$$g^{11}[33, 1] = -r\sin^2\theta = \begin{Bmatrix} 1 \\ 3\,3 \end{Bmatrix}$$

$$g^{22}[33, 2] = 1/r^2\,(-r^2\sin\theta\cos\theta) = -\sin\theta\cos\theta = \begin{Bmatrix} 2 \\ 3\,3 \end{Bmatrix}$$

$$g^{22}[21, 2] = r/r^2 = 1/r = \begin{Bmatrix} 2 \\ 2\,1 \end{Bmatrix}$$

$$g^{33}[31, 3] = r \sin^2\theta/r^2 \sin^2\theta = 1/r = \begin{Bmatrix} 3 \\ 3\,1 \end{Bmatrix}$$

$$g^{33}[32, 3] = r^2 \sin\theta \cos\theta/r^2 \sin^2\theta = \cos\theta/\sin\theta = 1/\tan\theta = \cot\theta = \begin{Bmatrix} 3 \\ 3\,2 \end{Bmatrix}$$

Note that multiplication of the "square" three-index symbol by the metric tensor has the effect of raising the last index and replacing square brackets by curly brackets (braces). In all, there are eight non-vanishing symbols of each kind since, for example, $\begin{Bmatrix} 2 \\ 2 \quad 1 \end{Bmatrix} = \begin{Bmatrix} 2 \\ 1 \quad 2 \end{Bmatrix}$. It should also be noted that, as we should expect, in Cartesian coordinates in which $g_{ik} = \delta^i_k$, all of the three-index symbols vanish and the formula for covariant differentiation reduces to that of ordinary differentiation. It is also useful to note that it is easily shown that the covariant derivatives of g_{ik}, g^{ik} and δ^i_k are all zero. The ambitious reader may wish to obtain the pleasure of proving these results for himself.

Finally, it should be remarked that just as it is possible to form the total derivative from the set of partial derivatives of a given function, provided that the various independent variables are all functions of a single variable or parameter such as time, so a counterpart exists in the case of covariant differentiation. The total derivative of a tensor is called its *intrinsic derivative*. For example, suppose we are given the covariant derivative of a contravariant vector $V^j_{;q}$. Then its intrinsic derivative with respect to t is given by

$$\frac{DV^j}{Dt} = (V^j_{;q})\frac{dX^q}{dt} = V^j_{,q}\frac{dX^q}{dt} + \begin{Bmatrix} j \\ q\,r \end{Bmatrix} V^r \frac{dX^q}{dt}$$

$$= \frac{\partial V^j}{\partial x^q}\frac{dX^q}{dt} + \begin{Bmatrix} j \\ q\,r \end{Bmatrix} V^r \frac{dX^q}{dt} = \frac{dV^j}{dt} + \begin{Bmatrix} j \\ q\,r \end{Bmatrix} \frac{dX^q}{dt}. \tag{7.71}$$

Intrinsic derivatives of tensors of any type are a straightforward generalization of (7.71). The intrinsic derivative is an important concept, since if the intrinsic derivative of a vector along a curve is zero, then the vector undergoes parallel transport along that curve. However, we shall discuss this topic in much greater detail in Chapter 9, where it is related to the concept of *geodesic*.

The Riemann-Christoffel Tensor

As a further exercise in covariant differentiation, we shall now establish a very important result. Subject to some basic requirements on the continuity of functions and their derivatives, it happens to be the case that ordinary differentiation is commutative. That is to say, that when we perform a mixed partial differentiation, the order in which the derivatives are taken is of no importance. In other words, we do not distinguish between $A_{p,\,qr}$ and $A_{p,\,rq}$, since the two are normally equivalent. This is not generally true in the case of covariant differentiation. Moreover, it turns out that the difference between the two mixed partial derivatives is a tensor which is of fundamental importance in the geometry of Reimannian, manifolds. To derive this tensor, we shall simply form the two second partial derivatives and subtract the one from the other. At the same time we shall be afforded the opportunity of practising the formation of second covariant derivatives. We shall begin by forming an expression for $A_{p;qr}$, the second covariant derivative of A_p. We treat the expression as $(A_{p;q})_{;r}$. It should be noted that since the expression in brackets is a covariant derivative, it must be a rank two tensor of type $\begin{pmatrix} 0 \\ 2 \end{pmatrix}$ and may be treated accordingly. So we begin:

$$(A_{p;q})_{;r} = (A_{p;q})_{,r} - \begin{Bmatrix} i \\ p\,r \end{Bmatrix} A_{i;q} - \begin{Bmatrix} i \\ q\,r \end{Bmatrix} A_{p;i} \tag{7.72}$$

We now proceed to form the remaining covariant derivatives on the right-hand side of (7.72).

$$\left(A_{p,q} - \begin{Bmatrix} i \\ pq \end{Bmatrix} A_i\right)_{,r} - \begin{Bmatrix} i \\ p\,r \end{Bmatrix}\left(A_{i,q} - \begin{Bmatrix} j \\ i\,q \end{Bmatrix} A_j\right) - \begin{Bmatrix} i \\ q\,r \end{Bmatrix}\left(A_{p,i} - \begin{Bmatrix} k \\ p\,i \end{Bmatrix} A_k\right). \tag{7.73}$$

We now eliminate all parentheses in the following two steps.

$$A_{p,qr} - \left(\begin{Bmatrix} i \\ p\,q \end{Bmatrix} A_i\right)_{,r} - \begin{Bmatrix} i \\ p\,r \end{Bmatrix} A_{i,q} - \begin{Bmatrix} i \\ p\,r \end{Bmatrix}\begin{Bmatrix} j \\ i\,q \end{Bmatrix} A_j - \begin{Bmatrix} i \\ q\,r \end{Bmatrix} A_{p,i} + \begin{Bmatrix} i \\ q\,r \end{Bmatrix}\begin{Bmatrix} k \\ p\,i \end{Bmatrix} A_k$$

$$= A_{p,qr} - \begin{Bmatrix} i \\ p\,q \end{Bmatrix} A_{i,r} - A_i \begin{Bmatrix} i \\ p\,q \end{Bmatrix}_{,r} - \begin{Bmatrix} i \\ p\,r \end{Bmatrix} A_{i,q} + \begin{Bmatrix} i \\ p\,r \end{Bmatrix}\begin{Bmatrix} j \\ i\,q \end{Bmatrix} A_j - \begin{Bmatrix} i \\ q\,r \end{Bmatrix} A_{p,i}$$

$$+ \begin{Bmatrix} i \\ q\,r \end{Bmatrix}\begin{Bmatrix} k \\ p\,i \end{Bmatrix} A_k. \tag{7.74}$$

To form the other second covariant derivative, it is not necessary to repeat the computation. We simply interchange q and r in (7.74). This will give us

$$A_{p,rq} - \begin{Bmatrix} i \\ p\,r \end{Bmatrix} A_{i,q} - A_i \begin{Bmatrix} i \\ p\,r \end{Bmatrix}_{,q} - \begin{Bmatrix} i \\ p\,q \end{Bmatrix} A_{i,r} + \begin{Bmatrix} i \\ p\,q \end{Bmatrix}\begin{Bmatrix} j \\ i\,r \end{Bmatrix} A_j - \begin{Bmatrix} i \\ r\,q \end{Bmatrix} A_{p,i}$$

$$+ \begin{Bmatrix} i \\ r\,q \end{Bmatrix}\begin{Bmatrix} k \\ p\,i \end{Bmatrix} A_k. \tag{7.75}$$

We now subtract (7.75) from (7.74) which yields

$$A_i \begin{Bmatrix} i \\ p\,r \end{Bmatrix}_{,q} - A_i \begin{Bmatrix} i \\ p\,q \end{Bmatrix}_{,r} + \begin{Bmatrix} i \\ p\,r \end{Bmatrix}\begin{Bmatrix} j \\ i\,q \end{Bmatrix} A_j - \begin{Bmatrix} i \\ p\,q \end{Bmatrix}\begin{Bmatrix} j \\ i\,r \end{Bmatrix} A_j. \tag{7.76}$$

We are entitled to replace the index i in the first two terms of (7.76) since it is a dummy index. This then yields

$$A_j \begin{Bmatrix} j \\ p\,r \end{Bmatrix}_{,q} - A_j \begin{Bmatrix} j \\ p\,q \end{Bmatrix}_{,r} + \begin{Bmatrix} i \\ p\,r \end{Bmatrix}\begin{Bmatrix} j \\ i\,q \end{Bmatrix} A_j - \begin{Bmatrix} i \\ p\,q \end{Bmatrix}\begin{Bmatrix} j \\ i\,r \end{Bmatrix} A_j \tag{7.77}$$

$$= \left(\begin{Bmatrix} j \\ p\,r \end{Bmatrix}_{,q} - \begin{Bmatrix} j \\ p\,q \end{Bmatrix}_{,r} + \begin{Bmatrix} i \\ p\,r \end{Bmatrix}\begin{Bmatrix} j \\ i\,q \end{Bmatrix} - \begin{Bmatrix} i \\ pq \end{Bmatrix}\begin{Bmatrix} j \\ i\,r \end{Bmatrix}\right) A_j \tag{7.78}$$

Now $A_j(= A_p)$ is an arbitrary tensor. Moreover, the expression as a whole is the difference of two tensors and must, therefore, be a tensor. Hence, the expression in parentheses is a tensor, which we shall symbolize as R^j_{pqr}. Note that the index i of (7.78) does not appear since it only occurs as a repeated or dummy index, indicating that a summation is to be performed. R^j_{pqr} is the celebrated Riemann–Christoffel tensor. It is a function of the metric tensor and of its first and second derivatives. Its significance is as a precise measure of the curvature of the manifold in the region where it is evaluated.

In the four-dimensional space-time manifold, the Riemann–Christoffel tensor would have 256 distinct components. However, due to various symmetries there are, in fact, only 20 distinct components. To know the values of these is to know everything about the local geometry of the manifold. As we shall see in a subsequent chapter, the Riemann–Christoffel tensor supplies the clue to the formulation of the law of gravitation of general relativity.

Before leaving this arduous chapter, I have a few final words for the reader. Many of you will have found these pages extremely difficult. You probably read the text and simply skimmed over the calculations. This is perfectly in order for a first reading in which you wish to familiarize yourself with the basic concepts. However, it is not really enough. I urge you to reread the material while armed with pencil and paper and to give every effort to reproducing all of the calculations. Even though you are not a scientist, there is nothing in this chapter which you cannot master, provided that you mastered Chapter 1. Many books on the philosophy of relativity place no more burden on the reader than to have him learn to mouth such terms as affinity, metric, Christoffel symbol, etc. However, while it is true that to know how to calculate the values of a Christoffel symbol is not to understand philosophy, I am firmly convinced that one really does not have a "feel" for such things until one has done at least a little rudimentary work with them. And the "feel" is a necessary condition for satisfactory philosophizing about a subject in contrast to the mere waving of hands.

Bibliography for Chapter 7

ADLER, I., *A New Look at Geometry*, New American Library, Toronto, 1967. A popular and comprehensive survey of various branches and types of geometry.

BERGMANN, P. G., *The Riddle of Gravitation*, Scribner, New York, 1968.

COOLIDGE, J. L., *A History of Geometrical Methods*, Dover, New York, 1963. A somewhat dated but thorough technical development.

LAWDEN, D. F., *An Introduction to Tensor Calculus and Relativity*, Methuen, London, 1962. A favourite for use in upper-level undergraduate courses. Possibly the best book with which to begin.

MCQUISTAN, R. B., *Scalar and Vector Fields: A Physical Interpretation*, Wiley, New York, 1965. A detailed account of the application of classical vector analysis to physical fields. Tensors are not employed.

O'NEILL, B., *Elementary Differential Geometry*, Academic Press, New York, 1966. A detailed account of the elementary part of the subject from an advanced, contemporary viewpoint. More for the use of mathematicians than of physicists.

PETROV, A. Z., *Einstein Spaces*, Pergamon Press, Oxford, 1969. An advanced treatise on the mathematics of general relativity. Not for the beginner but useful in so far as it deals almost exclusively with manifolds of Lorentz signature.

SCHRÖDINGER, E., *Space-Time Structure*, Cambridge University Press, Cambridge, 1963. A charming and insightful treatment of the mathematics of space-time manifolds. Definitely recommended.

SPAIN, B., *Tensor Calculus*, Oliver & Boyd, Edinburgh, 1953. A brief account suited to the needs of physicists.

SYNGE, J. L., and A. SCHILD, *Tensor Calculus*, University of Toronto Press, Toronto, 1949. A clear and comprehensive treatment for the advanced undergraduate.

WEATHERBURN, C. E. *An Introduction to Riemannian Geometry and the Tensor Calculus*, Cambridge University Press, Cambridge, 1963. A good standard work.

WREDE, R. C. *Introduction to Vector and Tensor Analysis*, Dover, New York, 1972. My subjective favourite. It is sufficiently rigorous but not unduly so.

8

The Path to General Relativity

The Classical Theory of Gravitation

WE have already had first-hand experience of the desirability of gaining insight into classical mechanics prior to studying relativistic mechanics. A comparison of the two is methodologically and philosophically rewarding as, for example, when it is brought to bear on the problem of the commensurability of theories. Furthermore, since physics came of age, in the seventeenth century, it has undergone a fairly smooth development, so that an understanding of its earlier stages is a virtual necessity for one who wishes to gain insight into its more recent aspects. The theory of general relativity, whatever it may encompass, is primarily a theory of gravitation. Consequently, to appreciate its significance and to understand its physical content, it is mandatory first to acquire a basic understanding of the classical or Newtonian theory of gravitation. We shall begin by attempting to gain some intuitive insight into the principal factual commitment of Newton's theory—namely, that the force of gravitation that is usually associated with a body's falling to the ground is also responsible for the motion of the moon and the planets. Then we shall attempt to reformulate the theory with the aid of more sophisticated mathematical tools. It is this formulation of the theory which provides the direct link with the Einsteinian theory of gravitation.

The first decisive step towards a satisfactory theory of gravitation was taken by Galileo, who discovered what is probably the most significant feature of gravitational phenomena—namely, that the acceleration of a body due to a gravitational force is independent of the mass of that body. Provided that appropriate allowance is made for the variability of air resistance on different bodies, it is found that given similar initial conditions, they all fall to earth at the same speed. The rate of acceleration is approximately 32.2 ft/sec^2. The distance traversed is $\frac{1}{2}at^2$, where a is the acceleration and t is the duration of the fall. These are the facts revealed by the celebrated, albeit apocryphal, story of Galileo's dropping of cannon balls from the tower of Pisa.

The other key precursory step was taken by Kepler, who, on the basis of the painstaking observations of his predecessors, formulated his celebrated three laws of planetary motion. (1) All planetary orbits are elliptical, with the sun located at one focus of the ellipse. (2) The line connecting the sun and the orbiting planet sweeps equal areas in equal times. (3) The ratio of the square of the period of a planet to the cube of its distance from the sun is a constant. It was left to Newton to synthesize the apparently diverse phenomena of locally falling bodies and orbiting planets. Newton undertook to show that the trajectory of a planet or earth satellite is determined by the same law that governs the falling of a body in the vicinity of the earth's surface.

The basic intuition underlying Newton's theoretical construction is as follows. It is well known that when a body is projected in a horizontal direction above the surface of the earth, say from a mountain peak, it will follow a parabolic path and eventually hit the earth. The curvilinear trajectory is the resultant of the two components of the projectile's motion: the horizontal component which was imparted to it when first projected or thrown and the vertical component which is due to its falling towards the earth. Now, since the surface of the earth actually curves away from the horizontal, the situation is readily visualized in which the horizontal (tangential) velocity of the projectile is given a critical initial value which enables it to follow a closed curve around the earth. This is the basis of the technology of artificial satellite launching.

These considerations may be refined by the following quantitative approximations.

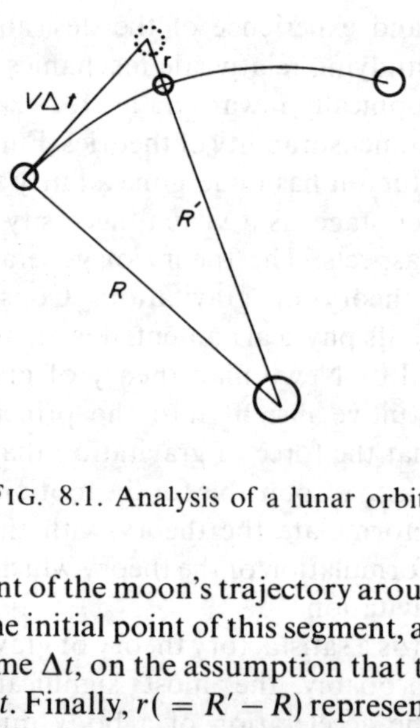

FIG. 8.1. Analysis of a lunar orbit

Figure 8.1 represents a segment of the moon's trajectory around the earth. R is the distance between earth and moon at the initial point of this segment, and R' is the distance between the two after an interval of time Δt, on the assumption that the moon travels in a straight line, covering the distance $v\Delta t$. Finally, $r(= R' - R)$ represents the actual displacement of the moon from its horizontal path towards the earth. On the basis of the Pythagorean theorem we may write the initial relationship

$$R'^2 = R^2 + (v\Delta t)^2 \tag{8.1}$$

from which by definition of R' we get

$$(R+r)^2 = R^2 + (v\Delta t)^2. \tag{8.2}$$

Expanding (8.2), we then have

$$R^2 + 2Rr + r^2 = R^2 + v^2\Delta t^2. \tag{8.3}$$

By some elementary algebraic manipulation we may rewrite (8.3) as

$$r + r^2/2R = v^2\Delta t^2/2R \tag{8.4}$$

For a suitably small interval of time, r is very much smaller than $R\,(r \ll R)$ and the second term on the left-hand side of (8.4) may be disregarded, so that

$$r = v^2 \Delta t^2 / 2R = \tfrac{1}{2}(v^2/R)\Delta t^2. \qquad (8.5)$$

Now since r represents the distance which the moon "falls" during the interval Δt, we may relate (8.5) to Galileo's law of falling bodies by equating the factor v^2/R with the acceleration of the body. In this way Newton, on the basis of a knowledge of the moon's velocity and distance from the earth, was able to calculate its acceleration. Comparing this result with the acceleration of a body in the immediate vicinity of the earth, he arrived at an extremely important result. He discovered that the acceleration of the moon was very much less than that of an object near the surface of the earth and, more precisely, that the force or acceleration falls off inversely as the square of the distance. That is to say, that gravitational force is proportional to $1/R^2$.

Furthermore, in accordance with his second law of motion, Newton knew the acceleration imparted to a body by a given force to be inversely proportional to the mass of that body. For a given force, the effect of doubling the mass would be to halve the acceleration. At first glance, this would appear to be incompatible with Galileo's discovery that all bodies undergo the same gravitational acceleration independently of their respective mass. However, the apparent discrepancy is easily resolved by assuming that as the mass of the body increases, the body will experience a proportionately greater force. Hence the inertia of the body or its tendency to resist a change of motion will be precisely compensated by the increased gravitational force. Thus, Newton concluded that the force of gravitational attraction is directly proportional to the mass of the body. Accordingly, the gravitational force function relating two mutually attracting bodies is given by

$$\mathbf{F} = Gm_1 m_2 \hat{r}/R^2 \qquad (8.6)$$

where $\hat{r}$ is the unit vector in the direction of R, the line joining the two bodies, and G is a factor of proportionality called the gravitational constant.

The Newtonian law of gravitation incorporates a truly remarkable feature. In the law of motion, $\mathbf{F} = m\mathbf{a}$, the mass is that property of a body which is manifested by its resistance to change of motion. We may call it the body's *inertial mass*. On the other hand, the role of mass in (8.6) is quite different. It is associated with the body's capacity to be a source of gravitational force. Accordingly, we may call it the *gravitational mass*. From a strictly logical standpoint, the inert mass and the gravitational mass (which could also be called gravitational charge by analogy with electric charge) might be independent quantities. Nevertheless, extremely accurate experiment has revealed a precise proportionality between them which, for a suitable choice of units, amounts to numerical equality.

Within the context of Newtonian mechanics such an equality is left as a most remarkable coincidence. Every body has two distinct properties, an inertial mass, which makes it resist the action of a force, and a gravitational mass, which makes it respond to the influence of a gravitational force. Moreover, these two countervailing tendencies are so precisely adjusted, the one to the other, that the net result is that all bodies fall to earth with the same acceleration. The Achilles' heel of Newtonian gravitation theory is that the coincidence went unexplained—to be accepted as a brute fact. However, Einstein always held fast to the view that no example of natural harmony should be regarded as standing without need of rational interpretation. Such an attitude to nature would seem to be methodologically required if only on the basis of considerations of statistical likelihood.

The situation is reminiscent of the use of the Fitzgerald contraction to explain the null result of the Michelson–Morley experiment. Nature eschews conspiracies!

Before proceeding to the next level of sophistication, let us pause to consider an important feature of the gravitational force. Newton's law as formulated in (8.6) is an example of what is called a *central force* law. A central force between two bodies is simply a force whose magnitude is a function of the distance between the two bodies and whose direction is along the line which joins them. A superficial examination of (8.6) reveals that the force of gravity is indeed central. Since (8.6) is a vector equation, the left-hand and right-hand sides must agree in direction. Hence, the force must be parallel to the unit vector $\hat{r}$ which lies along the line joining the two bodies. Secondly, the magnitude of the force is obviously given as a function of R, which is the distance between them.

A central force is an instance of a more general class of forces known as *conservative* forces. Any physical system which is governed only by forces of this kind is said to be conservative. In any such system, the total energy is a constant; that is to say, that it is conserved. An interesting property of any conservative system is that the work expended in moving an object from any point A to any other point B is independent of the path along which the object is constrained to move. Let an object be constrained to move along a differential element of path $d\mathbf{S}$. From the definition of work given in Chapter 3, the work expended will be $\mathbf{F}\cdot d\mathbf{S}$. The total work done in causing the object to be moved the finite distance AB will simply be the sum of all such differential increments of work. Thus, we may define the work done in moving a particle from A to B as

$$W = df \int_A^B \mathbf{F} \cdot d\mathbf{S} \tag{8.7}$$

An integral of this kind, i.e. along a path, is called a *line integral*.

If the force is conservative, it is a very easy matter to calculate the value of such an integral, since it may be computed along the most mathematically tractable path as, for example, a straight line. We may render this notion more comprehensible by considering motion in some radially symmetric field. In Fig. 8.2, we interpret B to be a point at a certain height above the point A in a gravitational field. In elevating a particle from A to B, one must perform positive work against the field. If the jagged path is followed, such work is done along each vertical segment. However, no work is done along the horizontal

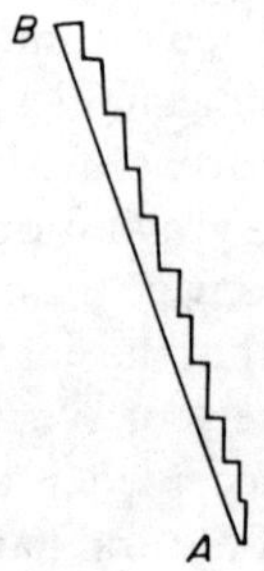

FIG. 8.2. Comparison of two paths in a conservative field

segments (to a first approximation). It is easy to show that the work required to push the object along the jagged path is exactly equal to that which is required to push it along the straight path. It is then possible to generalize this result to any arbitrary path. We end up with the result that the work required to move the particle from A to B simply depends on the positions of A and B.

Consider, now, the situation in which an object is raised to a fixed point above the surface of the earth, e.g. from A to B. The object is then allowed to fall along the same path from B to A. For the first part of the trajectory, work must be put into the system, represented by the formula:

$$\int_A^B \mathbf{F} \cdot d\mathbf{S}.$$

In the second part of the trajectory, work is taken out of the system, represented by

$$\int_B^A \mathbf{F} \cdot d\mathbf{S}.$$

But

$$\int_A^B \mathbf{F} \cdot d\mathbf{S} = - \int_B^A \mathbf{F} \cdot d\mathbf{S} \tag{8.8}$$

It follows that the total work is zero. However, since both left-hand and right-hand sides of (8.8) are independent of the path, it would not have been necessary for the object to have followed the same path for both legs of the trajectory in order that (8.8) be valid. All that is required is that the object follow a *closed* path. That is to say, that if it starts at A, it must end at A. If follows that the work done by a conservative force in moving an object in a closed path is zero.

The Field-Theoretic Standpoint

Up to this point our treatment has been squarely in the Newtonian tradition, in which the gravitational force is treated as a form of action-at-a-distance. In other words, we have not postulated the existence of an intervening medium through which the gravitating bodies interact. It would, in fact, be possible to suppose that if there were only one body there would be no gravitational force but that the latter springs into existence as soon as a second or test body is brought sufficiently close to the first one to be affected by it. Nevertheless, the considerations which emerged at the end of the previous section virtually cry out for reinterpretation in an entirely different conceptual framework.

The cosmological tradition of which Newtonian mechanics was a part is one which extends back to the atomism of Greek antiquity. In this tradition, all natural phenomena are to be explained as the outcome of the motion of material particles in an otherwise empty space or void. Gravitation, on this view, is nothing more than the way in which material objects move relative to each other. However, it is apparent that the explanatory power of such an approach is quite impoverished. To return to our previous considerations, since the motion of a test body under the influence of a more massive body turns out to depend on nothing but the position of the test body *vis-à-vis* the latter, it is natural and, indeed, almost unavoidable to suppose that at every point in the space surrounding the primary body, there is a "condition" which is just a function of position. This condition of each point in space is what causes the test particle to be accelerated from one place to another. However, one would suppose that even in the absence of a test particle, the condition would persist. The motion of the test body is merely a visible manifestation of the invisible condition. Finally, we suppose that the source of the condition is the gravitating mass. Thus we arrive at the notion of a *potential* field which is a real physical presence, although an immaterial one in the sense of its not consisting of atoms or molecules of matter. That, of course, is not to deny it all physical properties. It should not be supposed that the invisible or intangible is the same as the occult. However, it does involve a fundamental extension of our conception of the boundaries of what constitutes physical reality beyond the possibly simplistic view of mechanistic materialism.

It is therefore postulated that at each point in space there is a *gravitational potential*, which generally varies from point to point. The adjacent points which happen to have the same potential constitute equipotential surfaces. Motion under the influence of gravity is then treated as a case of a body's being accelerated from a region of higher potential to one of lower potential. Specifically, the body will move in the direction of the greatest rate of change of the potential. The reader should endeavour to picture the analogous situation of a relief map in which the contour lines which encircle the hills and mountains play the role of equipotential lines. The trajectory of a smooth object which is placed at the top of a bulge and then permitted to slide will be along the lines of steepest slope or gradient. Your geometric intuition should reveal to you that this steepest path is the one which intersects all of the equipotential contour lines at right angles. All of this leads us to recall the mathematical notion of the gradient of a scalar function. We learned that if a scalar field Φ is given in a region of space, then the greatest rate of change of Φ is given by $\nabla\Phi$. Accordingly, we have postulated the existence of precisely such a scalar field, the field of gravitational potential. In fact, we showed intuitively, what is also easily demonstrated mathematically, that the existence of a scalar field, at least formally, is a necessary consequence of the force's being conservative.

The force to which a body is subject in a gravitational field is the outcome of two factors, the intensity of the pre-existing gravitational field and the mass of that body. However, we are presently concerned with the field, in itself, apart from the effects which it has on bodies. To this end, therefore, we introduce a new quantity, by definition, which is obtained by dividing the force at a point by the mass at that point. This quantity, which will be independent of the mass, we call the gravitational field intensity or strength.

$$\boldsymbol{F} = df\, \mathbf{F}/m \tag{8.9}$$

Obviously it is closely related to the force and may be thought of as the force experienced by a body of unit mass. From Newton's law of gravitation, it is apparent that the field intensity whose source is a body of mass M will be given by $-GM\hat{r}/R$. The minus sign is characteristic of an attractive force. It serves to indicate that the intensity of the field points in the direction opposite to the source of that field. We are now able to introduce the gravitational potential Φ, which is related to the gravitational intensity by

$$\boldsymbol{F} = -\nabla\Phi, \text{ where } \Phi = -GM/R \tag{8.10}$$

Just as ∇ operates on a scalar field in a manner analogous to the multiplication of a scalar and a vector, so it also operates on a vector field in a manner analogous to the "dot" or scalar multiplication of two vectors. The resultant of this operation is called the *divergence*. Formally the divergence of the field vector **A** is given by

$$\nabla \cdot \mathbf{A} = \left(\frac{\partial}{\partial x} \hat{x}^1 + \frac{\partial}{\partial y} \hat{x}^2 + \frac{\partial}{\partial z} \hat{x}^3 \right) \cdot (A_x \hat{x}^1 + A_y \hat{x}^2 + A_z \hat{x}^3) = \frac{\partial}{\partial x} A_x + \frac{\partial}{\partial y} A_y + \frac{\partial}{\partial z} A_z \tag{8.11}$$

As a simple physical interpretation of the divergence we choose the following example. The velocity field associated with the flow of an incompressible fluid, such as water, is an instance of a vector field. Every point in the fluid has associated with it a definite velocity arrow. We consider a small volume Δv in a region through which the water is flowing. Since the water is practically incompressible, it is apparent that the amount of water entering the volume must be equal to the amount which is leaving it. Therefore the *net* flow through the surface of the volume must be zero.

Now, by the familiar limit process, we may imagine the volume ΔV to be shrinking to a point. When we consider the net flow from the differential element of volume dV, we speak of the *divergence* of the vector field at that point. Clearly, if there is no *source* of water at the point of interest, the divergence of the field at that point will be zero. The assertion that the divergence of a vector field is everywhere zero expresses an important mathematical or physical property, as the case may be, of that field. It is to say, that the field has no sources (or sinks).

We may now apply these ideas to the vector field $\boldsymbol{F}$, the gravitational intensity. Since matter is the only source of the field, it follows that in so-called *free-space*, the divergence of $\boldsymbol{F}$ must vanish, whereas within matter it will be non-vanishing. This is expressible by the following field equation

$$\nabla \cdot \boldsymbol{F} = -4\pi G \rho \tag{8.12}$$

where ρ is the density of the matter and G is the gravitational constant. We now combine (8.9) with (8.12) to get

$$\nabla \cdot (-\nabla \Phi) = -4\pi G \rho$$

or

$$\nabla^2 \Phi = \frac{\partial^2 \Phi}{\partial x^2} + \frac{\partial^2 \Phi}{\partial y^2} + \frac{\partial^2 \Phi}{\partial z^2} = 4\pi G \rho. \tag{8.13}$$

(8.13) is possibly the most important equation of mathematical physics. It is known as Poisson's equation. In matter-free space, it reduces to what is called Laplace's equation. In fact, the operator $\nabla \cdot \nabla = \nabla^2$ is called the Laplacian. Poisson's equation is valuable in providing a direct link between the gravitational potential and the material source. While it is somewhat beyond our scope to explain the full mathematical significance of Poisson's equation, it is helpful to think of it as expressing the difference between the value of the potential at a given point and its average value in the region which surrounds it. The salient point is that it is a field equation from which all of the information contained in Newton's law of gravitation is derivable.

It may seem to the reader that little has been accomplished through the formulation of Poisson's equation beyond a more sophisticated mathematical formalism, which happens, incidentally, to be handier to use than Newton's law in all but the simplest of applications.

Two remarks should be made in this regard. In the first place, there might be some merit in this objection were it the case, as some empiricists suppose, that a physical theory is nothing more than a piece of machinery for grinding out empirical predictions. However, we regard this view as mistaken. When we adopt the formalism of a field theory, it is because we wish to posit the existence of a field as an explanatory concept. In particular, we do so because we find the field concept to make more sense than the alternative of action-at-a-distance. The correlative notions of field intensity and potential represent putative conditions of space.

In the second place, that the field concept is more than an idle metaphor follows from its making an empirical difference. Having posited the thesis that gravitational interactions take place via an intervening field, we are inevitably led to consider the velocity with which an influence may be propagated through the field; that is, if we are paying more than mathematical lip-service to the concept. In the action-at-a-distance version of gravitational theory, such influences are treated as instantaneous. But an instantaneous velocity of propagation is scarcely compatible with the field-theoretic standpoint. In any case, it is obviously ruled out by the theory of special relativity, which holds that no transfer of energy can exceed the velocity of light. Suppose, therefore, that we wish to handle the rather complex problem of determining a planetary orbit around the sun, while taking account of the perturbations that are produced by the gravitational influence of a second planet. Now, action-at-a-distance theory would consider the instantaneous configuration of the three bodies. But the more subtle field-theoretic approach would consider how the planetary orbit *now* is affected by the second planet not in its present location but in the one which it occupied when the present field disturbance was propagated. In any case, special relativity teaches us that the instantaneous configuration of the three bodies is not a well-defined notion. So we repeat that field theory is not a mere metaphor which is introduced to interpret a mathematical formalism. In any event, it is obvious that the starting-point for Einstein's modification of the theory of gravitation must be from the standpoint of field theory. As we shall see when we explore the theory of general relativity, the physical significance of the field is even more striking than in the case of classical gravitation theory.

Mach's Principle

In virtue of the philosophical richness of the theory of general relativity, one who wishes to give an introductory level exposition is faced with the not inconsiderable task of sorting out its philosophical presuppositions, heuristic principles and physical axioms. To be sure, the enumeration of the last of these together with the derivation of the principal theorems is the simplest and most precise procedure. However, it has the disadvantage of being nearly unintelligible to the beginner as well as running the risk of being, at least in part, arbitrary. Accordingly, we shall follow an order of presentation which closely parallels that of Einstein in his fundamental paper of 1916, *The Foundation of the General Theory of Relativity*.

We begin, therefore, as Einstein did, by invoking what has generally come to be known as *Mach's Principle*. The basis for this principle is what Einstein refers to as "an inherent epistemological defect" of both classical and special-relativistic physics. Einstein endeavours to illustrate what he has in mind by means of the following example. Let us suppose that two fluid bodies, S_1 and S_2, of similar size and composition are so remote from all other matter in the universe including each other that neither is subject to any external gravitational influences. Suppose that their relative configuration is such that from the standpoint of an observer who is at rest with respect to S_1, S_2 is rotating with constant angular velocity about the line which joins them. Furthermore, if appropriate geometric measurements revealed that S_1 is a sphere while S_2 is ellipsoidal, it would be concluded from the classical standpoint that the difference between them must be the outcome of inertial forces.

The reader will of course be aware of such forces. They are associated with various accelerating frames of reference. For example, when we ride an ascending lift, we are conscious of a force which tugs us in a downward direction. When we are negotiating a curved path in a car, we experience the sensation of being pulled away from the centre of curvature. The latter is due to what is called the centrifugal force, and it is this which causes a rotating sphere to be deformed into an ellipsoid. Accordingly, a Newtonian observer would conclude that S_1 must be at rest with respect to a Galilean or inertial frame of reference, while S_2 is accelerating with respect to that same frame. But the Galilean frame is usually viewed as one which is unaccelerated with respect to space.

According to the physicist-philosopher of the late nineteenth century, Ernst Mach, such an approach is unacceptable since absolute space is not an observable. Absolute space, as an explanatory principle of inertial forces, has all of the defects which were associated with the aether of classical electrodynamics. In particular, it is yet another example of a material entity which affects bodies but is not reciprocally affected by them. Mach argued that all motion should be understood as relative motion. You will recall that the most immediate consequence of special relativity is that uniform motion is relative. The position of Ernst Mach is that accelerated motion is no more absolute than motion of the uniform variety. Moreover, just as the relativity of uniform motion requires the equivalence of a large class of reference frames, so the relativity of motion in general, by the same token, must require the equivalence of all possible frames of reference. But this is essentially what the principle of general relativity asserts—namely, that the laws of nature, properly formulated, retain their form when referred to arbitrary frames of reference.

The thrust of the argument is that hitherto there has been no adequate explanation of the *cause* of the difference between inertial and non-inertial reference frames. Special relativity simply postulates the equivalence class of Lorentz frames while failing to provide any explanatory account of their existence. In this regard, absolute space is, as Einstein puts it, merely a "*factitious* cause". Mach resolved the problem by claiming that the description of a system such as that of S_1 and S_2 is essentially incomplete. As the situation stands, it is meaningless to assert that the one body rotates with respect to the other. However, if the system be extended to include the distant or "fixed" stars, no matter how remote they might be from the standpoint of classical gravitation theory, an explanation of the ellipsoidal shape of S_2 may be provided in terms of inertial forces which arise as a result of motion relative to the fixed stars.

Properly or otherwise, Mach regarded motion with respect to the fixed stars as a sounder basis of explanation than motion with respect to a fixed space. Accordingly, what is known as Mach's principle is the claim that all inertial effects arise from the mutual interactions of physical bodies. It follows that the equivalence class of inertial reference frames is determined by the actual configuration of the matter in the universe at a particular time.

Evidently, it is fair to treat Mach's principle as a meaningful claim about physical reality. Whether, however, it could actually be upheld would require a precise quantitative formulation of it, something which Mach failed to provide. In any event, Einstein attributed his fundamental insight that no frame of reference plays an *a priori* privileged role to the influence of Mach's principle. Moreover, he endeavoured with a degree of success that is at least questionable to fashion the theory of general relativity in a manner that would incorporate it.

In very general terms, the incorporating of Mach's principle in the framework of a physical theory would mean, *inter alia*, that the theory would have to avoid the attribution of any absolute properties to space or spacetime. The structure or geometry of spacetime would be entirely *a posteriori* and dependent on the distribution of matter in the universe. Every student of general relativity knows that Einstein certainly took some major steps in that direction. However, a corollary of Mach's principle would seem to be that in a universe free of all matter, time and space would cease to have any structure and, since spacetime is not a substance, cease, in fact, to exist. The "void", at any rate, would be completely amorphous. In this regard, at least, the laws of general relativity appear to be incompatible with Mach's principle, as we shall eventually discover.

The Principle of Equivalence

Of much greater heuristic significance than Mach's principle, and considerably less controversial, is the principle of equivalence, for which Einstein deserves to be uniquely credited. He appears to have reached it as early as 1907, and by 1911 had published a paper in which the principle is utilized to predict, at least qualitatively, certain of the empirical consequences of the still-to-be-formulated theory of general relativity. Unlike Mach's principle, the principle of equivalence is factual in origin rather than epistemological. Specifically, it is introduced to clear up a curiosity of Newtonian mechanics to which we alluded in an earlier section. We saw that classical mechanics requires that one distinguish between inert and gravitational mass. The former, be it recalled, is the property in virtue of which a body resists a change of motion, while the latter is that in virtue of which it is accelerated by a gravitational field as well as being the source of such a field. The truly remarkable feature of the two properties is that their ratio is equal to one and the same universal constant for all bodies, regardless of their size or density. This being so, a suitable choice of units permits us to ascribe the same number to the two types of mass of a given body, i.e. to make the aforesaid ratio equal to unity. Thus, in classical mechanics there exists what amounts to a stray or theoretically isolated principle, that of the equality of gravitational and inert mass. The direct and well-known consequence of this is that all bodies are accelerated in precisely the same way by the same gravitational field, a fact emphasized by Galileo.

As we have already noted, Einstein was motivated by the conviction that nothing in nature is arbitrary. Accordingly, he took the position that the two kinds of mass could be equal only if they were, in fact, one and the same. However, this claim of numerical identity for two seemingly independent properties of matter entailed a profound revision of the conceptual framework in which they occur.

Einstein accounts for the close link between inertia and gravitation in the following way. Suppose that K is an inertial frame of reference with respect to which a certain body is in uniform motion. Let $\bar{K}$ be accelerating uniformly with respect to K. The aforementioned body which is not influenced by external forces will, of course, accelerate with respect to $\bar{K}$. However, it is obvious that the actual acceleration will be wholly independent of the composition or state of the body itself, since it is due solely to the acceleration of the system of reference, $\bar{K}$. From a classical standpoint, in fact, the acceleration of the body would provide a basis for an observer at rest in $\bar{K}$ to be able to detect the acceleration of his own reference frame. Einstein now points out that such a viewpoint is incorrect. Although the acceleration of the object is independent of its intrinsic nature, one may not conclude that it is merely the kinematic consequence of choosing $\bar{K}$ as a frame of reference. Specifically, this is because the very same type of acceleration could have been imparted to the body by a suitably chosen gravitational field. Hence, the observer in $\bar{K}$ could claim to be attached to an inertial frame by holding the body of interest to be under the influence of a gravitational field. In short, inertial effects, which are the result of the choice of reference frame, are of the same kind as those which are due to gravity. So an observer who occupies a frame of reference which is apparently inertial and who suddenly experiences a series of phenomena which might lead him to suspect that it has ceased to be inertial would always have at his disposal the possibility of postulating the existence of a suitable field of gravity as the cause of the phenomena. The *principle of equivalence* is simply the claim that inertial fields and gravitational fields, together with their effects, are indistinguishable. The same name, however, also refers to the closely related principle of the equivalence of gravitational and inertial mass.

For example, suppose that one is in a lift in which a weight is suspended from the ceiling by an extension spring. If the gravitational field acting on the weight were suddently to increase, the spring would be observed to undergo an increase of length. This would be an indication that the weight has become heavier. However, precisely the same extension of the spring would be observed if the lift were to undergo a sudden upwards acceleration. But in this case, the lengthening of the spring would be accounted for by the action of a field of inertia instead of one of gravity.

How would the situation differ if inert and gravitational mass were distinct? In the first place, if the two properties were independent, it would be quite normal for two bodies to have equal gravitational mass but different inertial mass. Under the action of a gravitational field they would experience the same amount of force, but the one with greater inertial mass would accelerate at a lower rate than the other. Thus, bodies of different size and composition would fall at different rates. Similarly, the occupant of the lift would be able to ascertain whether the elongation of the spring were the result of gravity or inertia, since the latter would cause all similarly constituted springs to stretch by the same amount, whereas under the action of gravity the stretching would be proportional to the gravitational mass of the suspended weight.

As a final illustration, let us consider what would happen in a lift which is falling freely through space. At the dawn of the age of space travel, we know that all the bodies in the lift would appear to be weightless. For example, suppose you were standing on an ordinary domestic scale. The indicator would register zero, since the equality of gravitational and inertial mass would guarantee that you and the scale would always fall at the same speed, so that you would exert no pressure on it. The Newtonian physicist would not allow that the objects had really become weightless, since that would be tantamount to the admission that the gravitational field had disappeared. He would claim that the apparent weightlessness was due to the fact that the gravitational field is precisely balanced by an inertial force, which, in turn, is explained by the equivalence of the two kinds of mass.

Characteristically, Einstein took the position that if no experiment can reveal a difference between the two kinds of mass, then it is redundant to postulate that two masses exist. Accordingly, the basis of the principle of equivalence is that gravitational and inertial mass are not merely equal but identical! It is the immediate consequence of this postulate that is often identified, in practice, with the principle of equivalence. Since gravitational mass is that property of a body which responds to the gravitational field while inertial mass may be thought of as that property of a body which responds to an inertial field, it follows that if the two properties are really one, then the two fields are really one, in the sense that they must be of the same kind. We have already examined several cases where the effects of a gravitational field are indistinguishable from those of a field of inertia. It would now appear that the explanation of this similarity is much deeper than had been envisaged in classical physics.

Now the so-called inertial forces, such as the centrifugal force, of classicial physics had a special status which made them distinct from all other forces. Specifically, such forces arise when one employs a "wrong", that is a non-inertial, frame of reference. Thus, in the description of a physical system, all inertial terms may be eliminated by performing an appropriate coordinate transformation corresponding to a shift to an inertial reference frame. What is suggested by the principle of equivalence is that a gravitational force should be capable of being handled in the same way. That is to say, that a gravitational field, like an inertial field, has a purely mathematical existence, being a function of the choice of coordinates. However, as we shall now explain, it is possible to be misled by such considerations.

It does not follow that any real gravitational field can be duplicated by a field of inertia and therefore "transformed away". In other words, it would not be possible to duplicate any arbitrary gravitational field by an appropriate acceleration of a frame of reference. For example, the gravitational field of the earth is radial and falls off with the square of the distance. Such a field could not be reproduced in its entirety by an accelerating frame. However, if we consider a sufficiently small region of the field, it will be found to be approximately uniform, with parallel lines of force, and therefore indistinguishable from the sort of field which arises from a uniform, straight-line acceleration. Conversely, Einstein's lift, which is a frame in free fall, is an inertial frame in a certain small neighbourhood in the sense that no inertial terms arise when physical systems are described with its aid. But beyond that neighbourhood, the effects of gravitational fields would appear. In short, the equivalence of gravitational and inertial fields is only a *local* one. From an ontological standpoint, this restriction of the principle of equivalence is important. However, since the fundamental laws of general relativity are partial differential field equations, they tell us what is taking place in the neighbourhood of a point. Accordingly, the local equivalence of gravity and inertia is still of great methodological significance.

On the other hand, that gravitational and inertial fields are not, in general, globally equivalent is also of great importance. When that is overlooked, one is apt to make too facile an interpretation of the principle of general relativity. The most popular and commonplace interpretation of it is to the effect that it is a straightforward generalization of the special principle. Whereas the latter holds that there is a privileged class of Lorentz frames which are all equivalent for the formulation of the laws of nature, the former is held to assert that all physically possible frames of reference are equivalent for the formulation of the laws of nature. Now such an interpretation is correct from one standpoint but not from another. As we shall see in the next section, it is correct in the sense that the laws of general relativity are so formulated that if they hold in one frame of reference they will hold in any other. But this is often taken to imply that whereas special relativity precludes the possibility of detecting the absolute velocity of a reference frame, general relativity entails the impossibility of detecting an accelerated motion.

Let us briefly recapitulate the train of reasoning. The most precise measurements reveal that there is no genuine distinction between gravitational and inertial mass. It follows that gravitational and inertial forces are of the same kind. In particular, it is well known that the acceleration of a body is independent of its mass when under the sole influence of either. Therefore, the acceleration of a test body with respect to a particular frame of reference could be explained either as the result of an acceleration of the frame of reference or as the effect of a gravitational field. The two accounts are equivalent descriptions. But since accelerated motion is now deprived of its absolute significance, it follows that there should be no essential difference between a Lorentz frame and any other frame which is accelerating with respect to it. All reference frames are equivalent. It is fairly obvious that the general thrust of the argument is inspired by the claim of Mach that all motion, whether uniform or accelerated, is purely relative.

If there is a single flaw in this line of argument, it is in the assumption that there is no difference between an inertial field and a field of gravity. If this were of global validity, Mach's principle would be well on the way to achieving its ultimate vindication. However, as the matter stands, it is possible to distinguish the two types of field. That is not to say that Mach's principle is false. The type of field which cannot be transformed away may be due to the interaction of a body with the rest of the matter in the universe. But this remains to be shown. However, what cannot be gainsayed is this: The theory of general relativity is a field theory rather than an action-at-a-distance theory. Consequently, its laws are concerned, in the first instance, with what is taking place *locally*. But locally, at least, a gravitational field is *virtually* like an inertial field. The reason for the qualification will eventually emerge. Hence, from the local standpoint all frames of reference are equivalent, and the laws of nature should reflect this.

A truly striking feature of the principle of equivalence is its heuristic significance. The critical philosopher of science is apt to be suspicious of heuristic claims on behalf of highly general principles. Generally, it will turn out that they are only applicable *ex post facto*, revealing the wisdom of hindsight. The Leibnizian principle of sufficient reason would appear to be vulnerable to such a criticism. The principle of equivalence appears to be an exception in so far as relativity physicists do habitually employ it as a guide to further discoveries in the field of gravitational phenomena. Its success as a heuristic principle is almost certainly explicable by its being a factual or constitutive principle of nature. Very simply, in order to ascertain, at least qualitatively, the effects of gravitation on a given kind of phenomenon, one has merely to consider how that phenomenon would be described when referred to an appropriately accelerating frame of reference. Then, in virtue of the equivalence of the inertial and gravitational descriptions, one immediately determines how the phenomenon will be affected by a gravitational field.

The most celebrated example of the application of this principle is probably Einstein's prediction of the bending or deflection of light-rays by a sufficiently strong gravitational field. Imagine that a light-ray is beamed from one wall of a lift to the opposite one. Now, if the lift is made to ascend very rapidly the effects which take place inside it are indistinguishable from those that would be produced by a sudden increase in the field of gravity. Now it is not too difficult to show, and is intuitively fairly clear, that the beam of light will not hit the opposite wall at the point one would expect if the lift were stationary or moving uniformly, but at a somewhat lower point. In other words, the light ray is curved when referred to the accelerating frame. It follows on the basis of the principle of equivalence that light-rays must be curved or deflected from their straight path by a gravitational field. This consideration enabled Einstein, several years before he had formulated his general-relativistic law of gravitation, to predict that starlight will be deflected as it passes the massive sun.

The Principle of General Covariance

The ideal of science, like that of any cognitive venture worthy of the name, is objective knowledge. Every scientist is subject to the imperative that his theories must be as independent as humanly possible from the idiosyncracies of race, temperament, culture and so forth. Obviously, the patterns of reality have an objective existence to which all such factors are irrelevant. *Inter alia*, the patterns of reality are clearly independent of the language in which they are expressed. The principles of physics are the same whether written in English, German or Swahili. Equivalently, when a physical law is translated from one language to another, it should neither gain nor lose anything of an essential nature.

The systems of coordinates which we employ to describe the world in space and time are analogous to impoverished but orderly languages. The basic function of a quadruple of space-time coordinates is to label or name an event. The reader is already aware that there are uncountably many ways of labelling events or of coordinatizing the space-time manifold. Since the various systems of coordinates have a purely conceptual existence, the description of nature which is carried out with their aid should not depend in any essential way on their particular properties. The nature of the world, including the space-time manifold, must be indifferent to our choice of oblique or orthogonal coordinates or to that of rectilinear or curvilinear ones. Just as objective knowledge precludes "English laws" and "German laws" so it should also preclude "spherical laws" and "rectangular laws". In some sense or other, the formulation of a law of nature should be "coordinate free". Accordingly, when a coordinate transformation is carried out on a space-time description of the world, nothing essential to that description should be gained or lost. Thus, a law's independence from the coordinate system is equivalent to its retaining its "essential form" under an arbitrary coordinate transformation. The requirement that a law be so constructed that it be unaffected by a mathematically admissible coordinate transformation is the prima facie meaning of the *principle of general covariance*.

Formally, if a given law be represented as a function F of certain quantities q_i in a given system of coordinates S, then under a transformation to the system $\overline{S}$ in which the q_i become $\overline{q}_i$, the law should continue to hold in the form $F(\overline{q}_i)$. The transformation from S to $\overline{S}$ is subject only to the restrictions that it be one-one, continuous and differentiable. The technical term for such a transformation is *diffeomorphism*. The principle of general covariance is, therefore, equivalent to the claim that the fundamental laws of nature are invariants of the group of coordinate diffeomorphisms.

It is a matter of historical record that Einstein attached profound importance to this principle in the context of general relativity. In fact, many writers continue to identify the principle of general covariance with the principle of general relativity. We shall eventually argue that such an identification is difficult to sustain. However, it is instructive to begin by approaching the matter from the original Einsteinian standpoint.

Einstein offers two distinct arguments for the imposition of general covariance on the fundamental laws of physics. The first may be called the argument from geometry. The *spatial* geometry of both classical and special relativity physics is Euclidean. The Euclidean assumption is reflected in the way in which coordinate numbers are assigned to points. Specifically, if $P_1(x_1, y_1, z_1)$ and $P_2(x_2, y_2, z_2)$ are two points in space, the values of their Cartesian coordinates must satisfy the Euclidean distance function, which is given by

$$d(P_1, P_2) = ((x_2 - x_1)^2 + (y_2 - y_1)^2 + (z_2 - z_1)^2)^{1/2} \tag{8.14}$$

The numbers which are assigned to the various points along the three orthogonal axes are required to conform to this function in the following sense. To say that a particle is located at x_1 on the X-axis of some reference frame is roughly equivalent to saying that a rigid measuring rod of unit length could be laid off between the origin and the location of the particle exactly x_1 times. Moreover, if a second particle were located at $x_2 = 2x_1$, it would follow that the rod could now be laid off precisely twice as many times as before. It is obvious that this state of affairs which is characteristic of special relativity is far removed from the spirit of general covariance.

From the standpoint of general covariance, the following coordinatization of the plane would be acceptable.

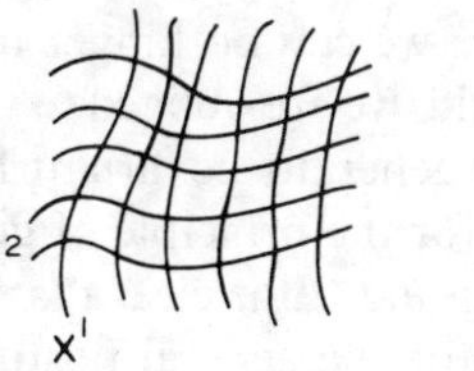

FIG. 8.3. Coordinatization of the plane by arbitrary general coordinates

However, it is apparent that were one to employ such a system, the customary relation between coordinate values and standard measuring procedures would not obtain. Now the reader may very well object at this point that since that is the case, then such coordinate systems along with the principle of general covariance which countenances them should be rejected. The point which Einstein would make in reply is that once one has accepted the equivalence of all frames of reference the Euclidean character of space, presupposed by our standard measuring procedures, breaks down. Accordingly, there would be, in his view, a factual basis for general covariance.

Einstein's point may be illustrated by comparing measurements carried out in a rotating frame $\bar{K}$ with those carried out in an inertial frame K. In particular, we let $\bar{K}$ be a rotating circular disk having a common $\bar{Z} = Z$-axis with K. Measurements are carried out in $\bar{K}$ by means of a measuring rod which is very small compared to the radius of the disk. These measurements determine that the radius of the disk is r, while its circumference is C.

However, from the standpoint of the inertial frame K, the measuring rod must undergo a Lorentz contraction when placed along the circumference. If ω is the angular velocity of $\bar{K}$ with respect to K, then the linear or tangential velocity of a rod at a distance r from the centre of the disk will be ωr. Consequently, a rod of unit length will undergo a contraction to the length $(1-\omega^2r^2/c^2)^{\frac{1}{2}}$ from the standpoint of K. Accordingly, since the number of times that the rod has been laid off along the circumference must be the same with respect to both K and $\bar{K}$, the length of the circumference with respect to K must be $C(1-\omega^2r^2/c^2)^{\frac{1}{2}}$. But from the formula for the circumference of a circle we may write

$$C(1-\omega^2r^2/c^2)^{\frac{1}{2}} = 2\pi r \tag{8.15}$$

It follows immediately from (8.15) that in the rotating reference frame the ratio of circumference to diameter of a circle of radius r is given by $\pi/(1-\omega^2r^2/c^2)^{\frac{1}{2}}$, which is greater than π. We conclude that the geometry of the rotating disk is non-Euclidean. It is also clear that for similar reasons the rates of clocks would vary with their distance from the axis of rotation.

We have been dealing with a relatively simple and well-behaved example of a non-inertial frame. In general, an accelerated frame may describe a very complicated trajectory with respect to an inertial frame, with ensuing complications in the geometry. Consequently, in virtue of the principle of equivalence, which entails the equivalence of all frames of reference, it follows that there is no longer any justification for assuming the correctness of Euclidean geometry. But once the Euclidean postulate is dropped, we are obliged to abandon the old restrictions on the choice of coordinate systems, which merely reflected our belief in the validity of the Euclidean metric. Since, in principle, one metric geometry is as good as any other, it follows that one set of coordinates must be as good as another. More accurately, since we can no longer make *a priori* assumptions about the space-time geometry of the world, we are obliged to formulate the laws of physics in terms of *arbitrary coordinates*, i.e. in generally covariant form.

Einstein's second argument for the principle of general covariance is very much more brief and elegant than the foregoing: "It is clear that a physical theory which satisfies this postulate will also be suitable for the general postulate of relativity. For the sum of *all* substitutions in any case includes those which correspond to all relative motions of three-dimensional systems of coordinates" (1916, p. 117).

By a "substitution", Einstein simply means a coordinate transformation. He is here arguing that the principle of general relativity puts all frames of reference on a par. Thus, unlike both the classical and the special principles of relativity, the general principle must allow for transformations between mutually accelerating frames. Such transformations are non-linear. However, a generally covariant formulation of a law will accommodate such non-linear frame transformations since it accommodates, within certain mathematical limits, any coordinate transformation whatsoever. Thus, the principle of general covariance is here treated, loosely speaking, as the mathematical realization or reflection of the principle of general relativity.

The attentive reader may have been struck by a significant shift in the course of this discussion. We began by depicting the requirement of general covariance as an *a priori* desideratum concerning the formal language in which the fundamental laws of nature are to be expressed. However, we subsequently justified it on the basis of such presumably *a posteriori* considerations as the non-Euclidean character of accelerating frames and the claim that all frames of reference are equivalent with respect to the fundamental laws of nature. Unfortunately, this dichotomy merely reflects general uncertainties concerning the status of the principle. While there is universal agreement that general covariance plays an essential role in general relativity, as does Lorentz covariance in the special theory, there is no standard view as to the precise nature of that role.

In his fundamental paper of 1916, Einstein clearly believed that the principle of general covariance had factual content, since it apparently led to new physics. In brief, if we require a field theory of gravitation to satisfy the principle, we are led to field equations which yield testable consequences, such as the precession of planetary orbits, which are not derivable from Newton's theory of gravitation. However, it was pointed out as early as 1917 by the mathematician, Kretschmann, that it is possible to express any theory, including Newtonian mechanics, in a generally covariant form. This, in turn, led Einstein formally to renounce his belief in the factuality of general covariance. However, there is ample textual evidence that, consciously or otherwise, he actually continued to treat the principle as though it had factual content.

More recently, the philosopher of physics, Mario Bunge, has argued that all covariance principles are devoid of factual content. He contends that when one analyzes the foundations of a physical theory, one must draw a sharp distinction, as in formal logic, between the laws which belong to the object language of the theory, that is to the theory proper, whose referents are the objective patterns of nature, and the laws which belong to the metatheory, those laws which refer to the structure of the theory. According to Bunge, the claim that classical mechanics is Galilei covariant is not a statement about the world but about the structure of the theory. Now since one and the same theory may be expressed in many different forms, having different covariance properties, there would certainly appear to be considerable merit in Bunge's claim that covariance principles are factually empty metatheorems. (Here, "factually empty" does not mean "vacuous").

On the other side of the ledger, the physicist, J. L. Anderson, has forcefully argued in an important and original treatise on relativity physics that covariance principles do have factual content, in the sense that when a particular covariance requirement is imposed on a theory which is to contain only certain specified geometric objects and whose laws are to be of a specific differential order, then the types of possible laws which may belong to that theory are severely delimited and may even be uniquely determined. The Kretschmann position can only be maintained if we are permitted to add additional elements in the form of geometric objects, *ad libitum*, to the theory. But since those additional elements are not required to stand for any particular physical entity, they are merely a manifestation of a certain mathematical legerdemain.

Anderson's position would appear to gain support from the established fact that there are certain types of covariance properties of physical laws which lead, in turn, to conservation laws, the latter being indisputably factual. These covariance properties are called *symmetries*. For example, if a physical theory satisfies a certain mathematical requirement, which we shall neglect to specify, and its laws are independent of position, i.e. possess translational symmetry, one may deduce from them the law of conservation of linear momentum. If they are independent of direction, i.e. possess rotational symmetry, then they entail conservation of angular momentum. Moreover, if they possess time-translation symmetry, they will entail the fundamental law of the conservation of energy. It is easy to see that each of the aforementioned symmetries corresponds to a coordinate transformation. For example, to say that a law possesses rotational symmetry is equivalent to saying that its form will not be affected by a pure rotation of the coordinate system in terms of which it has been expressed.

In spite of all this, the position of Bunge is logically incontrovertible. However, while we may agree that the principle of general covariance is metatheoretical, it is less apparent that Bunge is correct in subsequently identifying general covariance with general relativity. If general covariance is devoid of physical content, it would then follow that the principle of general relativity must be similarly devoid. In fact, one leading relativity theorist, V. Fock, has gone so far as to argue that the very notion of general relativity is a contradiction in terms. According to Fock, Einstein's theory is simply a generally covariant theory of gravitation which does not satisfy any relativity principle of greater generality than the special principle. Clearly, these questions which bear on the meaning and very existence of a principle of general relativity are among the most fundamental ones that may be posed. For this reason, we shall postpone a full treatment of them to a later chapter. In any event, we may conclude by pointing out that all authorities are in agreement concerning the necessity, for whatever reason, of formulating the theory in generally covariant form. We shall now proceed to an outline of the theory itself.

Bibliography for Chapter 8

EDDINGTON, A. S., *Space, Time and Gravitation*, Harper, New York, 1959 (1st edn., 1920). A good popularization of general relativity by an outstanding authority.

EINSTEIN, A. (1952). See reference in Chapter 3.

GAMOW, G., *Gravity*, Doubleday, New York, 1962. A semi-popular account of classical gravitation and a brief look at general relativity. Highly readable.

WALD, R. M., *Space, Time and Gravity*, University of Chicago Press, Chicago, 1977. Based on a series of popular public lectures. Includes an up-to-date look at the cosmological implications of general relativity.

9

An Outline of General Relativity

THE reader will appreciate that since these are not the pages of a treatise on physics, it would be beyond our scope to do more than open the door to this somewhat complex and conceptually challenging theory. The aim of the present chapter is to provide just enough material to achieve two basic ends. The first is to capture sufficient of the flavour of the theory to provide a basis for a critical appreciation of some of the main philosophical problems which arise from it. The second is to provide the type of introduction to the subject which may enable the more ambitious student to gain access both to the easier treatises on the subject as well as to the more sophisticated philosophical literature.

To these ends, we shall restrict the exposition to the case of matter-free space—that is, to the case of space which surrounds matter and is modified by it as distinct from the condition in the interior of a gravitating mass such as a cloud of particles. Moreover, we shall follow what are essentially the lines of the "classical" version of the theory, presented to the public by Einstein in 1916. In fact, however, we believe this to be hardly a real restriction at all, since the theory has exhibited remarkable powers of survival, so that, apart from mathematical sophistications, the most commonly accepted refinements are little more than natural extensions of Einstein's original position. Accordingly, the reader should not be concerned that he is spending effort on a system of outmoded ideas. Finally, a more serious limitation is in our decision to omit all considerations of the cosmological consequences of general relativity. In the first place, we felt that their discussion would involve more subtle mathematical matters than our reader could realistically be expected to follow. In the second place, the cosmological aspect of general relativity is far from complete, and we suspect that it would be rash to treat it as a basis for fruitful philosophical speculation at this point in time.

You will have gathered from the preceding chapter that a key, if not the most central, idea of general relativity is that the traditional presuppositions concerning the geometry of space and of spacetime are to be abandoned in favour of the view that the geometry of spacetime is not fixed but is dependent on the distribution of matter. Just as matter is regarded in Newtonian theory as a source of the gravitational field, so now it is treated as a source of the "geometric field" or, as physicists put it, of the *metric field*. Apart from any other considerations, the *a posteriori* and variable character of the geometry suggests the desirability of formulating such a theory in a generally covariant form.

We have seen that the character of Newtonian space renders the Cartesian coordinate system a particularly suitable device for the labelling of points. Moreover, the Cartesian vector is an especially useful geometric object for the expression of laws of nature which presuppose a background of Euclidean space. In the arena of special relativity, the Minkowski coordinate system was found to be particularly well adapted for the coordinatization of the flat spacetime which is presupposed by that theory. Similarly, the geometric object called the four-vector was of particular value as it is a vector with respect to a rotation of the Minkowski coordinate system, the latter being the model of a Lorentz frame transformation. Accordingly, an equation in four-vector form is automatically Lorentz covariant and, therefore, a prospective candidate for qualification as a special-relativistic law of nature.

Moving to general relativity, we find that spacetime reveals no particular uniformity or symmetry, at least in general, which could lead to a pre-eminently well-adapted coordinate system. Hence, it is now appropriate to employ general coordinates which, to a large extent, are geometrically uncommitted, thus treating spacetime as a general Riemannian manifold. Correspondingly, we shall need geometric objects which permit the formulation of laws of nature which are covariant with respect to tranformations between arbitrary general coordinate systems. As the reader knows, the Riemannian tensor is precisely such an object, which is why we spared no effort in learning the fundamentals of tensor analysis.

In addition to assuming the manifold to be Riemannian, we impose a further condition which is of fundamental importance. A distinguishing feature of a Riemannian manifold is that in a sufficiently small neighbourhood of any point, it will appear to be flat, as, for example, in the case of a sufficiently small area on the surface of the earth, which approximates a plane surface. In the case of the space-time manifold, this means that it must be possible to coordinatize any small volume by the type of coordinate system employed in special relativity. The system could be that of Minkowski, with $x^4 = ict$. However, the imaginary time coordinate is of no particular value in general relativity, so we shall let our Greek indices run from 0 to 3 with $X^0 = ct$ serving as the time coordinate. Accordingly, we require that in the neighbourhood of any point the metric tensor may be transformed to the form:

$$g_{\mu\nu} = \begin{pmatrix} 1 & 0 & 0 & 0 \\ 0 & -1 & 0 & 0 \\ 0 & 0 & -1 & 0 \\ 0 & 0 & 0 & -1 \end{pmatrix} \tag{9.1}$$

Hereafter, we shall denote this particular matrix of metric coefficients by $L_{\mu\nu}$ and speak of Lorentz coordinates where previously we have spoken of Minkowski ones.

There is a well-known theorem of Riemannian geometry, which is really a theorem about symmetric matrices, that whatever the metric coefficients $g_{ik}(i, k = 1, 2, \ldots, n)$ may be at a given point of a Riemannian manifold, there will always exist a reversible transformation which will carry the g_{ik} into a diagonal matrix whose elements are a fixed sequence of 1, -1 and 0. Clearly, $L_{\mu\nu}$ is one such instance. Moreover, while the transformation to this "canonical" form will vary from place to place, depending on the nature of the g_{ik}, the resulting canonical matrix is unique for each manifold and thus serves to identify its family. The sequence of plus and minus signs which occurs, e.g. $(+ - - -)$, is, the reader will recall, to be called the *signature* of the manifold, although this is a minor departure from standard usage. In general relativity, it is always assumed that the latter example is the signature of the space-time continuum.

A coordinate system such as the one which corresponds to the canonical matrix $L_{\mu\nu}$ is an example of what is termed a *geodesic* coordinate system. Physically, a geodesic coordinate system may be interpreted as the coordinatization of a frame of reference which is in a state of free fall, so that there is an absence of gravitational effects. In such a system, a material particle will obey Newton's first law. Accordingly, such systems are the closest counterpart to the inertial or Lorentz frames of special relativity. However, it must be emphasized that they are, in general, only of local significance. Their local character has precisely the same basis as the purely local equivalence of gravitational and inertial force fields. Thus, one cannot extend the axes of such a system indefinitely. At the origin of a geodesic frame it will be found, as previously noted, that free particles obey Newton's first law. However, they will be found to accelerate at a certain distance from the origin due to the inhomogeneities of the gravitational field.

In spite of this limitation, the existence of geodesic coordinates is a great aid to mathematical computation. A tensorial law which holds when expressed in terms of one coordinate system must be valid with respect to any other system. Thus, if such a law is found to hold when expressed in terms of geodesic coordinates it will be generally covariant. But geodesic coordinates are the easiest to work with since the constancy of the g_{ik} associated with such a system entails the vanishing of the Christoffel symbols, which are formed from the partial derivatives of the g_{ik}. Hence, in such systems ordinary differentiation replaces covariant differentiation.

Finally, in this regard, we should note that if there is a *single* transformation which carries the metric tensor into its canonical form over the entire manifold, then that manifold is flat. In any such manifold, the variability of the g_{ik} would merely be symptomatic of the choice of coordinates. For example, the g_{ik} of the familiar spherical coordinate system are non-constant even though they coordinatize a flat manifold. But the basic mathematical relationship between special and general relativity is that the latter presupposes a geometric background in which the former enjoys local validity. In other words, it is assumed that the laws of special relativity would be found to hold in the absence of gravitational forces.

We now proceed to elaborate Einstein's identification of gravitation with geometry from several standpoints. We have already learned that the metric of a general Riemannian manifold is a straightforward generalization of the theorem of Pythagoras in arbitrary or general coordinates. This homogeneous quadratic form as applied to the space-time manifold of physics is given by

$$(ds)^2 = g_{\mu v} dX^{\mu} dX^{v} \; (\mu, v = 0, 1, 2, 3). \tag{9.2}$$

The physical justification for (9.2) is that it is known to hold in special relativity in the form:

$$(ds)^2 = L_{\mu v} dX^{\mu} dX^{v}. \tag{9.3}$$

But to say that (9.3) holds in the case of special relativity is equivalent to saying that it holds when expressed in terms of geodesic coordinates. (9.2) is simply the generalization of (9.3) in terms of arbitrary or general coordinates. But since it is a tensor equation it is valid in any frame of reference and is a suitable metric for general relativity. As already noted, the significant difference between the $g_{\mu v}$ of (9.2) and the $L_{\mu v}$ of (9.3) is that the former are variable functions of position in spacetime whereas the latter are constants.

From the classical standpoint, (9.2) would be regarded as an inconvenient outcome of our having chosen an awkward coordinate system. In particular, (9.2) will be especially clumsy when the coordinates are attached to an accelerating frame of reference in which "inertial" forces arise which seemingly have a distorting effect on our measurements. However, from the standpoint of general relativity, the interpretation of (9.2) is quite different. Now there are no avoidable inertial forces, only the perfectly natural gravitational forces. The functional variation of the $g_{\mu\nu}$ from point to point is associated with the presence of a gravitational field. In the stronger gravitational fields, the values of the $g_{\mu\nu}$ vary more rapidly. It is, therefore, natural to suppose that the $g_{\mu\nu}$ may play a role in a general-relativistic theory of gravitation which is analogous to that of the Newtonian scalar potential Φ in classical gravitational theory.

We are now in a position to understand in what sense Einstein may be said to have "geometrized" gravity. (9.3) is the condition for both a flat space-time manifold and the gravity-free condition in which the laws of special relativity are assumed to be accurate. Similarly, (9.2) is to be interpreted as the expression of both the local curvature of spacetime and the presence of a gravitational field. However, as we shall now see, the theory of gravitation is "geometrized" in a much stronger sense than the foregoing.

You will recall that Newton's first law of motion asserts that a free particle, that is one on which no forces are acting, will follow a spatial trajectory which is a straight line. Now a straight line may be thought of as a curve with a constant velocity. Hence, it is a path with zero acceleration. Thus, the trajectory of a free particle in Newtonian dynamics is given by

$$\frac{d^2X^i}{dt^2} = 0. \tag{9.4}$$

To some degree, this law may be regarded as geometric in so far as it asserts that the natural or "unforced" trajectory is a geometrically distinguished path—namely, a Euclidean straight line. However, it is not fully geometric since it is not an "intrinsic" or coordinate-free description. We shall shortly prove this claim analytically. However, for the moment, the following intuitive reminder will suffice to indicate our meaning. Suppose that a ball is rolling in a frictionless manner along the floor of a room and that it is always equidistant from one of the walls. This would seem to be sufficient to characterize the ball's trajectory as "straight" or in accordance with (9.4). However, if the room, which is taken as a frame of reference, happens to be rotating, the path of the ball will appear to be curvilinear to an observer who is outside the room. In so far as the ascription of straightness to the trajectory is dependent on the chosen frame of reference, it cannot be regarded as expressing a geometric matter of fact. As you know, the natural and immediate generalization of (9.4) in special relativity is

$$\frac{d^2X^\mu}{d\tau^2} = 0. \tag{9.5}$$

Although (9.5) is Lorentz covariant, it suffers from the very same defect as (9.4) in so far as it is dependent on the choice of coordinates.

In the context of general relativity we shall seek an expression which is generally covariant and, therefore, coordinate free. That is to say, that the general-relativistic counterpart of (9.5) should be a tensorial equation. As you know, in virtue of the linear and homogeneous character of its transformation law, if a tensor vanishes with respect to one coordinate system, it will vanish with respect to every other. Hence, it would be fortunate indeed, if the quantity on the left-hand side of (9.5) were a tensor. However, we shall now show that such is not the case. We begin with the fairly obvious assumption that the four-velocity of special relativity is a tensor.

$$V^{\mu} = df\frac{dX^{\mu}}{d\tau}. \tag{9.6}$$

Its transformation law will, therefore, be the usual one for a contravariant vector.

$$\overline{V}^{\alpha} = \frac{\partial \overline{X}^{\alpha}}{\partial X^{\mu}} V^{\mu}. \tag{9.7}$$

We now ascertain the transformation law of the four-acceleration by differentiating (9.7) with respect to τ. By the Leibniz law for the differentiation of a product, we get

$$\frac{d\overline{V}^{\alpha}}{d\tau} = \frac{\partial \overline{X}^{\alpha}}{\partial X^{\mu}}\frac{dV^{\mu}}{d\tau} + \frac{\partial^2 \overline{X}^{\alpha}}{\partial\tau\partial X^{\mu}} V^{\mu} \tag{9.8}$$

This clearly shows that

$$\frac{dV^{\mu}}{d\tau} = \frac{d^2 X^{\mu}}{d\tau^2}$$

is not a tensor with respect to the group of diffeomorphisms in virtue of the presence of the second (inhomogeneous) term on the right-hand side of (9.8).

The four-acceleration, which failed to satisfy our requirements, was obtained by taking the ordinary derivative of the four-velocity with respect to the proper-time. However, it is possible to define a generally covariant counterpart to this operation, which is called *intrinsic differentiation.* As one might expect, the intrinsic derivative of a vector is closely related to the already familiar notion of the covariant derivative.

The reader may recall that we briefly acquainted ourselves with this notion in Chapter 7 where it was compared with the total derivative of elementary calculus. However, since it is so fundamental in general relativity, we shall now consider it in greater detail. The intrinsic derivative of a vector is formally defined as the inner product of the covariant derivative of that vector with a unit vector. Thus, if A^{μ} is an arbitrary vector, and U^{v} is a unit vector in any direction, then $A^{\mu}; vU^{v}$ is the intrinsic derivative of A^{μ} with respect to U^{v}. Intuitively, it gives the rate of change of A^{μ} in the direction of U^{v}. Now suppose that A^{μ} is a vector field along a timelike curve $X^{v}(\tau)$. We now wish to explore the concept of the intrinsic derivative of A^{μ} in the direction of the curve X^{v}. At any point on the curve, the direction of the curve is simply the unit tangent vector $dX^{v}/d\tau$. Thus, the intrinsic derivative of A^{μ} is $A^{\mu}; vdX^{v}/d\tau$. Now if this quantity is zero, it follows that as A^{μ} is "transported" along the curve, it has a constant magnitude and direction. Thus, we have arrived, by an alternative route, at the notion of the parallel-transport of a vector along a curve in terms of the vanishing of the intrinsic derivative.

$$\frac{DA^{\mu}}{D\tau} = A^{\mu}; v\frac{dX^{v}}{d\tau} = \left(\frac{\partial A^{\mu}}{\partial X^{v}} + \begin{Bmatrix} \mu \\ v\,\rho \end{Bmatrix} A^{\rho}\right)\frac{dX^{v}}{d\tau} = 0$$

or

$$\frac{DA^{\mu}}{D\tau} = \frac{dA^{\mu}}{d\tau} + \begin{Bmatrix} \mu \\ v\,\rho \end{Bmatrix} A^{\rho}\frac{dX^{v}}{d\tau} = 0. \tag{9.9}$$

We are now in a position to introduce the notion of the "straightest curve" or "curve of minimum (or maximum) length" or *geodesic* as that path whose tangent vector forms a parallel vector field. Intuitively, such a curve is one which has a constant velocity, i.e. a constant tangent vector or vanishing acceleration. This is obviously a generalization of the concept of straight line which occurs in the context of Euclidean space, in which the tangent vector to a straight line lies on the line and would be transported parallel to itself by being slid along that line. Thus, a geodesic is a curve such that the intrinsic derivative of its tangent vector along the curve vanishes. This is the generally covariant version of the notion of a curve of vanishing acceleration. The equation of such a curve is obtained by the direct substitution in (9.9) of $dX^\mu/d\tau$ for A^μ.

$$\frac{D}{D\tau}\left(\frac{dX^\mu}{d\tau}\right) = \frac{d^2X^\mu}{d\tau^2} + \begin{Bmatrix} \mu \\ \upsilon\rho \end{Bmatrix} \frac{dX^\upsilon}{d\tau}\frac{dX^\rho}{d\tau} = 0 \tag{9.10}$$

Since the intrinsic derivative is the inner product of two tensors, it, too, is a tensor. Consequently, (9.10) must be regarded as an intrinsic characterization of the concept of geodesic. In other words, what constitutes a "shortest path" on a given surface is a matter of "geometric fact", having nothing to do with the choice of coordinate system. Incidentally, we may point out that this consideration provides an elegant resolution of the twin paradox. It is simply a geometric fact that the path of T_r is a geodesic in spacetime, whereas that of T_m is not. Moreover, such a geodesic path in spacetime is always a *maximum* rather than a minimum interval.

Returning to physics, we remark that (9.10) is one version of what Einstein called the *geodesic hypothesis*. Physically interpreted, it expresses the truly remarkable conjecture that the path of a free particle is that of a geodesic in spacetime. The reader will note the obvious facts that for $g_{\mu\upsilon} = L_{\mu\upsilon}$, the second term on the left-hand side vanishes and one is left with the special relativistic law of inertia, and that if we then replace μ by i, we have Newton's first law. Now although it is reasonable to expect such correspondences, the reader should not be misled into attaching undue importance to them. The fact is that (9.10) has immensely more physical content than either of its counterparts. It embodies much of what we were struggling to express without the aid of mathematics in the preceding chapter.

In the first place, it should be noted that the term which involves the Christoffel symbol is to be interpreted as representing all the forces of gravitation and inertia, the two being lumped together in general relativity. But the reader may object that a gravitational term has no place in an equation which is supposed to depict what goes on in the absence of external force. The best answer that can be offered to this reasonable observation is that the heart and soul of the theory of general relativity rests on the assumption that gravitation is not a force in the usual sense. The rectilinear motion that is associated in Newtonian mechanics with a free mass point and which we attribute to "inertia" is the standard or normal motion of a body. It is what a body does when nothing causes it to do otherwise. Now Einstein is claiming that, in reality, the "natural" or "normal" or standard trajectory of a particle when nothing is causing it to go astray, i.e. when it is moving freely, is rather more complicated than Newton had envisaged. On this view, it is no longer deemed appropriate to ask why a body falls to the ground, since this is the natural motion of a body when no force interferes with it. Similar considerations apply to the planetary orbits. The path of a planet is now conceived to be the natural path which it will follow when no force intervenes to make it deviate from geodesic motion.

If the geodesic hypothesis is correct, then the mystery of the equality of inertial and gravitational mass simply cannot arise. The point has already been somewhat belaboured in earlier chapters that all bodies will be accelerated in the same way in a gravitational field if and only if the inertial and gravitational masses are equal. But according to the geodesic hypothesis, every body is bound to follow a geodesic path as its "natural" motion. Accordingly, since gravitational acceleration is nothing but geodesic motion, it follows that no explanation is called for. That is not to say, of course, that it would be methodologically unsound to speculate on the origin of geodesic motion but merely that it occupies a fundamental place in the theory and, therefore, does not stand in need of explanation in that context.

Although few would deny the place of the geodesic hypothesis among the most aesthetically satisfying principles in all of physics, many will initially find it to be somewhat preposterous. However, it is usually the case that the counterintuitive character of the principle derives from the tendency to interpret it spatially instead of in terms of spacetime. For example, the beginner is likely to object that although he is prepared to accept the idea that space departs in some degree from Euclidicity, there is abundant evidence that such a departure is not particularly great, at least within the limits of the solar system. How, then, it will be asked, can it be possible that the planetary orbits are geodesics? The answer, of course, is that they are not geodesics of three-dimensional space but of spacetime. Due to the hyperbolic metric, i.e. signature $(+ - - -)$, what constitutes a straightest possible path is far different from what we associate with the Euclidean metric.

Another objection which is based on the same misunderstanding is to the effect that a theorem of Riemannian geometry asserts that through a point in a given direction there will pass one and only one geodesic. It must, therefore, follow from the geodesic hypothesis that if two bodies pass through a given point in the same direction, they are bound to follow similar trajectories. But this is demonstrably false. For example, two balls may be thrown in the same direction and from the same point, but trace quite different parabolic paths before falling to the ground. In this case the explanation is that even though the balls travelled in the same spatial direction, so long as their initial velocities differed, they would be travelling in different directions, hence along different geodesics, in *spacetime.* In fact, as early as the mid-nineteenth century, Riemann had struggled unsuccessfully to construct a geometric theory of gravitation. It was he who first seriously conjectured that the geometry of physical space might be a dynamic function of the distribution of matter. However, his efforts were to no avail largely because he was working with space instead of spacetime.

So far, we have in the geodesic hypothesis something which is akin to Newton's law of motion. It tells us, in effect, how a body will move in any given field of $g_{\mu\nu}$. The claim is simply that it will follow a geodesic path. However, the latter fails to provide one with sufficient information to predict actual trajectories in the absence of knowledge concerning the specific distribution of the $g_{\mu\nu}$'s in spacetime. This the geodesic hypothesis, of course, does not deal with. Thus, the geodesic hypothesis stands in need of supplementation by something akin to the Newton–Poisson equation of classical gravitation. A solution of the latter gives Φ, the gravitational potential, as a function of the spatial coordinates, X^i. What we need, now, is an analogous equation for the $g_{\mu\nu}$. Obviously, however, the field law of general relativity is bound to be much more complex than the Newton–Poisson law or Laplace's equation if only because Φ is a scalar, while $g_{\mu\nu}$

is a symmetric tensor of the second rank, having ten independent components.

Einstein utilised three heuristic principles in the tortuous path to the field law of gravitation. Firstly, it should be generally covariant. Secondly, it should approximate the Newton–Poisson field law in the limit of low velocities and weak fields. Finally, it should admit the laws of special relativity in the special case of vanishing gravitation, i.e. $g_{\mu v} = L_{\mu v}$. Now since the Newton–Poisson equation is of second differential order in Φ, it is reasonable to seek an equation which is of second differential order in $g_{\mu v}$. In fact, we have already encountered one such equation in the form:

$$R^{\alpha}_{\beta\gamma\delta} = 0 \tag{9.11}$$

You may recall that the Riemann–Christoffel curvature tensor involves the first derivatives of the Christoffel symbols, which, in turn, involve the first derivatives of the $g_{\mu v}$. Hence, (9.11) is a second order equation in the $g_{\mu v}$. Clearly, however, it fails to meet our requirements since it corresponds to a perfectly flat space or vanishing of the metric affinity (Christoffel symbol). Thus, while it would certainly admit the flat-space solution of special relativity, it is far too restrictive in ruling out solutions corresponding to non-zero curvature or the presence of a gravitational field.

As you know, one of the basic operations of tensor algebra is that of contraction, which consists in equating an upper and lower index of a tensor and then performing the indicated summation. In the case of the Riemann–Christoffel tensor, this would amount to the following.

$$R^{\alpha}_{\beta\alpha\delta} = R_{\beta\delta} = 0 \tag{9.12}$$

The resulting tensor is known as the Ricci tensor. (9.12) is less stringent than (9.11) in that its validity does not require the flatness of space. The latter is certainly sufficient but is not necessary for (9.12). That is to say, that (9.12) admits as solutions $g_{\mu v}$ which correspond to curved as well as to flat spaces. But, by our earlier discussions, the curved-space $g_{\mu v}$'s correspond to the presence of gravitational fields.

Obviously, in the highly compact notation of tensor analysis, (9.12) scarcely looks like a law of gravitation or, for that matter, like any sort of differential equation. Accordingly, our immediate task must be to provide some form of intuitive physical interpretation for our abstract mathematical formalism. Only then can we determine whether (9.12) is a useful model of gravitation. This is not an altogether easy matter in the limited scope of this chapter. However, our task will be facilitated if we can show how the metric tensor, the geodesic law and the Ricci tensor combine to yield, as a first approximation, the classical law of gravitation in its field-theoretic form. To this end, throughout the following discussion, we shall make several simplifying assumptions. We know that Newtonian gravitational theory is very accurate for the field of the earth. Moreover, it turns out that, to a first approximation, the earth may be treated as an inertial system. In particular, the laws of electromagnetism and light propagation are not visibly affected by terrestrial gravitational influences. Consequently, we conclude that the metric tensor, $g_{\mu v}$, which we have tentatively interpreted as the counterpart of the familiar gravitational potential, must deviate only by very small quantities from the Lorentz metric tensor in the case of weak fields such as that of the earth. In the second place, we shall assume that the field is static, which is to say that it does not vary with respect to time, again in conformity with our gross experience of the earth's gravitational field. Finally, we shall assume that our "probe" or test particle, whose behaviour measures the field, is moving very slowly in comparison to the velocity of light.

The assumption that the metric departs from its Lorentz form but very slightly permits us to decompose the metric tensor in the following manner.

$$g_{\mu v} = L_{\mu v} + \varepsilon h_{\mu v} \tag{9.13}$$

where $L_{\mu v}$ is the familiar Lorentz metric tensor, diag (1, −1, −1, −1), and the second term on the right-hand side represents the slight deviation from flat spacetime, so that $\varepsilon \ll 1$. Accordingly we shall drop all terms which are of second order in ε, on the assumption that they would be too small to have any significant physical influence.

We may now write the equation of motion employing coordinate time in place of proper-time. This is justifiable in the Newtonian approximation in which $dt/d\tau \to 1$. However, it is obvious that we must now abandon any pretense of achieving a coordinate-free treatment. But, in any case, since all that follows depends crucially on the assumption that the velocity of a test particle is very small, we are clearly engaging in a frame-dependent account. Without further ado, we write the law of motion in the form

$$\frac{d^2X^\mu}{dt^2} + \begin{Bmatrix} \mu \\ v\rho \end{Bmatrix} \frac{dX^v}{dt}\frac{dX^\rho}{dt} = 0. \tag{9.14}$$

We now consider the factor $dX^v/dt\, dX^\rho/dt$ which occurs in (9.14). We may break it up into the two parts: $dX^i/dt\, dX^j/dt$ and $dX^0/dt\, dX^0/dt$. The first part is simply $|\mathbf{V}|^2$, where $\mathbf{V}$ is the velocity of the test particle, while the second part is equal to c^2, since $X^0 = ct$. But since the velocity is taken to be very small as compared to c, it follows that the spatial part of the quadratic factor is negligible as compared with the temporal part. Therefore, we may make the significant simplification of setting $v = \rho = 0$. Now we have

$$\frac{d^2X^\mu}{dt^2} + \begin{Bmatrix} \mu \\ 0\,0 \end{Bmatrix} \frac{dX^0}{dt}\frac{dX^0}{dt} = 0. \tag{9.15}$$

We now rewrite (9.15) in the form

$$\frac{d^2X^\mu}{dt^2} + \begin{Bmatrix} \mu \\ 0\,0 \end{Bmatrix} c^2 = 0. \tag{9.16}$$

We may now turn our attention to the problem of constructing an explicit formulation of the Christoffel symbol of (9.16). The reader should recall that one must first work out the Christoffel symbol of the first kind which, in this case, is

$$[0\,0, \lambda] = \frac{1}{2}\left(\frac{\partial g_{\lambda 0}}{\partial X^0} + \frac{\partial g_{\lambda 0}}{\partial X^0} - \frac{\partial g_{00}}{\partial X^\lambda}\right). \tag{9.17}$$

We now exploit our assumption that the field is not varying with time. This entails that the time derivatives of the $g_{\mu v}$ will vanish, so that we are left with

$$[0\,0, \lambda] = -\frac{1}{2}\frac{\partial g_{00}}{\partial X^\lambda}. \tag{9.18}$$

By the same token, we should replace the space-time coordinate variables X^λ by the spatial coordinate variables X^i to get

$$[0\,0, i] = -\tfrac{1}{2}\frac{\partial g_{00}}{\partial X^i}. \tag{9.19}$$

The Christoffel symbol of the second kind must be

$$\begin{Bmatrix} j \\ 0\,0 \end{Bmatrix} = -\frac{g^{ij}}{2}\frac{\partial g_{00}}{\partial X^i}. \tag{9.20}$$

In virtue of the nearly Lorentz metric, (9.20) may be rewritten as

$$\begin{Bmatrix} j \\ 0\,0 \end{Bmatrix} = -\tfrac{1}{2}(L^{ij}+\varepsilon h^{ij})\left(\frac{\partial L_{00}}{\partial X^i}+\frac{\partial \varepsilon h_{00}}{\partial X^i}\right). \tag{9.21}$$

Now, since we drop the term of order ε^2 and use the fact that L^{00} is constant, (9.21) is immediately simplified to give

$$\begin{Bmatrix} j \\ 0\,0 \end{Bmatrix} = -\tfrac{1}{2}L^{ij}\frac{\partial \varepsilon h_{00}}{\partial X^i}. \tag{9.22}$$

But L^{ij} is simply $-\delta^{ij}$. Therefore,

$$\begin{Bmatrix} j \\ 0\,0 \end{Bmatrix} = \tfrac{1}{2}\frac{\partial \varepsilon h_{00}}{\partial X^j}. \tag{9.23}$$

We have now to insert this result in the equation of motion (9.16).

$$\frac{d^2X^i}{dt^2} = -\frac{c^2}{2}\frac{\partial \varepsilon h_{00}}{\partial X^i}. \tag{9.24}$$

This is an eminently satisfying result, since it may be interpreted as the equation of motion in a Newtonian gravatitional field, provided that we relate the Newtonian potential Φ to h_{00} by

$$\Phi = \frac{c^2}{2}\varepsilon h_{00}. \tag{9.25}$$

From (9.25), there immediately follows

$$\varepsilon h_{00} = 2\Phi/c^2. \tag{9.26}$$

and, therefore,

$$g_{00} = 1 + 2\Phi/c^2. \tag{9.27}$$

We have now achieved our first aim, which was to provide a physical interpretation of the $g_{\mu\nu}$. In fact, it has emerged that at low velocities only the 00-component of the metric tensor is of significance and that it may be related in a very direct way to the classical gravitational potential. Attention should also be given to the purely practical point concerning the development from (9.21) to (9.24). This justifies the generally adopted convention that when dealing with "nearly Lorentz" metrics, $L^{\mu\nu}$ is to be used in place of $g^{\mu\nu}$ to raise indices. This, of course, tends to facilitate considerably various mathematical manipulations.

We now turn to the second phase of the task before us, which is to attach some direct physical significance to the field law, $R_{\mu\nu} = 0$. Once again, we shall use the various simplifying assumptions that were exploited in the interpretation of $g_{\mu\nu}$. The Ricci tensor is defined in terms of the Riemann Christoffel tensor as follows:

$$R_{\beta\delta} = R^{\alpha}_{\beta\alpha\delta} = \begin{Bmatrix} \alpha \\ \beta\,\delta \end{Bmatrix}_{,\alpha} - \begin{Bmatrix} \alpha \\ \beta\,\alpha \end{Bmatrix}_{,\delta} + \begin{Bmatrix} \varepsilon \\ \beta\,\delta \end{Bmatrix}\begin{Bmatrix} \alpha \\ \varepsilon\,\alpha \end{Bmatrix} - \begin{Bmatrix} \varepsilon \\ \beta\,\alpha \end{Bmatrix}\begin{Bmatrix} \alpha \\ \varepsilon\,\delta \end{Bmatrix}. \tag{9.28}$$

In the case of a weak gravitational field, the third and fourth terms may be dropped since they are of second order in ε. The resulting equation

$$R_{\beta\delta} = \begin{Bmatrix} \alpha \\ \beta\,\delta \end{Bmatrix}_{,\alpha} - \begin{Bmatrix} \alpha \\ \beta\,\alpha \end{Bmatrix}_{,\delta} \tag{9.29}$$

represents an enormous simplification since its linearity greatly facilitates the discovery of solutions. In fact most of the practical results of general relativity stem from the so-called "linearized field theory". We shall now rewrite (9.29) in terms of the Christoffel symbols of the first kind.

$$R_{\beta\delta} = \frac{L^{\alpha\rho}}{2}\frac{\partial}{\partial X^{\alpha}}[\beta\delta,\rho] - \frac{L^{\alpha\rho}}{2}\frac{\partial}{\partial X^{\delta}}[\beta\alpha,\rho]. \tag{9.30}$$

Note that we have used $L^{\alpha\rho}$ in place of $g^{\alpha\rho}$ for reasons that have already been stated. We must now expand the Christoffel symbols. But this will be done in terms of the $h_{\mu\nu}$ since the $L_{\mu\nu}$ are constant and, therefore, have vanishing derivatives.

$$R_{\beta\delta} = \frac{L^{\alpha\rho}}{2}\frac{\partial}{\partial X^{\alpha}}\left(\varepsilon\frac{\partial h_{\beta\rho}}{\partial X^{\delta}} + \varepsilon\frac{\partial h_{\delta\rho}}{\partial X^{\beta}} - \varepsilon\frac{\partial h_{\beta\delta}}{\partial X^{\rho}}\right) - \frac{L_{\alpha\rho}}{2}\frac{\partial}{\partial X^{\delta}}\left(\varepsilon\frac{\partial h_{\beta\rho}}{\partial X^{\alpha}} + \varepsilon\frac{\partial h_{\alpha\rho}}{\partial X^{\beta}} - \varepsilon\frac{\partial h_{\beta\alpha}}{\partial X^{\rho}}\right). \tag{9.31}$$

Performing the indicated differentiations, we get

$$R_{\beta\delta} = \frac{L^{\alpha\rho}}{2}\varepsilon\left(\frac{\partial^2 h_{\beta\rho}}{\partial X^{\alpha}\partial X^{\delta}} + \frac{\partial^2 h_{\delta\rho}}{\partial X^{\alpha}\partial X^{\beta}} - \frac{\partial^2 h_{\beta\delta}}{\partial X^{\alpha}\partial X^{\rho}} - \frac{\partial^2 h_{\beta\rho}}{\partial X^{\delta}\partial X^{\alpha}} - \frac{\partial^2 h_{\alpha\rho}}{\partial X^{\delta}\partial X^{\beta}} + \frac{\partial^2 h_{\beta\alpha}}{\partial X^{\delta}\partial X^{\rho}}\right) \tag{9.32}$$

which simplifies to

$$R_{\beta\delta} = \frac{L^{\alpha\rho}}{2}\varepsilon\left(\frac{\partial^2 h_{\delta\rho}}{\partial X^{\alpha}\partial X^{\beta}} - \frac{\partial^2 h_{\beta\delta}}{\partial X^{\alpha}\partial X^{\rho}} - \frac{\partial^2 h_{\alpha\rho}}{\partial X^{\delta}\partial X^{\beta}} + \frac{\partial^2 h_{\beta\alpha}}{\partial X^{\delta}\partial X^{\rho}}\right) \tag{9.33}$$

It has been explained that for *low velocities* we need only retain the *OO*-component of such expressions. Consequently, setting $\beta = \delta = 0$, we write

$$R_{00} = \frac{L^{\alpha\rho}}{2}\varepsilon\left(\frac{\partial^2 h_{0\rho}}{\partial X^{\alpha}\partial X^{0}} - \frac{\partial^2 h_{00}}{\partial X^{\alpha}\partial X^{\rho}} - \frac{\partial^2 h_{\alpha\rho}}{\partial X^{0}\partial X^{\beta}} + \frac{\partial^2 h_{\beta\alpha}}{\partial X^{0}\partial X^{\rho}}\right). \tag{9.34}$$

We now exploit the assumption that the field is static or independent of time. This, as we already know, has the dual effect of eliminating all of the partial derivatives involving time and having the Greek indices replaced by Latin ones. Accordingly, we achieve the considerable simplification:

$$R_{00} = \frac{-\delta^{ar}}{2}\varepsilon\left(-\frac{\partial^2 h_{00}}{\partial X^{a}\partial X^{r}}\right) \tag{9.35}$$

or

$$R_{00} = \tfrac{1}{2}\varepsilon \frac{\partial^2 h_{00}}{\partial X^a \partial X^a} = \tfrac{1}{2}\varepsilon \nabla^2 h_{00}.$$

But

$$\varepsilon h_{00} = 2\Phi/c^2. \tag{9.36}$$

So by (9.36) and (9.12)

$$R_{00} = 1/c^2 \nabla^2 \Phi = 0. \tag{9.37}$$

which is a form of Laplace's equation. Thus, we have succeeded in deriving, *as an approximation*, Newton's free-space law of gravitation from the one conjectured by Einstein. Although this is certainly to be regarded as an impressive result, it should not be permitted to obscure the fact that we have made several simplifying assumptions along the way and that the basic law (9.12) is of almost mind-boggling complexity as compared with that of Newton. On the other hand, it should also be remembered that the greater complexity of Einstein's law is strictly algebraic. From a purely geometric standpoint, Einstein's law is actually simpler than Newton's.

We cannot terminate this part of the exposition without mention of a somewhat remarkable aspect of the mathematical structure of Einstein's theory. Following the example of Newtonian mechanics, we first constructed a law of motion, the geodesic hypothesis, and then a field law. This, indeed, was the procedure adopted by Einstein. However, it was soon shown that the law of motion is actually derivable from the field law. In fact, this is not as surprising as it first seems. The field law, as we have noted, is highly non-linear. This means, in physical terms, that the sum of two solutions is not generally a third solution. That is to say, that if two sources are present, the field which they produce in combination will not consist in the superposition of the two fields which would have been produced by each source in the absence of the other. In a linear field theory, such as electrodynamics, the effects of the various sources are mutually independent and additive. Since each source acts independently, the field law does not provide information as to the manner in which the sources interact. For this, a law of motion is required, which expresses how the sources are acted on by the fields of other sources. But in a non-linear field theory, the interaction via the field is revealed by the very non-linearity of the situation. For these reasons, once we had determined that Newtonian gravitation followed from the geodesic hypothesis, it was natural to suppose that it would also have been found to follow from the field law which entails the geodesic hypothesis.

Apart from its concordance with Newton's law of gravitation, the Einsteinian law has traditionally found support in three principal experimental areas. The most dramatic was its ability to account in a completely natural way for an irregularity in planetary orbits and in that of Mercury in particular. It had long been recognized that the perihelion of Mercury, the point at which Mercury is closest to the sun, is gradually advancing so that the planetary orbit is not quite closed. Newtonian mechanics proved incapable of providing a satisfactory explanation of the precession of the perihelion of Mercury. However, it follows directly from the principle of general relativity, thus strongly suggesting the latter's superiority. A second area is its ability to predict with a fairly high degree of accuracy the amount of bending or deflection of light-rays in the vicinity of a strong gravitational field. Finally, it was predicted that the frequency of a given light source would be slowed down by the effect of a gravitational field, so that in such a field the spectral lines of a given element would undergo a shift towards the red or low-frequency end of the spectrum. It is the last of these which we shall now examine in some detail. We do so for the reason that it, in particular, has attracted the attention of philosophers, who have interpreted it as indicating that time "slows down" in a gravitational field.

We begin by introducing the concept of a *standard clock*, which is any cyclical device which is assumed to measure equal intervals of proper-time along its trajectory. In short, it is a clock whose mechanism is impervious to the effects of acceleration. Clearly, such a clock is different from an ordinary mechanical clock which will be influenced by accelerative or gravitational forces. For example, one has only to consider what happens to a pendulum clock which is subjected to forces of this kind. It is normally assumed that the ideal of a standard clock is physically realized by an oscillating atom of a given substance, say caesium, having a characteristic frequency v_o, where the latter is the number of oscillations per unit of proper-time. On the basis of our hypothesis about standard clocks, it follows that the ratio of the rates of any two such clocks would be constant. Moreover, since we have defined this rate or frequency as the number of cycles per unit of *proper-time*, the aforementioned ratio is an invariant characterization of the relationship between such clocks.

Let us suppose that an oscillating atom is at rest at the point X_1^μ in a static gravitational field. For simplicity, we stipulate that it completes one period of vibration in the time $d\tau$. In this period, the clock will have moved along its trajectory to the point $X_1^\mu + dX_1^\mu$. However, since the clock is "at rest" relative to the spatial field coordinates we must have $dX_1^i = 0$. That is to say, that the clock has continued to occupy the same point in space but has, of course, "moved" in spacetime. Hence, we may write

$$(ds)^2 = g_{\mu\nu}\, dX^\mu\, dX^\nu = g_{00}(dX^0)^2 \tag{9.38}$$

and, dividing by c^2 and taking square roots,

$$d\tau = (g_{00})^{1/2}\, dt. \tag{9.39}$$

where g_{00} is evaluated at the location of the vibrating atom. It is tempting to interpret (9.39) as a mere generalization of the special-relativistic formula

$$d\tau = (1 - v^2/c^2)^{1/2}\, dt. \tag{9.40}$$

However, the latter relates the rates of two clocks which are attached to distinct frames of reference. Moreover, as we know, it is reciprocal. The former expresses a relationship between the proper-time and the coordinate time within a single frame of reference.

We must now establish an important lemma to the effect that in a static gravitational field, the coordinate-time interval corresponding to the period of oscillation will be a constant. Intuitively, this entails that whereas the self-measured proper-time period of a standard clock is only of local significance, the coordinate-time period is non-local, pertaining to the way in which time is "spread through space". Proceeding, now, to the proof of our lemma, let a coordinate clock at the location of the atom (which could be the atom, itself) record the coordinate-time interval $t_2 - t_1$, as the elapsed time between the beginning and end of a single cycle. Suppose that this takes place near the surface of the sun. Let there be a receiver located in a laboratory on the earth. This receiver will record the elapsed time between signals from the pair of events on the sun as $\overline{t}_2 - \overline{t}_1$. Since we are assuming the gravitational field to be static, at least for the duration of the signal transmission from the sun to the earth, it follows that the second signal will take the same time to reach its destination as the first one. Accordingly we have the following relationship:

$$\bar{t}_2 - t_2 = \bar{t}_1 - t_1. \tag{9.41}$$

The left-hand side represents the transit time of the second signal while the right-hand side is the transit time of the first one. Then there immediately follows

$$\bar{t}_2 - \bar{t}_1 = t_2 - t_1. \tag{9.42}$$

The left-hand side of (9.42) represents the coordinate-time interval of the clock period on the earth, while the right-hand side represents the same interval on the sun. Thus, our lemma is proved. So long as the field is static, this reasoning may be extended to the duration of any number of cycles, so that, in general, $\Delta t = \Delta \bar{t}$, independently of position in the field. On the other hand, such reasoning would not be applicable in the case of proper-time, which is a path-dependent quantity. For example, suppose that there are two identical clocks at the point A, such that $d\tau_1 = d\tau_2$. Let them traverse their respective space-time trajectories and be reunited at the point B. The fact is that the proper-time intervals for the two clocks will not generally be the same, since $\tau = \int_A^B d\tau$ depends on the specific path between A and B. Technically, we say that proper-time is not *integrable*. On the other hand, coordinate time is integrable. With respect to a given coordinate system, each event has a unique time coordinate, so that $\int_A^B dt$ is simply $t(B) - t(A)$. However, the latter is not an invariant or coordinate-free quantity.

We now introduce the notion of the *characteristic* or *proper* frequency of an atom, symbolized by v_o. An atom of a given substance will have a specific characteristic frequency which serves to identify the substance. This frequency is the number of vibrations per unit of proper-time. In fact, this is a formal consequence of the hypothesis that a vibrating atom is a standard clock. Accordingly, we claim that on the surface of the sun, in a period $\Delta\tau_1$ of its own proper-time, an atom of frequency v_o will undergo n vibrations in accordance with the relation

$$n = v_o \Delta\tau_1. \tag{9.43}$$

At the surface of the earth, the n waves will be received in the time period $\Delta\tau_2$ as measured by a standard clock at the point of reception. Hence,

$$n = v_e \Delta\tau_2. \tag{9.44}$$

There is, as already intimated, no reason to suppose that $\Delta\tau_1 = \Delta\tau_2$, since they measure the proper-time intervals along the distinct trajectories of an atom at the sun and one at the earth. However, the number of waves received must equal the number emitted. Thus, we may equate (9.43) and (9.44) to get

$$v_o \Delta\tau_1 = v_e \Delta\tau_2. \tag{9.45}$$

But in virtue of (9.39), (9.45) may be rewritten as

$$v_o (g_{00}(1))^{1/2} \Delta t = v_e (g_{00}(2))^{1/2} \Delta t. \tag{9.46}$$

But by our lemma (9.42) the coordinate-time intervals, Δt, are one and the same. Hence,

$$v_o (g_{00}(1))^{1/2} = v_e (g_{00}(2))^{1/2} \tag{9.47}$$

and
$$V_e = V_o \left(\frac{g_{00}(1)}{g_{00}(2)}\right)^{\frac{1}{2}} \tag{9.48}$$

This is the key relationship. It tells us that the frequency of an atom or, more generally, the rate of a standard clock which is V_o will undergo a change from the standpoint of a clock which is located at a different gravitational potential. In the case of the type of field we have been considering, it will be recalled that approximately

$$g_{00} = 1 + 2\Phi/c^2 \tag{9.49}$$

where Φ is the Newtonian potential. Let us take $g_{00}(1)$ to be the metric potential at the sun and $g_{00}(2)$ to be that at the surface of the earth. Since the Newtonian potential is a negative quantity, $g_{00}(1) < g_{00}(2)$. It follows that $v_e < v_o$. In other words, the frequency has shifted towards the red end of the spectrum. In the case of the sun and the earth, v_o/v_e is a mere 1.00000212. In the case of very dense stars, however, the red shift will be many times greater and readily accessible to experimental confirmation.

While the red shift is of importance as an experimental test of the theory of gravitation, we have already remarked that it has aroused philosophical interest due to its bearing on the concept of time. In effect, the vibrating atoms are highly accurate clocks and the red shift is a special case of the general principle that clocks at a higher gravitational potential run more slowly than those at a relatively lower potential. Two observers, one placed at the sun and the other at the earth, will be in agreement concerning the characteristic frequency of an atom of a given substance. That is to say, that they will be in agreement as to the number of beats per unit of proper-time. Yet the earth observer will find that the atomic clock on the sun is running slow as compared with the atomic clock on the earth. Conversely, the sun observer will find the atomic clock on the earth to be running fast—a blue shift. It must be emphasized that unlike the Lorentz time dilation, this effect is not reciprocal. Thus, it is possible, no doubt, to interpret it as the falsification of the long-held principle of the uniformity of time, which was affirmed by Newton. However, we suggest that such philosophic inferences be made with the utmost of caution. How can one meaningfully speak of the slowing down of time? This is to ascribe a certain rate of change or velocity to time. But if time has a rate of change, with respect to what independent variable is it changing? It cannot be the usual one, which is time itself. For then the rate of change of time would always be $dt/dt = 1$. I am able to comprehend the Lorentz time dilation phenomenon which is simply the kinematic realization of $dt/d\tau$, the rate of change of coordinate time with respect to proper-time. However, I would interpret gravitational time dilation as a different type of effect.

It is perhaps helpful to observe how gravitational time dilation is an inevitable consequence of a certain interpretation of the formal structure of the theory of general relativity. In brief, since the metric tensor is a tensor in spacetime and since it is employed to represent the gravitational potential, then as surely as the field intensity varies from region to region, so the g_{00} component of the metric will be a variable function of position. Therefore, it seems to me that the claim that time slows down in a gravitational field may be based on a semantic confusion. The strict semantic interpretation of $g_{\mu v}$ in general relativity is as the gravitational potential. However, if one allows a semantic ambiguity in the theory, one may also interpret $g_{\mu v}$ as the metric of physical spacetime. I do not wish, at present, to argue that it cannot be both but merely suggest that in arguing that the slowing down of time is a consequence of general relativity one is trading on the dual interpretation of $g_{\mu\nu}$. Nevertheless, my first duty as an expositor of the theory is to point out that the standard interpretation of the gravitational red-shift is indeed to the effect that time literally runs slow in regions of higher potential. Furthermore, it is customary to "explain" gravitational time dilation as a geometric phenomenon.

I must now confess that I fail to understand how geometry could explain a physical phenomenon. In my view the function of mathematical theories when employed in physics is not to explain but to describe. A physical explanation may employ the language of geometry but that does not make it a geometric explanation. By the same token, if a theory employs a purely algebraic or arithmetic formalism, it does now follow that the facts which it explains are explained algebraically or arithmetically. The following analogy may be helpful. Suppose that a small steel ball is released at the inner edge of a smooth dish. It is possible to explain the motion of the ball as it rolls to the bottom of the dish in terms of various mechanical forces which constrain it to follow a particular path. Now it may also be shown that the path of the ball is a *geodesic* of the curved surface. My question is whether the latter would constitute a geometric explanation of the ball's motion. My own view is that it would not and that, moreover, the situation in general relativity, while not strictly analogous, is not essentially different. To say that the path of the ball is determined, at least in part, by the shape of the dish is not to offer a geometric explanation but a physical one.

It is possible to derive the gravitational red-shift from the conjunction of the principles of the conservation of energy and equivalence. The former is independent of general relativity while the latter is, of course, the very basis for the so-called geometrization of gravity in the form of the geodesic hypothesis. Hence, any successful geometric theory of gravitational phenomena would be bound to predict the red-shift. A failure to do so would be in violation of either energy conservation or the very geometric programme of the theory. However, to admit that general relativity is, in a certain sense, a successful geometric theory of physics, i.e. one which incorporates the geodesic hypothesis, is not to say, in my opinion, that it provides geometric explanations of phenomena.

If it be granted that it is perhaps too facile to describe gravitational time dilation as a geometric phenomenon, how ought it then to be interpreted? A natural interpretation would, in the first place, treat it as a feature of the gravitational field. In particular, just as the field causes test particles to traverse given trajectories, so it also has an effect on the behaviour of clocks. Most relativists would object to this on the ground that a fundamental assumption of general relativity is the existence of clocks which are not affected by gravitational or accelerative forces. However, one must exercise caution in one's interpretation of this assumption. We have seen that what it comes to is the claim that standard clocks, such as vibrating atoms, have a fixed characteristic frequency, which is a function of proper-time. Thus, it amounts to the empirically confirmed claim that there exist classes of clocks which are not affected in a *differential* manner by gravitational forces. An observer who is at a given gravitational potential will affirm that his standard clock is unaffected by gravitational forces, since all physically suitable clocks in his immediate vicinity behave in the same way. This may be contrasted with the case of an ordinary pendulum clock and a shockproof chronometer, both of which are subjected to equal accelerations. In the latter case, the effect will clearly be differential, i.e. differently constructed clocks will be variously affected. We have seen that although proper-time is invariant, it does not have a fixed magnitude with respect to different potentials. In physical terms, this simply means that the rates of standard clocks differ at distinct potentials. This is a frame-independent fact, which is revealed by the gravitational red-shift, the latter being a phenomenon of the absolute kind. It is a great tribute to the success of general relativity that this phenomenon, among so many, may be incorporated in its geometric formalism. Nevertheless, we must not too hastily suppose that it may be treated as some sort of geometric feature of the world.

Before terminating this review of the basic ideas of general relativity, we must make passing reference to the matter-energy tensor, $T_{\mu\upsilon}$. The reader will recall that our discussion of the field equation was restricted to the case of matter-free space. The free space equation, $R_{\mu\upsilon} = 0$, is the counterpart of Laplace's equation for the gravitational potential. The counterpart of the more general Poisson equation is $G_{\mu\upsilon} = \kappa T_{\mu\upsilon}$, where $G_{\mu\upsilon}$ is just the divergenceless form of $R_{\mu\upsilon}$ and $T_{\mu\upsilon}$ represents every possible material and energy source of gravitation. The task of entering specific values of the various components of $T_{\mu\upsilon}$ is quite formidable, so we have preferred to ignore it. We merely mention it now because we shall have occasion to make casual references to it in a later chapter.

Our little survey of the basic ideas of general relativity is now complete. We intimated at the beginning of this chapter that we intended to do little more than to scratch the surface. However, the reader who has followed this exposition attentively and with understanding, albeit after several readings, should not suffer from undue modesty. It is a fact that he now knows as much about the subject as many physicists who have specialized in some other branch. Moreover, he has sufficient conceptual grasp of both the physics and the mathematics to undertake, should he be so inclined, the study of one of the many standard introductory treatises on the subject. But most important for the moment, he now knows enough about the theory to tackle realistically many of the interesting methodological and philosophical issues which arise from it.

Bibliography for Chapter 9

ADLER, R., M. BAZIN, and M. SCHIFFER, *Introduction to General Relativity*, McGraw-Hill, New York, 1965. An exceptionally lucid discussion of the mathematics of general relativity.

ANDERSON, J. L., *Principles of Relativity Physics*, Academic Press, New York, 1967. A sophisticated but beautiful exposition. Although generally too advanced for the beginner, he should attempt to read as much of it as he can.

BERGMANN, P. G. (1942). See reference in Chapter 6.

BERRY, M., *Principles of Cosmology and Gravitation*, Cambridge University Press, Cambridge, 1976. A unique treatment of general relativity for the average undergraduate. Tensors are scarcely employed. Considerable attention is devoted to the physics of black holes.

EINSTEIN, A., *The Foundation of the General Theory of Relativity* (1916). Reprinted in Einstein *et al.* (1923). See reference in Chapter 3.

EINSTEIN, A., (1953). See reference in Chapter 5.

MISNER, C. W., K. S. THORNE, and J. A. WHEELER, *Gravitation*, Freeman, San Francisco, 1973. The most monumental treatise on the subject. Uses both classical and contemporary mathematics, side by side. Arranged so that both advanced and intermediate students may benefit from it. Every student of relativity must at least look at it and get from it what he may.

MØLLER, C., *The Theory of Relativity*, Oxford University Press, London, 1952. A thorough treatment from a somewhat classical standpoint. Authoritative but a trifle unexciting.

OHANIAN, H. C., *Gravitation and Spacetime*, Norton, New York, 1976. I think it deals too harshly with the general equivalence of reference frames. However, it is a very well-organized textbook. A careful reading should prove highly rewarding.

10

Relativity and Covariance

WHEN one studies the "philosophy" of a physical theory, one may draw a rough and ready distinction between the internal and the external problems relating to that theory. Admittedly, it is sometimes difficult to know precisely where the line is to be drawn. However, there are certain questions which clearly pertain to the role and status of particular concepts within the theory while others pertain to the epistemological and ontological significance of the theory as a whole. It seems appropriate to begin with a treatment of one of the fundamental internal problems before turning our attention to questions concerning overall implications. This key problem concerns the role and, indeed, the very meaning of the principle of general covariance and its relation to the principle of general relativity.

Before joining the debate on the significance of general covariance, it is important to remark that there is even some formal ambiguity surrounding this notion. For example, a leading theoretician of general relativity, C. Møller, has characterized the requirement as follows:

". . . if the law in question is expressed by equations of the form

$$F(A, B, \ldots, \frac{\partial A}{\partial X^i}, \frac{\partial B}{\partial X^i}, \ldots) = 0$$

where $A, B, \ldots$ are physical quantities, we have in another arbitrary system of coordinates (X'^i) the same functional relation between the physical quantities in (X'^i), i.e.

$$F(A', B', \ldots, \frac{\partial A'}{\partial X'^i}, \frac{\partial B'}{\partial X'^i}, \ldots) = 0\text{''} \quad (1952, \text{ p. } 265).$$

It has already been remarked that a sufficient condition for the satisfaction of this requirement is that the equation be tensorial. For example, the tensorial expression for the interval (or distance) is

$$(ds)^2 = g_{ik}dX^i dX^k. \qquad (10.1)$$

When we subject (10.1) to an admissible but otherside arbitrary coordinate transformation, the resulting expression is of the form:

$$(d\bar{s})^2 = \bar{g}_{ik}d\bar{X}^i d\bar{X}^k. \qquad (10.2)$$

(10.2) certainly appears to have the same form as (10.1). However, the similarity of form in this case would appear to be a trivial consequence of a mere notational convention. Let us suppose that (10.1) is the expression for three-dimensional distance in Cartesian coordinates. As the reader knows, its explicit formulation in terms of its components will be

$$(ds)^2 = dx^2 + dy^2 + dz^2. \tag{10.3}$$

Let us suppose that (10.2) results from (10.1) in virtue of the familiar transformation to spherical coordinates. Then the explicit form of (10.2) will be

$$(ds)^2 = (dr)^2 + r^2(d\theta)^2 + r^2 \sin^2\theta(d\phi)^2. \tag{10.4}$$

Clearly, (10.4) does not have the same form as (10.3), notwithstanding the fact that (10.4) is equivalent to (10.2) and (10.3) is equivalent to (10.1). Thus, the first problem connected with the requirement of general covariance is that whether or not a given physical law will satisfy it is dependent on one's choice of notation.

The difficulty could be resolved if it were shown that one of the premises on which the argument is based—namely, that the choice of notation is merely conventional—is actually false. I believe that it is, in fact, possible to make a case for the claim that the tensor notation is objectively preferable to the *explicit* coordinate notation. In effect, we would argue that a physical law in tensorial form is to be interpreted as a universally quantified statement in which one of the quantifiers ranges over coordinate systems. The significance of any such expression would be that since it holds for all coordinate systems, its truth must be independent of any particular choice of coordinate system. What is essential in the semantic content of a physical law should not consist in part of information about the properties of a system of coordinates but should refer exclusively to physical reality. Using more sophisticated terminology, we would say that a physical law (of a space-time theory) must express an *intrinsic* or geometric relationship among physical quantities. Indeed, modern differential geometry, which was founded by Élie Cartan, employs a notation which goes one step beyond that of the classical tensor analysis of this book in so far as tensor indices and, *a fortiori*, coordinates are simply suppressed. All that appears is an expression of the essential geometric truth which has nothing to do with coordinates. In many ways, this approach is analogous to that of classical vector analysis. For example, an elementary law of electrodynamics asserts that the magnetic field has zero divergence, which is to say, that it has no sources. It may be expressed in the form

$$\nabla \cdot \mathbf{B} = 0. \tag{10.5}$$

This is a putative truth which makes no reference to coordinate systems. It is an example of an intrinsic statement. However, in order to *measure* the field one is obliged to select a frame of reference which has been appropriately coordinatized. Then the form of (10.5) may be vastly altered and complicated in accordance with one's more or less arbitrary choice. We conclude that the requirement of general covariance is to be interpreted as signifying that a law of nature must express a geometric relationship. The spirit of the requirement may be satisfied by the notation of classical tensor analysis provided that it be remembered that the tensor indices, which do refer to coordinate-dependent components, should not be interpreted to refer to any *particular* system of coordinates.

A somewhat related difficulty is that the covariance group of any law—that is, the group of transformations of which that law is an invariant expression—may be widened by the simple expedient of adding additional terms to that law. A simple illustration of this is provided by the case of Newtonian mechanics, which is usually regarded as being covariant with respect to the relatively narrow group of inhomogeneous Galilean transformations. As you know, this entails, *inter alia*, that a measurement which is performed in a uniformly rotating frame of reference will fail to be in conformity with $\mathbf{F} = m\mathbf{a}$, in virtue of the presence of inertial forces, in this case the centrifugal and Coriolis forces. However, it will always be possible to preserve the *form invariance* of the law of interest in a formal or vacuous way.

Consider the equation of motion of a point-particle with respect to a frame of reference which has both linear and angular acceleration with respect to an inertial frame. Let the position vector of the particle be $\mathbf{r}$ in the non-inertial frame. A_0 will be the linear acceleration of the origin of the non-inertial frame with respect to the inertial one and $\boldsymbol{\omega}$ is the angular velocity of the former with respect to the latter. Accelerations and velocities of the particle with respect to the non-inertial frame are enclosed in square brackets. A tedious but elementary calculation, the details of which are not important, yields the following "general" equation of motion.

$$m\left[\frac{d^2\mathbf{r}}{dt^2}\right] = \mathbf{F} - m\mathbf{A}_0 - 2m\boldsymbol{\omega} \times \left[\frac{d\mathbf{r}}{dt}\right] - m\left[\frac{d\boldsymbol{\omega}}{dt}\right] \times \mathbf{r} - m\boldsymbol{\omega} \times (\boldsymbol{\omega} \times \mathbf{r}) \qquad (10.6)$$

The second term on the right-hand side of (10.6) is due to the linear acceleration, while the succeeding terms are respectively the Coriolis, transverse and centrifugal forces. (10.6) is the expression for the motion of a particle with respect to an arbitrary frame of reference. Hence, its covariance group is much wider than the Galilean group. If one happens to choose an inertial frame, then all of the terms with the exception of the first will simply vanish and one will be left with the standard Galilei-covariant form of Newton's law of motion.

We must now face the question as to what is to prevent one from reformulating the law of motion along the lines of (10.6). To do so would certainly not affect the outcome of any computations, since as we choose "better behaved" frames of reference, the various inertial terms in (10.6) will simply vanish. There are various methodological stratagems which one might employ for the purpose of rejecting(10.6). The most obvious to the philosopher is by an appeal to simplicity. However, the criterion of simplicity no longer enjoys the respect with which it was once held. In the first place, there are so many diverse meanings of the term that it turns out in practice to be hopelessly vague. In the second place, it seems that nature is sometimes generous, in the sense of being uncomplicated, but not invariably so. In short, simplicity seems to be an instance of a textbook criterion, which is adapted to the wisdom of hindsight but is unreliable as a heuristic guide. An alternative approach to (10.6) would be to declare that any meaningful term in a physical law must be non-vanishing. Again there would seem to be little warrant for such a tactic. As every student of physics knows, there are laws in the form of equations of continuity which are of profound physical significance but which assert that a certain quantity vanishes identically.

The truth of the matter appears to have been recognized by Bunge, when pointing out that a covariance requirement is applicable only to the high-level or basic laws of a theory but not to all of their deductive consequences (1967, pp. 87 ff.). For example, while a certain differential equation may be an invariant of a group of coordinate transformations, its various solutions will not in general be so since they are tied to particular frames of reference. Now the difficulty with (10.6) is that each of the "inertial" terms contains a covert reference to a particular frame of reference. Thus the law cannot be deemed to express a fundamental or potentially global property of physical reality. What it does assert is a frame-dependent state of affairs. The shortcoming which we find in (10.6) is directly related to the frequently asserted methodological fiat that a law of nature should not contain any essential occurrence of a proper-name. It may be argued that (10.6) does contain hidden occurrences of the name of a frame of reference, at least in the sense of an "ambiguous name". (Cf. Suppes, 1957, for the significance of ambiguous names.) We conclude that whenever we speak of the covariance group of a theory, we must have in mind the covariance group of its fundamental laws. As our example illustrates, in the absence of some such proviso, the property of covariance with respect to a given group of transformations is trivialized.

We now come to a much more serious and perplexing issue. Since the very inception of general relativity, the question has been raised as to whether the principle of general covariance has any factual content. As we remarked in an earlier chapter, Einstein clearly thought that it did, on the ground that it led to new physics. That is to say, that in forcing a law into the mold of general covariance, one is obliged to modify it, and such modifications lead to new predictive consequences such as the correct value for the precession of the perihelion of Mercury. Kretschmann was the first to question the alleged factual status of general covariance on the ground that any theory including Newtonian mechanics can be expressed in generally covariant form. In fact, to make a Lorentz covariant theory generally covariant is an almost trivial undertaking which consists in replacing the Minkowski metric tensor by $g_{\mu\nu}$ and ordinary derivatives by covariant derivatives (commas replaced by semicolons). It is undeniable that the formal steps are extremely simple, given the appropriate mathematical instruments. However, it does not follow that the modification which thereby accrues to a physical law is devoid of physical significance. We have already seen that we arrive at many significant results, whether true or false is another matter, by taking a law of Newtonian mechanics and replacing the various three-vectors by four-vectors, etc. Few physicists would doubt the factual status of Lorentz covariance and Einstein perceived general covariance to be a generalization of the latter.

The most forceful critic of the Einsteinian view of general covariance is the eminent theoretical physicist, V. Fock. In his important treatise, *Space, Time and Gravitation*, he goes so far as to claim that there is no valid principle of general relativity but merely a generally covariant theory of gravitation. It is therefore appropriate to examine Fock's anti-Einsteinian viewpoint with considerable care. The thrust of his criticism is simply that the intimate link which Einstein supposed to exist between general covariance and general relativity is based on a confusion. However, what is striking about his rejection of the principle of general relativity is that it is based not on empirical grounds but on the strength of purely mathematical or *a priori* considerations.

On the one hand, Fock argues for a fairly conventional interpretation of a principle of relativity. Any such principle is simply interpreted to assert that if a certain sequence of events takes place in an inertial frame of reference, then a sequence in a second frame which is in uniform motion with respect to it which has the same initial state as the first sequence will be a copy of the latter in the sense that the same mathematical functions which serve to describe the first sequence will also describe the second. In short, the laws of nature are the same in the two reference frames. On the other hand, Fock argues less perspicuously that the very existence of a principle of relativity is a manifestation of what he calls the "uniformity" of spacetime.

By the uniformity of a space, Fock simply means its degree of symmetry. Thus, a space possesses uniformity if the interval or distance between two of its points is invariant with respect to a given group of transformations. He then cites the theorem that in an n-dimensional space, the largest group which can leave the separation invariant will be one which involves a maximum of $\frac{1}{2}n(n+1)$ parameters. If such a group exists, the space is said to have maximal uniformity.

The spacetime of physics is four-dimensional. Hence, its maximal uniformity requires the existence of a ten-parameter group. The homogeneity of spacetime or the equivalence of all points or events is related to the four-parameter transformation of displacing the origin of the coordinate system with respect to the three spatial and the one temporal axis. The isotropy of space or equivalence of all directions corresponds to the three-parameter group of spatial rotations. Finally, the equivalence of inertial frames corresponds to the three-parameter group of frame transformations representing the three components of relative velocity. Since Minkowski spacetime is of maximal uniformity, Fock concludes that the principle of special relativity, which is the physical embodiment, so to speak, of the Lorentz group, cannot possibly be further generalized.

Before proceeding, we should pause to clarify the notion of the uniformity of a manifold. In fact, the idea is quite simple. Consider the two-dimensional surface of a sphere. If it were of maximal uniformity, it would be invariant with respect to a three-parameter transformation group. Now the sphere is transformed into itself by a rotation about any central axis. But this is indeed a three-parameter group corresponding to three angles of rotation. We conclude that the spherical surface does possess maximal uniformity, i.e. symmetry. Now consider an arbitrary surface of revolution, which is a surface obtained by rotating an arbitrary curve about a fixed axis. The only transformation which carries it into itself is a rotation about the axis of symmetry. This corresponds to a one-parameter group. Hence, the surface of revolution has less uniformity than a spherical surface. Finally, an arbitrary or general surface has no uniformity.

Fock makes the point that the spacetime of the theory of gravitation is completely non-uniform. There is no transformation that carries the manifold into itself. Hence, there can be no principle of relativity associated with it. Hence, "general relativity" far from being a factual truth is actually a contradiction in terms! Einstein's theory of gravitation is far less relativistic than the theory of special relativity.

Fock's analytic argument is to the effect that the invariance of the space-time interval under the group of diffeomorphisms is only achieved by making $g_{\mu\nu}$ a dynamic variable. In the case of Minkowski spacetime, we may write

$$(ds)^2 = L_{\mu\nu}\,dX^\mu dX^\nu = \bar{L}_{\mu\nu}d\bar{X}^\mu d\bar{X}^\nu \tag{10.7}$$

where $L_{\mu\nu}$ is the same function of X^α that $\bar{L}_{\mu\nu}$ is of $\bar{X}^\alpha$. On the other hand, in the theory of gravitation the counterpart is

$$(ds)^2 = g_{\mu\nu}dX^\mu dX^\nu = \bar{g}_{\mu\nu}d\bar{X}^\mu d\bar{X}^\nu \tag{10.8}$$

where it is *not* in general the case that $\bar{g}_{\mu\nu}$ is the same function of $\bar{X}^\alpha$ that $g_{\mu\nu}$ is of X^α. In effect, invariance is achieved at the cost of non-uniformity. By the same token, any law which involves $g_{\mu\nu}$ could not, therefore, qualify as relativistic in the usual sense.

The requirement that Fock would impose on relativistic laws—namely, that they be functionally isomorphic under coordinate transformations of a given type—is actually very stringent. In fact, such isomorphism is in general only possible with respect to a group of *linear* transformations. Thus, Fock's mathematical interpretation of the principle of relativity implies that any such principle be associated with a group of linear transformations. For example, a comparison of (10.3) and (10.4) reveals that the non-linear transformation from Cartesian to spherical coordinates gives rise to non-isomorphic representations of spatial distance. Conversely, were we to employ general curvilinear coordinates, the Lorentz transformation would not be represented as a linear transformation. Now Fock would properly point out that in the context of special relativity, the non-linearity of the Lorentz transformation which results from the use of curvilinear coordinates is a trivial matter which has nothing to do with the structure of spacetime. But in gravitation theory, the non-constancy of the $g_{\mu v}$ is the unavoidable heart and soul of the matter. To put Fock's argument as succinctly as possible, a principle of relativity holds only if the mathematical form of physical laws is the same in a well-defined class of reference frames—the inertial frames. Einstein's theory of gravitation involves non-linear frame transformations with resultant changes in the mathematical form of laws. Hence, there can be no general principle of relativity associated with the theory of gravitation.

It should be reasonably obvious from what has been said that Fock is prepared to rule out general relativity on *a priori* grounds. However, while Fock's formal reasoning may be correct, it fails to establish the factual thesis that it is not the case that sequences of events may be mutual copies in frames of reference which are mutally accelerating. As to the factual issue, it is certainly true that over a *finite* region there is a distinct difference between a gravitational field in the conventional sense and an inertial field which results from the selection of an accelerating frame of reference. In the latter case, the Riemann–Christoffel tensor vanishes, which is not possible in the former case. However, the *basic* laws of general relativity are *local* laws. According to this theory, it is not possible to make a local distinction between the two kinds of fields. Locally, they are of a piece, in so far as the behaviour of a test particle is described by the components of the affinity or Christoffel symbol of the second kind. This law of motion is valid for both "real" and "pseudo" gravitational fields. Locally, they may be transformed away in either case. This, of course, is due to the non-tensorial nature of the Christoffel symbol. It is in this sense that Einstein's theory may be said to be generally relativistic. Similarly, the field equation holds true, according to the theory, in any physically realizable frame of reference. It is not the case that either the law of motion or the field law is valid for a distinguished set of reference frames but not for others. In this sense, too, the theory admits of a principle of general relativity.

The situation is not altogether different from that of the theory of electromagnetism. In the latter theory, a charge at rest will produce a pure electrostatic field. However, when viewed from a uniformly moving frame, the charge is in motion and, thus, constitutes an electric current which is the source of a magnetic field. Thus, there are cases in which magnetic fields may be made to come and go by a Lorentz transformation. On the other hand, there are cases in which the magnetic field cannot be transformed away. But in all cases, the very same laws of electrodynamics are employed to describe the field. However, the distinction between the electric components of the field and its magnetic components is not held to be absolute. The underlying reality is the coordinate-free electromagnetic field which will appear with one set of components or another in accordance with our choice of coordinate system. It seems to me that similar considerations might apply to the gravitational field. In one frame of reference certain components of the Christoffel symbol will be non-vanishing and represent an inertial field. In another frame of reference other components will represent a real gravitational field. Einstein's point is simply that there is no preferred set of representations and, consequently, no preferred set of reference frames.

There may be good reason for preferring the rest frame of the earth when describing the earth's gravitational field as there is good reason for describing the field of an isolated electric charge in its own rest frame, but neither choice pertains to a globally preferred frame of reference.

Although we reject Fock's rather intemperate dismissal of "general relativity", we do find ourselves obliged to admit that the fact that special relativistic theories, not to speak of Newtonian mechanics, can be recast in generally covariant form raises a perplexing conceptual problem. If Newtonian mechanics may be formulated in a generally covariant way, its laws, as reformulated, must hold with respect to all coordinate systems and, *a fortiori*, with respect to all frames of reference. Hence, Newtonian mechanics, in one of its versions, must admit a principle of general relativity! To avoid this unwanted conclusion would appear to call for a more careful analysis of the role of covariance and its relationship to relativity.

A distinctive line of attack was taken by Mario Bunge (1961). Every student of philosophy should be aware of the fundamental distinction between an object language and a metalanguage. Suppose that S is a statement which is formulated in the language L. If one now wishes to make a statement about S or about L as a whole, one must resort to a second language L_m which is the metalanguage with respect to which L is the object language. But, in turn, a statement about L_m would belong to a different metalanguage, with respect to which L_m would then be an object language. For our purposes, however, we shall treat the two levels of discourse as absolute. The object language will be the language of physics in which we formulate theories which refer to the objective world. The metalanguage will be that language in which we speak about scientific theories and their various constituents such as laws. Bunge is particularly interested in statements about laws, which he calls metalaw statements. This is a very wide class of statements which includes such disparate elements as "Maxwell's equations are invariant under space inversion" ("Laws of Physical Laws", *Am. J. Phys.* **29**, 519) and "Electricians employ Kirchhoff's laws" (*ibid.*).

Of particular interest to philosophers and physicists is the subset of *lawlike* metalaw statements, for which Bunge has coined the expression *metanomological statements.* He proceeds to subdivide these into various categories but the details are not essential to our present needs. The key point of Bunge's analysis is that it is of fundamental importance to maintain the sharp distinction between laws about the world and laws about scientific theories, i.e. our modes of representing the world.

"As can be seen, the mere awareness of the existence of metanomological statements constitutes an effective reminder that not every scientific statement refers to facts" (*ibid.*, p. 521).

Bunge exploits this analysis to make the profoundly important point that the various principles of covariance which occur in physical theories strictly belong to the corresponding metatheory. For example, the statement that the laws of Newtonian mechanics are Galilei covariant is not a statement about the physical world but one which refers to certain mathematical features of a theory which we have formulated to describe the world. Hence, it would be inappropriate, for example, to perform an experiment to test Newton's laws for Galilei covariance. That they possess such a property is to be determined with the aid of paper and pencil. In short, a covariance principle is a mathematical metatheorem.

We find Bunge's view concerning the status of covariance principles to be both attractive and compelling. However, we disagree with his further contention that principles of relativity are likewise metanomological. Bunge asserts, we believe erroneously, the equation: "general covariance = general relativity". However, it is no easy task to formulate a principle of relativity in such a way that it is clearly a part of the object theory. Still, we would contend that a relativity principle is experimentally testable. Moreover, any such principle is committed to the existence of at last two frames of reference, whereas a mathematical metatheorem is committed to nothing beyond the realm of mathematics.

It is certainly helpful to apply Bunge's thesis on covariance to our present predicament. It obviously helps to lessen the force of the dilemma raised by the generally covariant formulation of Newtonian mechanics. We may now argue that the latter implies nothing about the hypothetical world which that theory purports to describe but only refers to the mathematical formalism in which one has chosen to represent it. However, it would be simplistic to suppose that our problem has been resolved. In spite of our agreement with the contention that covariance principles are formal rather than factual, there still remains the fact of a close relationship between Galilei covariance and classical relativity, Lorentz covariance and special relativity and general covariance and general relativity. Bunge's thesis merely points the way to a possible solution. The precise relationship between a covariance principle and its associated relativity principle remains to be elucidated.

The essence of the problem at this stage is as follows. A covariance principle asserts that the basic laws of a theory have a mathematical form which is unaffected by any mathematically admissible coordinate transformation. On the other hand, a relativity principle is an assertion to the effect that the physical content of the basic laws is independent of the frame of reference and will therefore be unaffected when a transformation is carried out from one frame of reference to another. Formally, the scope of a covariance principle is much wider than that of a relativity principle. While there are uncountably many coordinate systems, there are only finitely many frames of reference. A working distinction between a coordinate transformation and a frame transformation is that the transformation coefficients in the latter case will vary with time. Thus, a *pure* coordinate transformation in two dimensions representing a rotation of axes through the angle θ will have the following transformation matrix.

$$C_{ij} = \begin{pmatrix} \cos\theta & \sin\theta \\ -\sin\theta & \cos\theta \end{pmatrix}. \tag{10.9}$$

However, if one wishes to represent a transformation to a second reference frame which is in rotational motion with respect to a given frame, the transformation matrix might be

$$C_{ij} = \begin{pmatrix} \cos\theta t & \sin\theta t \\ -\sin\theta t & \cos\theta t \end{pmatrix}. \tag{10.10}$$

While the transformation corresponding to (10.9) is physically trivial, (10.10) models a physically significant situation. Accordingly, we adopt the rather obvious conclusion that *formally* speaking frame transformations are a proper subset of coordinate transformations. It was precisely such reasoning that led Einstein to favour a generally covariant formalism, i.e. the tensor calculus. He held that if any law is invariant with respect to the entire group of coordinate diffeomorphisms it must, *a fortiori*, be covariant with respect to the group of frame transformations which is a subgroup of the diffeomorphisms. In short,

Einstein appears to have held that general covariance *implies* general relativity. By the same token, it could well be concluded that Lorentz covariance implies special relativity and that Galilei covariance implies classical relativity. However, if we accept this, we are thrown back on our original difficulty. For what we have just said is at variance with our earlier contention that covariance is a metatheoretical property. There is simply no way that a matter of fact is deducible from a mathematical metatheorem. Hence, if relativity principles are deducible from covariance principles either the former must be formal or the latter factual.

In fact, we accept neither horn of this dilemma. We shall try to show that the argument which led to it is plausible but incorrect. It should by now be obvious to the reader that the covariance or transformation properties of a law statement have little to do with the semantic interpretation or physical content of that law. For example, any well-formed four-vector equation, regardless of its meaning, may be declared to be Lorentz covariant. However, when one is concerned with a fully interpreted physical law as distinct from its mathematical skeleton, one must bear in mind that the selection of a particular type of geometric object to represent a certain physical quantity involves a definite factual commitment. Thus, a quantity which has been treated as a scalar in pre-relativistic physics may turn out to be a component of a four-vector in relativistic physics. Hence, that a law has certain covariance properties does not entail that it must satisfy a particular principle of relativity. Such an entailment would only be warranted if, *per impossibile*, the law were incorrigible. What is, however, the case is that when a law has been endowed with certain covariance properties, the existence of an appropriate principle of relativity may be strongly suggested. It is then the task of experimental physics to determine whether such a relativity principle can be confirmed. An example may help to clarify our point. From the *true* statement that Newtonian mechanics is Galilei covariant, it does not follow that the laws of mechanics obey the principle of classical relativity, since the latter is factually false, as revealed, for example, by the subadditive character of the three-velocity. By the same token, from the *true* statement that Cartan's formulation of Newtonian mechanics is generally covariant, it does not follow that there is a principle of general relativity which is satisfied by that theory, although it is the case that Newtonian mechanics becomes a *candidate* for inclusion in general relativity. That is to say, that a general-relativistic Newtonian mechanics is not a logical impossibility. It is not the case that Einstein was justified in claiming that his theory of gravitation was generally relativistic because he had succeeded in expressing it in generally covariant form. On the contrary, because he believed that a correct theory of gravitation must be generally relativistic, he supposed himself to be justified in writing it in tensorial form. That his theory of gravitation is generally covariant is undeniable; that it does, in fact, satisfy a principle of general relativity is for experiment to decide.

Although it is undeniable that the tensor calculus, in one or another form, is admirably suited as a mathematical instrument for expressing the laws of a generally relativistic theory, we suspect that undue preoccupation with general covariance has led to philosophic excess in two directions. On the one hand, there are those who seem to treat it as a philosopher's stone while overlooking its purely formal character. On the other hand, there are those who regard it as so trivial that they are led to reject the principle of general relativity with which they may tend to equate it. We wish now to argue for what we would commend to the reader as a more balanced view of general covariance which should lead to neither type of excess.

A generally covariant equation is one which has its form preserved under any mathematically admissible transformation. By now it should be unnecessary to emphasize that this is a claim which may only be applied meaningfully to mathematical expressions but not to the world. Nature cannot be subjected to coordinate transformations. Indeed, it is not even meaningful to speak of transformations of the space-time manifold. All that may be transformed is our coordinate-dependent description of that manifold.

However, it remains true to say that one must not lose sight of the fact that generally covariant equations purport to represent reality. In particular, one must remember that our coordinate systems are intended to model physical frames of reference. Likewise, many coordinate transformations are intended to model physically significant frame transformations. Moreover, what is mathematically trivial may be physically significant once the mathematics has received a certain interpretation. For example, one of the basic equations of electrodynamics in the flat or gravity-free manifold of special relativity may be written in the following Lorentz covariant form.

$$F^{\mu\nu}{}_{,\sigma} + F^{\nu\sigma}{}_{,\mu} + F^{\sigma\mu}{}_{,\nu} = 0. \tag{10.11}$$

We render (10.11) generally covariant by the "trivial" operation of replacing the commas (partial derivatives) by semicolons (covariant derivatives). But the resulting equation now implicitly contains several Christoffel symbols which relate to the gravitational field. Thus, the modified equation indicates how the electromagnetic field is influenced by gravitation. Now it would be a profound mistake to suppose that the generally covariant equation must be true if (10.11) is true. However, it is a significant physical hypothesis, which is of greater generality than (10.11). Hence, while the achieving of general covariance is mathematically trivial, the results which accrue therefrom may be highly suggestive. Much of the basic confusion *vis à vis* covariance stems from the tendency to overlook the great difference between an item of pure mathematics and a physical hypothesis which is expressed in mathematical form.

The tendency to overlook the physical interpretation of mathematically formulated laws has also resulted in the opposite tendency—namely, to over-emphasize the role of general covariance. The fact is that the tensor formalism is remarkably permissive. The range of transformations which are permitted by it extends far beyond what any physicist would dream of carrying out, since their physical interpretation would conflict with various basic physical requirements. Accordingly, once the laws of relativity physics have been given a generally covariant formulation, they must be supplemented by the so-called reality conditions which limit the possible space-time transformations. In short, what is permitted by the syntax of the theory, i.e. its mathematical structure, may be ruled out as dynamically unacceptable.

The reality conditions, pertaining to standard types of reference frames, which must always be satisfied by the transformed coordinate variables are the following. The X^i must be spatial coordinates while X^0 is temporal. Furthermore, for two points, $P_1(X^\mu)$ and $P_2(X^\mu)$: if $P_1(X^i) = P_2(X^i)$ and $P_1(X^0) \neq P_2(X^0)$, then $P_1(\overline{X}^0) \neq P_2(\overline{X}^0)$. Moreover, the temporal order of P_1 and P_2 is fixed. For an infinitesimal temporal interval, the points will have the temporal coordinates X^0 and $X^0 + dX^0$ such that $g_{00}dX^0dX^0 > 0$, from which it follows that $g_{00} > 0$ and $\overline{g}_{00} > 0$. Similarly, it can be shown that $g_{ik}dX^idX^k < 0$ and $\overline{g}_{ik}d\overline{X}^id\overline{X}^k < 0$. Any transformation which violated one or more of those conditions would be prohibited on physical grounds. Such restrictions prohibit any transformation which models a transition to a *physically unrealizable* frame of reference. A physically realizable frame of reference is one which, among other things, is not necessarily rigid but is such that no material point of it is moving with respect to another with a velocity in excess of *c*. Now if "general relativity" is of real significance, it must surely mean that all physically realizable frames of reference are, in some sense, equivalent.

We have used the qualification "in some sense" because we are obliged to admit that there is an inescapable difference between a general transformation and a Lorentz one. In particular, the transformation coefficients of the latter are constant, so that the transformation affects the entire system in the same way. On the other hand, the coefficients of the general transformation are functions of the coordinates. Consequently, their effect on one part of the system will be different from that on another part. This again, suggests that the general equivalence of frames is something which requires a *local* interpretation.

The most careful and balanced account of the entire issue is the one provided by J. L. Anderson in his *Principles of Relativity Physics*. At the very outset, Anderson points out that from the indistinguishability of the various points of the space-time manifold it follows that there can be no unique coordinatization of the manifold and that no result of coordinatization should depend on the particular set of coordinates that is adopted. This he calls the requirement of *coordinate covariance*. He mentions it but once and clearly regards it as both obvious and trivial. Yet it is worth remarking that it is precisely this sense of covariance that has been analyzed by Fock and other writers.

Every space-time theory of physics must specify the geometric objects which are to be associated with the physical quantities of interest. The components of the geometric objects may assume various sets of values. Any such set of values is called a *kinematically possible trajectory* (*kpt*). The laws of the theory determine the subset of the *kpt* which may be realized in fact. Any member of this subset is called a *dynamically possible trajectory* (*dpt*). A given group of transformations *G* is said to be a covariance group of the theory if two conditions are satisfied. Firstly, the *kpt* must be a faithful realization of the group. That is to say, that each element of the group, i.e. each individual transformation, must associate a *kpt* with a *kpt*. Secondly, and more important, each *dpt* must be associated with another *dpt* of the theory. This means that the covariance group of a theory depends not only on the type of geometric objects but also on the way in which those objects interact or are coupled. Furthermore, whether *G* is a covariance group of a theory *T* is something which is testable. Given a certain *dpt* one may carry out transformations by changing the position or motion of the observer to determine whether the corresponding transformed *dpt* actually exists. Moreover, Anderson is able to show that such a notion of covariance places a considerable restriction on the possible dynamic laws which are compatible with a certain set of geometric objects. Finally, the *dpt* of *T* are partitioned into equivalence classes. Two *dpt* are equivalent if and only if they are associated by some element of *G*. It is such equivalence classes of *dpt* that correspond to the *intrinsic states* of the system.

Although this form of covariance, which might be called the *covariance of laws* as distinct from the physically trivial coordinate covariance, has some physical significance, it does not reach as far as that which Anderson terms the *invariance of laws*. The notion of an invariance group of a theory includes what we have called a relativity principle. Roughly speaking, while covariance preserves the *form* of a law, invariance preserves its *content*. Formally, Anderson divides the geometric objects into those that are affected by the state of the system, which he calls *dynamical objects*, and those that influence the system but are not affected by it. The latter are the absolute objects of the theory. For example, ds, t and δ_{ik}

are absolute objects of Newtonian mechanics; c and $L_{\mu\nu}$ are absolute objects of special relativity. Moreover, the absolute objects would appear in every equivalence class of *dpt*. Anderson defines the *symmetry group of a physical system* as the largest subgroup of the covariance group which is simultaneously the symmetry group of its absolute objects. An element of a group G is a symmetry of a geometric object $Y(x)$ if $Y'(x)$ is its transform under that element and $Y'(x) - Y(x) = 0$. The totality of such elements will constitute the symmetry group of Y. As a vivid instance of the distinction between covariance and symmetry, the Lorentz covariance of electrodynamics could be imposed on the basic laws even though c were not a constant, so long as the transformed equations looked like their original under a Lorentz mapping. However, Lorentz invariance would go further in requiring that c be a Lorentz invariant of the system, i.e. a symmetry of the system.

Anderson draws two important conclusions from the foregoing analysis. The first is that the general group of coordinate transformations, what we have called the group of diffeomorphisms, which he calls the manifold mapping group, is the covariance group of all space-time theories. However, that is not to say that general covariance is thereby trivialized. The argument of Kretschmann to the effect that any lawlike statement of physics may be made generally covariant is dismissed by Anderson on the ground that it involves the introduction of mathematical quantities which are not observables, i.e. which cannot be measured. This agrees with our own view except that Anderson states it as an empiricist criterion. We would contend that every element of a theory, whether or not it is observable, must have a physical interpretation. With either of these restrictions, it may be shown that the possible laws of a space-time theory are delimited by the requirement of general covariance. At the same time, it can be shown that the inhomogeneous Galilean group is the symmetry group of Newtonian mechanics and that the Lorentz group is the symmetry group of special relativistic physics.

The second conclusion, which is the one of immediate concern to us, is that the manifold mapping group, the diffeomorphisms, is the symmetry group of general relativity. That is possible simply because general relativity contains no absolute objects. In particular, $g_{\mu\nu}$ is not an absolute object of Einstein's later theory, but a dynamic one which is affected by the distribution and action of matter. We are at last in a position to understand what makes Einstein's theory of general relativity distinct from other space-time theories. It is not that its laws are generally covariant but that its covariance group is coextensive with its symmetry group. Moreover, this is the case not because its absolute objects are unaffected by any element of the group of diffeomorphisms but simply because it lacks any absolute object. Since this is the case, it is impossible to single out a class of privileged reference frames, i.e. those frames with respect to which measurements of absolute objects are unaffected. In this sense, Einstein's theory certainly warrants the name of *general relativity*. It boils down to there being no class of reference frames in which certain quantities (absolute objects) play a pre-eminent role.

Since the ultimate goal of the present chapter is the formulation of the principle of general relativity in terms of the concrete (and controversial) notion of the general equivalence of reference frames, we shall first prepare the conceptual ground with a review of the concept of reference frame in general and inertial frame in particular.

We take as our conceptual starting-point the notion of the total set of reference frames $\{K\}$. As the reader knows, according to the theory of Newtonian mechanics, there is a distinguished subset $\{I\} \subset \{K\}$, the so-called inertial frames. The latter are distinguished by the fact that Newton's fundamental laws of motion hold with respect to them. For example, $\mathbf{F} = m\mathbf{a}$, where $\mathbf{F}$ is a known function, is found to hold when measurements are carried out in any such frame. Furthermore, the set $\{I\}$ is found to be a basis for a group with respect to the Galilean velocity mappings. It is important to note that the linearity of these mappings follows from the mathematical structure of Newton's laws—namely, their being second-order differential equations. If it were the case that forces produced changes in acceleration instead of changes in velocity then an actual acceleration of a frame of reference would not preclude the latter from serving as an inertial frame, in which case the set of mappings between the augmented set of inertial frames would not be linear functions of the coordinates. Parenthetically, it may be remarked that Aristotelian physics maintained that a force produces a change in position. That is to say, that, in modern terminology, the Aristotelian law of motion is of first differential order. If the world were of the form conceived by Aristotle, which is surely a real possibility, a fixed, immutable space would have played a pre-eminent role as a frame of reference. There would be an objective difference both ontologically and epistemically in any such world between rest and uniform motion. The crucial point to keep in mind at this juncture, however, is that the *linearity* of the transformation is not an *a priori* feature but depends on the form of the theory which, in its turn, depends on the putative structure of the world.

Turning now to special relativity, we may define its inertial frames $\{L\}$ in various ways. The obvious one would be as frames in which $F^{\mu} = m_0 dU^{\mu}/d\tau$ is found to hold. Another would be any frame in which the velocity of electromagnetic propagation is the universal constant c. What is of immediate importance is that with respect to the Lorentz mappings $\{L\}$ constitutes a basis for the realization of a group. The question may be raised as to whether $\{I\}$ is identical with $\{L\}$. Is an inertial frame of classical mechanics also an inertial frame of electrodynamics? This question could be put aside with the simple observation that since classical mechanics is false, the inertial frames of classical mechanics are non-existent. However, a consideration of one aspect of the question may still be instructive. It should be clear that the couples $\langle \{I\}, \gamma \rangle$ and $\langle \{L\}, \lambda \rangle$, where γ and λ are respectively the Galilean and Lorentz transformations, are certainly distinct. The result of performing two successive Lorentz mappings is not the same as that of performing two Galilei mappings. There are more or less sophisticated ways of demonstrating this point. However, an elementary and direct way follows from the relativistic law for the addition of velocities. Let I_1 and I_2 be inertial frames with relative velocity $\mathbf{v}$. A particle has velocity $\mathbf{w}$ with respect to I_2. Hence the rest frame I_3 of that particle must have the velocity $\mathbf{v} + \mathbf{w}$ with respect to I_1. Now consider the same arrangement, *mutatis mutandis*, involving L_1 and L_2. In this case, the rest frame L_3 of the particle will have the velocity

$$\frac{\mathbf{v} + \mathbf{w}}{1 + \dfrac{vw}{c^2}}$$

with respect to L_1. Clearly, even though I_1 and I_2 were numerically identical to L_1 and L_2, respectively, L_3 would not be identical to I_3. This simply serves to illustrate that the status of inertial frames, like that of the transformation equations, is a factual matter. It only remains to be said that the linearity of the Lorentz transformation, like that of its Galilean counterpart, follows from the structure of the world and is not given *a priori*.

We now turn to the question as to whether there is a distinguished subset of $\{K\}$, in the context of general relativity, which plays a role analogous to that of $\{I\}$ and $\{L\}$. Many writers have contended that the locally Lorentz or geodesic frames do in fact perform such a role. However, there seems to be little justification for this rather widely held view. Various types of coordinate systems have been constructed to yield mathematically simple descriptions of physical systems, just as Cartesian coordinates are particularly handy in many classical contexts. Any such system of coordinates must be associated with certain measuring procedures whereby numbers are associated with points and events. This is the practical import of the association between a coordinate system and a frame of reference. The latter is often depicted as an idealized scaffolding. But the core significance of a frame of reference is an appropriate set of measuring instruments which have been anchored to a laboratory, radar station or other vehicle.

From what has just been said, it follows that a geodesic frame is little more than a mathematically convenient idealization. Strictly speaking, such frames are only geodesic at their origin. Furthermore, it seems generally to be overlooked that such frames are approximated in Newtonian gravitational theory, in the form of artificial satellites, as much as in general relativity. In spite of these objections, there would be an interesting analogy between geodesic frames and inertial frames if the former happened to be related by some restricted, and preferably linear, set of transformation equations. However, such is not the case. A pair of geodesic frames may be related by any element of the group of diffeomorphisms, depending on the structure of the gravitational field. That is to say, that, in general, two Einstein cages will undergo a mutual acceleration due to the inhomogeneity of the field.

To be consistent in the use of the term "inertial frame", we must require that an inertial frame of general relativity be any frame in which the basic laws of motion hold. But in the case of general relativity, such a frame is one in which

$$\frac{d^2 X^\mu}{d\tau^2} + \begin{Bmatrix} \mu \\ \nu\rho \end{Bmatrix} \frac{dX^\nu}{d\tau}\frac{dX^\rho}{d\tau} = 0$$

is valid. But as we know, this is the equation of a geodesic of spacetime. Since it expresses an intrinsic geometric "fact", it is valid for all systems of coordinates and, *a fortiori*, for all frames of reference. Those who perceive geodesic frames to be privileged in any but a superficial sense are failing to think geometrically. The trajectory of a test particle either is or is not a geodesic and no manipulation of coordinate systems or choice of reference frame can or should have any effect on the matter. Accordingly, we are led to conclude that there is no privileged subset of $\{K\}$ in general relativity. Thus, we are led to assert that there is a general equivalence of reference frames in that theory. One has the choice of claiming either that all frames are inertial or that no frames are; the matter is trivially terminological.

The most telling argument against the general equivalence of frames is to the effect that there is an objective difference between a frame of reference with respect to which a real gravitational field is detected and an accelerating frame with respect to which a pseudo-gravitational field is detected. It is argued that in the one case the Riemann–Christoffel tensor will vanish, whereas in the other it will not. Now since the vanishing or non-vanishing of a tensor, even *locally*, is an intrinsic matter, it follows that there must be an intrinsic difference between the two frames such that the relativity of motion cannot be sustained. In fact, all that this objection demonstrates is that Einstein's original Machian justification for the general equivalence of reference frames was somewhat confused. His argument is to the effect that an observer is unable to distinguish between the effects of a force which is due to an acceleration and those of a force which is due to a gravitational source. This was then coupled with Mach's rather obscure conjecture that the former, like the latter, are due to the distribution of matter. Einstein was, of course, aware that his argument only applied locally. Nevertheless, it seemed to confirm Mach's empiricist claim that all motion, both uniform and accelerated, is non-absolute. Now the force of the objection which presently confronts us is that there are even *local* differences between two such frames—namely, that the Riemann–Christoffel tensor is zero in the one but not in the other. But if this be so, in what sense can it be possible to "transform away" or to "create" a gravitational field, even locally?

This whole subject seems to be a source of conceptual confusion, so let us examine it with appropriate care. Einstein maintained that in a frame of reference which is in free fall in a gravitational field, that is to say which is moving along a geodesic, the effects of gravitation will disappear, at least locally. In mathematical terms, this simply means that with respect to such a so-called geodesic frame, the metric tensor will have constant components at the origin and these may be transformed to $L_{\mu\nu}$. It should be noted that $g_{\mu\nu}$ is not transformed away but is merely reduced to a particular canonical form. However, this being so, the Christoffel symbol *will* be transformed away, since it is constructed from the first derivatives of the $g_{\mu\nu}$. There is no cause for concern at that possibility since the Christoffel symbol, or metric affinity, is not a tensor. What, then, of the Riemann–Christoffel tensor. It is constructed from both the first and the second derivatives of the $g_{\mu\nu}$. It may be said to be a measure of the relationship between the $g_{\mu\nu}$ at a point and the $g_{\mu\nu}$ in a four-dimensional differential volume containing that point. As such, it is a measure of the local inhomogeneities of the $g_{\mu\nu}$ field. Thus, it need not vanish. The beginner may be puzzled by the fact that the second derivative of the $g_{\mu\nu}$ does not vanish even though the first derivative does. However, a simple intuitive analogy should serve to dispel any confusion on this point. Consider a particle which is smoothly accelerating and decelerating. At a particular time it decelerates to zero velocity so that the first derivative of the path vanishes. Nevertheless, so long as the velocity is changing in a smooth way, it will have a non-vanishing acceleration at that point, which is to say that the second derivative does not vanish. If you are still not convinced consider the following numerical example. Suppose that the time-track of the particle is $y = t^5 - t^3 - 2t$. Then

$$\frac{dy}{dt} = 5t^4 - 3t^2 - 2; \; \frac{d^2y}{dt^2} = 20t^3 - 6t.$$

At
$$t = 1, \; \frac{dy}{dt} = 0, \; \frac{d^2y}{dt^2} = 14.$$

It is possible to show that, in general, when $g_{\mu\nu,\lambda} = 0$, $R^{\alpha}_{\beta\gamma\delta}$ will simply reduce to $R^{\alpha}_{0\gamma 0}$. Moreover, in the Newtonian approximation, this may be further reduced to $\nabla^2\Phi$, where Φ is the scalar potential. In conclusion, although the gravitational force may be reduced to zero at a point or along a geodesic, this should not be construed as signifying that the field is literally transformed away but merely that certain of its *effects*, as expressed by the frame-dependent Christoffel symbol, have been made to disappear. The only effect on the Riemann–Christoffel tensor is that certain of its components are made to vanish.

Now that these fundamental points have been clarified, we may return to the original objection. What it amounts to, once again, is that strictly there is no equivalence between a gravitational field and a pseudo-gravitational field. If a real gravitational field exists in a given region, then it will be "seen" to exist from the standpoint of any observer regardless of his frame of reference. On the other hand, there are frames of reference with respect to which certain non-intrinsic effects of the gravitational field will disappear. In such frames, the Christoffel symbol will vanish locally along a geodesic. However, it is important to note that when the Christoffel symbol vanishes in the equation of motion, the resulting equation is still one of geodesic motion, albeit of the type that one would normally associate with flat Minkowski spacetime.

The situation which we have just described is not essentially different from that which is found in special relativity. The special principle does not require that the world look exactly the same from the standpoint of all inertial observers. We may refer again to the frame-dependent *appearance* of the electromagnetic field, which is just one of many obvious examples. What it does require is that the *laws of nature* be the same for all inertial observers. The intrinsic properties of the world cannot vary from one observer to another. That is just to say that the geometric properties are completely objective. But what our earlier discussion revealed is precisely that the laws of general relativity are the same for all possible observers. The intrinsic properties do not vary. Specifically, the Riemann–Christoffel tensor either vanishes or does not with respect to all frames of reference, and, in particular, the Einstein field equation will hold in every frame if it holds in any. The constancy of $g_{\mu\nu}$ at the origin of a geodesic frame is merely superficial. I conclude, therefore, that the principle of the general equivalence of frames makes perfectly good sense in the context of general relativity, even though Einstein sought to justify it on the basis of an incorrect heuristic argument—namely, Mach's principle. What this comes down to, therefore, is that the general equivalence of frames does not entail the general relativity of motion.

The general equivalence of frames has been attacked on various grounds. As we have seen, Fock's criticism implies that a transformation between equivalent frames should be linear. But our earlier discussion would seem to dispose of this on the ground that it is based on the simple prejudice that there are *a priori* grounds for holding that the coordinate transformations relating equivalent frames of reference must be linear. That they are linear in the case of the two restricted principles of relativity is to be regarded as a physical accident. If it could be shown that there is an *a priori* necessity for the laws of motion to be of the second-differential order, our thinking on this matter would doubtless stand in need of revision. A more serious objection is that it is allegedly impossible to establish any *a priori* connection between the space-time description of a physical system and an accelerating frame of reference as one does in the case of an unaccelerating frame. This is due to the various stresses suffered by an accelerating frame, which depends on the particular constitution of that frame. In other words, our measuring instruments will be affected to a more or less unpredictable extent when subject to inertial forces. In fact, Einstein himself was aware of this difficulty, which was one of the motivations for employing general coordinates in regard to which no *a priori* assumptions are required concerning the actual coordinate numbers to be associated with events by this or that frame of reference. The requirement of practical rigidity of the reference frame is abandoned. But that is not to say that space-time descriptions with respect to arbitrary frames are beyond one's control. In the first place, atomic clocks are only negligibly affected by accelerations. Thus, atomic clocks furnish us with a practical method of measuring timelike separations. Moreover, it can be demonstrated that if one is given the timelike separations among events, it is possible to deduce therefrom the spacelike separations among them. A related but somewhat more complex difficulty concerns the matter of associating a natural coordinate system with an accelerating frame. However, such coordinate systems have been constructed which exhibit no pathological properties provided that one does not attempt to extend them to distances which are beyond the order of magnitude of the acceleration. The details are of no particular importance to us. However, an exceptionally clear discussion of the method of construction is to be found in Misner, Thorne and Wheeler (1973, pp. 327 ff.).

Having argued that the general equivalence of frames makes physical and mathematical sense, we may proceed to a formal statement of the principle of general relativity, which is similar in spirit to our formulation of the special principle but necessarily rather more complex. What follows may hardly be claimed to be exhaustive. Furthermore, we do not claim that it is the uniquely correct one. However, it is hoped that it succeeds in capturing most of what is essential.

$Fx = x$ is a physically realizable reference frame.
$Lx = x$ is a fundamental law of nature.
$Hxy = x$ holds with respect to y.
$Dx = x$ is a diffeomorphic mapping.
$Mxyz = x$ maps y diffeomorphically onto z.

$$(\exists u)(v)\{Fu \wedge (Lv \rightarrow Hvu) \wedge (x)(y)[(Dx \wedge Fy \wedge Mxuy) \rightarrow (z)(Lz \wedge Mxvz \rightarrow Hzy)]\}.$$

This statement may be best understood as consisting of two principal parts. The first main conjunct ensures that the principle is not vacuous by requiring the existence of a reference frame with respect to which all fundamental laws hold. The second main conjunct is strictly universal apart from the occurrence of an existentially bound variable. It asserts that any diffeomorphism which maps the given frame onto a diffeomorphic image will simultaneously carry each fundamental law onto a diffeomorphic image which holds with respect to the latter frame.

We shall now make some general remarks concerning the above formulation. Firstly, it should be noted that the principle of general covariance is neither necessary nor sufficient for the principle of general relativity. It is not necessary, since general relativity is restricted to the set of physically admissible frame transformations. It is obviously not sufficient, since it may be satisfied by a law which does not hold, i.e. is not factually the case, in any frame of reference. Nevertheless, general covariance certainly continues to lurk in the background, since it is the easiest and only practical way of arriving at laws which satisfy the principle of general relativity, just as the easiest formal route to the construction of laws which satisfy the special principle is by way of manifest, i.e. four-vector, Lorentz covariance. Indeed, it is difficult to see how the principle of general covariance could have been incorporated while retaining the status of general relativity as an object law of physics. Once again, I would like to repeat the message that when doing the philosophy of physics it is of crucial importance to maintain a sharp conceptual distinction between the mathematical formalism and the physical content of the theory which is under investigation. In the second place, any reference to inertial frames in the form of so-called Einstein cages or geodesic frames is conspicuously absent. It seems to me that despite their obvious mathematical appeal they play no essential theoretical role. A crucial feature of our formulation is that it refers to all physically possible reference frames indifferently. Hence, a main point of the principle is that privileged reference frames are eliminated. In the absence of this feature, it is hard to see what significance could be attached to the "geometrization" of physics.

Finally, we note that the principle refers to *all* fundamental laws. It does not single out the laws of gravitation. In short, we have deliberately kept it sufficiently open that it may incorporate the laws of electrodynamics, which is relatively easy, as well as the quantum-theoretic laws which seems to continue to be a hope for the future. In this regard, we might mention that a distinction is sometimes drawn between the *weak* and *strong* principles of equivalence. The former asserts that locally the effects of gravitation are equivalent to accelerative effects. It is appropriate to those interpretations of the theory of general relativity which treat it as nothing but a theory of gravitation. The strong principle is the very general claim that no local phenomena, gravitational or otherwise, display distinct properties when under the influence of gravitational forces from those displayed as a result of accelerative forces. We obviously are opting for the strong interpretation, which is the only one of particular epistemological significance. Moreover, the distinction seems to be of dubious merit. For example, if the local appearance of the electromagnetic field were to differ depending on whether it was subject to gravitational or inertial forces, this would constitute a prima facie difference between the local effects of gravitation and acceleration. In other words, if the strong principle fails to hold, it is difficult to see how the weak one could be sustained. In any event, we remind the reader that the distinction only applies to the effects of such forces and is not to be construed as the claim that the two kinds of forces, gravitational and pseudo-gravitational, are indistinguishable.

Bibliography for Chapter 10

ANDERSON, J. L. "Covariance, Invariance, and Equivalence: A Viewpoint", *General Relativity and Gravitation*, **2**, 161 (1971).

ANDERSON, J. L. (1967). See reference in Chapter 9.

BUNGE, M. (1967). See reference in Chapter 2.

BUNGE, M. (1961). "Laws of Physical Laws", *Am. J. Phys.* **29**, 518 (1961).

EARMAN, J., "Covariance, Invariance, and the Equivalence of Frames", *Foundations of Physics*, **4**, 267 (1974). A sophisticated argument against the general equivalence of frames.

FOCK, V., *The Theory of Space, Time and Gravitation* (2nd edn.), Macmillan, New York, 1964. A mathematically advanced treatise.

FRIEDMAN, M., "Relativity Principles, Absolute Objects and Symmetry Groups", in SUPPES, P. (ed.) (1973). See reference in Chapter 5. An attempt to clarify and refine Anderson's viewpoint.

HAVAS, P., "Foundation Problems in General Relativity", in *Delaware Seminar in the Foundations of Physics* (ed. M. BUNGE), Springer-Verlag, Berlin, 1967.

HAVAS, P., "Four Dimensional Formulations of Newtonian Mechanics and their Relation to the Special and General Theory of Relativity", *Rev. Mod. Phys.* **36**, 938 (1964).

MISNER, C. W., K. S. THORNE, and J. A. WHEELER (1973). See reference in Chapter 9.

MØLLER, C. (1952). See reference in Chapter 9.

SUPPES, P. (1957). See reference in Chapter 1.

11

Spacetime and Geometry

THE principal aim of the present chapter is to attempt to resolve the fundamental question as to the actual information concerning the geometric structure of the world, i.e. the structure of spacetime, that is conveyed to us through the theory of general relativity. In what sense are we being informed about the structure of physical reality when we are told that spacetime is a pseudo-Riemannian manifold of variable curvature? It can hardly be overemphasized that the question is a deep one whose treatment may lapse into unedifying obscurity when one is less than completely clear as to its precise meaning. It is significant that expert opinion on the mutual relevance of general relativity and geometry ranges all the way from J. A. Wheeler's celebrated slogan: "Physics is geometry", to the no less remarkable comment of J. L. Anderson that ". . . Einstein succeeded actually in eliminating geometry from the space-time description of physical systems by letting the gravitational field take over all its functions" (1967, p. 329).

When two eminent authorities offer such apparently diverse interpretations of so fundamental an issue one may suspect that the problem cannot be resolved by simple recourse to the facts of science but requires, at least as a prolegomenon, a serious conceptual analysis. Accordingly, we shall approach the solution by stages, beginning with an account of the relationship between pure and applied mathematics.

The Distinction Between Pure and Applied Mathematics

The controversy surrounding the problem of the relationship between mathematical theory and physical reality has had a long and varied history from the extreme realism of the Pythagoreans, the somewhat obscure epistemological dualism of Plato to the transcendentalism of Kant and finally to a certain point of view which is current among contemporary philosophers. I shall refer to the latter viewpoint as *formalism*, for the sake of a convenient label, since it has been inspired by certain aspects of the movement of that name which was initiated by the great German mathematician, David Hilbert. The aim of Hilbert and his followers was to establish various metamathematical properties of mathematical theories, such as consistency and completeness, on the basis of purely formal, syntactical features of such theories. This programme may be carried out by first disinterpreting the various primitive, descriptive predicates and terms of the theory of interest. That is to say, that they are replaced by purely symbolic placeholders and variables. Thus, after disinterpretation a relation such as "lies between" will be replaced by an abstract ternary placeholder whose significance is purely syntactical. The resulting theory is an instance of what logicians call an uninterpreted or formal calculus. The construction of such calculi is admirably suited to the purpose of investigating the properties of theories as determined by their logical syntax. Many contemporary philosophers have utilized the notion of an uninterpreted calculus to locate the demarcation between pure and applied mathematics. In brief, the uninterpreted calculus is taken to exemplify a system of pure mathematics. Accordingly, when the predicates and terms are supplied with a semantic interpretation or model, the resulting system is taken to be a theory of applied mathematics. The factual truth or falsity of such a theory will depend in part on the model which one adopts. Thus, one and the same system of pure mathematics may be true or false depending on the choice of model. By the same token, seemingly incompatible systems of pure mathematics may be rendered simultaneously true by supplying them with different interpretations. Obviously, such an approach provides an excellent framework for conventionalism.

Although I do not object in principle to the inclusion of uninterpreted calculi under the general heading of *pure mathematics*, I would argue that in the present context it is misleading to do so. In its usual sense, the distinguishing feature of a pure mathematical theory is not the absence of referents but their purely conceptual character. An excellent example is afforded by point set topology, whose basic referents are various subsets of a universal set, which are characterized as open subsets.

From our standpoint, the characterization of a formal system as a mathematical theory is pernicious only because the proponents of formalism then treat an interpreted mathematical system, pure mathematics in the usual sense, as an instance of applied theory. Our claim is simply that the demarcation between pure and applied mathematics depends on the *kinds* of entities referred to by the descriptive vocabulary of the system. Whereas a pure theory refers to such conceptual entities as point sets, an applied theory refers to what may very broadly be characterized as substantial entities, or idealizations of the latter, such as the points of a rigid body, paths of light-rays, events, etc.

To clarify the relationship between pure mathematical theory and the application of such a theory we shall now construct a rudimentary illustration. Let Σ be a basic set consisting of all the rational points, P_i, on the Euclidean plane. The plane is coordinatized by a rectangular system of coordinates (X, Y). We define a relation R_1 on Σ such that $P_j R_1 P_k$ if and only if the X-coordinate of P_j has a lower value than that of P_k, and we say that P_j is a *predecessor* of P_k. We then introduce a second relation R_2 such that $P_j R_2 P_k$ if and only if neither $P_j R_1 P_j$ nor $P_k R_1 P_j$. We then say that the two points are *equivalent*. We proceed to construct a set of axioms for $\langle \Sigma, R_1, R_2 \rangle$.

(i) $(P_i R_1 P_j) \wedge (P_j R_1 P_k) \rightarrow (P_i R_1 P_k)$	(transitivity of R_1)
(ii) $(P_i R_1 P_j) \rightarrow \sim (P_j R_1 P_i)$	(asymmetry of R_1)
(iii) $P_i R_2 P_i$	(reflexivity of R_2)
(iv) $[(P_i R_2 P_j) \wedge (P_j R_2 P_k)] \rightarrow (P_i R_2 P_k)$	(transitivity of R_2)
(v) $P_i R_2 P_j \rightarrow P_j R_2 P_i$	(symmetry of R_2)
(vi) $(P_i R_1 P_k) \vee (P_j R_1 P_i) \vee (P_i R_2 P_j)$	(trichotomy law).

We now provide a simple theorem whose elementary proof may be left to the inclination of the reader.

(vii) $[(P_i R_1 P_j) \wedge (P_j R_1 P_k)] \rightarrow \sim (P_i R_2 P_k)$.

It may be parenthetically noted that although (i) to (vii) are obvious truths about Σ, they are clearly not truths of logic.

Let us now consider how a scientist might employ such a "mathematical theory" to make claims about some aspect of the world. Suppose that a psychologist is interested in the nature of human preference. As a first approximation to a theory of individual preference, he constructs a psychological model whereby Σ, or rather a proper subset of Σ, is interpreted as a set of commodities, R_1 as the relation of preference and R_2 as that of indifference. The six postulates are now interpreted as statements about individual preferences. For example, the factual counterpart of (ii) is to the effect that if one commodity is preferred to a second by some individual, then the second will not be preferred to the first by the same individual, while the counterpart of (vi) may be interpreted as asserting that an individual is able to rank any two commodities in the order of preference. Having a mathematical theory of preference at his disposal, albeit a most rudimentary one, the psychologist is able to employ it as a basis for prediction. The theory is, therefore, testable. In the course of examining the preferences of a particular test subject, he finds that the individual indicates a preference for steak over chicken, is indifferent with respect to steak and fish while preferring chicken to fish. Since this is contrary to (vii), the psychologist may conclude that the theory is false and stands in need of further refinement.

The picture that emerges from the foregoing example is of applied mathematics as an attempt to *coordinate* a given conceptual structure or mathematical model with the structure of some aspect of the world. The extent to which the conceptual structure fits or reflects the structure of reality will seldom if ever be complete. Moreover, in most cases, the physical model will contain additional postulates which supplement the mathematical ones. Moreover, in certain cases, no mathematical model will be found which allows of a convenient physical reinterpretation or coordination. In that event, a new mathematical model may be invented. The theory of Minkowski space is a case in point.

The point which now deserves to be emphasized is that there is little that is mysterious or problematic about all of this. It would indeed be mysterious had it turned out that such coordinations of physical and mathematical model were always successful. One would then know before the fact that deductively obtained predictions are bound to be correct. One would then be confronted, like Kant, with the perplexing problem of constructing a philosophy to account for *synthetic a priori* knowledge. At the other extreme, it is a logical possibility that every such coordination be a complete failure. This would suggest either that the world is so complicated as to be mathematically intractable or that our mathematical resources stand in need of further refinement and enrichment. On the other hand, it might simply indicate that the creative imagination of the scientist has, so far, failed to penetrate physical reality to the point of discovering a suitable physical model with which to coordinate an existing mathematical structure. In fact, this seems to be the situation from which the social and biological sciences are just now emerging.

We have earlier remarked that general relativity has been characterized as the geometrization of physics. Unfortunately, this familiar dictum is more dazzling than illuminating. If it is intended to signify that Einstein succeeded in replacing a physical theory, namely Newtonian gravitation, with a geometric one, it is certainly false. This is borne out by the fact that Einstein had a partial grasp of the physical model before being acquainted with the theory of Riemannian manifolds. On the other hand, it is suggestive of the particularly close fit that obtains between physical and geometric model. The latter is due to the contingent fact of the equivalence of gravitational and inertial mass, which is the basis for the geodesic hypothesis. In brief, general relativity, from the standpoint of applied mathematics, is the coordination of a physical model of the gravitational field as determined by matter with a mathematical model of a four-dimensional pseudo-Riemannian manifold of variable curvature such that ideal gravitational test particles traverse timelike geodesics and photons traverse null geodesics. The success of this coordination, i.e. the closeness of the fit, possibly exceeds that of any other scientific theory. Consequently, the geometric machinery is a particularly apt instrument for the prediction of physical consequences. Nevertheless, one cannot simply read the physics from the geometric formalism. For example, the spacelike geodesics play no less a *mathematical* role than the timelike ones. Accordingly, the impossibility of spacelike motion does not follow from the geometry but has to be inserted as a physical hypothesis. General relativity vindicates Pythagoras no more than Kant.

The Geometric Conventionalism of Henri Poincaré

Mathematics in the modern sense may be said to have originated in the work of Euclid. His monumental achievement was the invention of the axiomatic method. Prior to this time, the various "truths" of geometry had been treated as isolated or mutually independent. The essence of the axiomatic method in geometry is to demonstrate that the truth of all geometric theorems depends solely on the truth of a proper subset of them, the *axioms*, from which all of the former are derivable. Thus, Euclid succeeded, albeit imperfectly, in reducing the science of geometry to the set of axioms. The problem of the confirmation of the system of geometry was accordingly reduced to the much simpler problem of the confirmation of the axioms. But from Euclid's standpoint, this was no problem at all since the axioms or initial assumptions were deemed to be self-evident or intuitively certain principles.

However, to several of his successors it still appeared that one shadow of doubt was cast on the credibility of Euclid's enterprise. The famous Euclidean parallels postulate was thought to lack the self evident character of the remaining axioms. As the reader probably knows, the parallels postulate may be expressed as the claim that through any point in a plane of a given straight line there is one and only one straight line which does not intersect the given line. It appeared to Euclid's successors that such an assumption marred the perfection of his system in so far as it involves not simply line segments but the entire Euclidean plane extended to infinity. Can it be self-evident that there are lines of infinite length which do not eventually intersect? Accordingly, one of the central themes of post-Euclidean geometry was the attempt to derive the parallels postulate from the remaining axioms. If this could be accomplished, the dubiety of the parallels postulate would cease to be a problem since its truth would then be guaranteed by the truth of the other axioms.

An important step in this general development was first taken by the Italian mathematician, Saccheri (1667–1733), who perceived that the desired theorem should be derivable by the *reductio ad absurdum* or indirect method. Specifically, by assuming a postulate which is inconsistent with the parallels postulate, one should, in conjunction with the other axioms, be able to derive a contradiction. By showing that the remaining axioms together with the denial of the parallels postulate are mutually inconsistent, one demonstrates that the parallels postulate is a deductive consequence of the former.

Such an approach was carefully pursued in the early nineteenth century by several mathematicians and most notably by Lobatschewski, who produced a system of geometry whose axiomatic basis included, in place of the parallels postulate, the postulate that there are at least two lines which do not intersect a given line. This system included many exotic and unexpected theorems. For example, in *hyperbolic* geometry, as it is called, the angle-sum of a triangle is less than two right-angles. Moreover, the angle-sum depends on the area of the triangle. As the area decreases, the angle-sum approaches the Euclidean limit, so that a vanishingly small triangle would be Euclidean. However, despite the strangeness of many of its theorems, the new system was shown to be as internally consistent as the geometry of Euclid. Thus was born non-Euclidean geometry. Its results were soon generalized by Riemann, along the lines which were detailed in Chapter 7, who succeeded in showing that the geometries of Euclid and Lobatschewski were merely two of the infinitely many possible forms of metric geometry.

Riemann's generalization may be interpreted as showing that a system of metric geometry is determined by the choice of metric function $g_{ik}\, dX^i dX^k$, of which there are in principle infinitely many. Equivalently, it may be interpreted as the claim that a system of geometry is determined by a standard of congruence. To say that a line segment AB is congruent to a line segment CD is to affirm that a rigid rod which coincides with AB will, after transportation, also coincide with CD. However, one's mathematical predictions as to the partitioning of point pairs into equivalence classes of congruent intervals depend on the choice of g_{ik}. Moreover, the standard of rigidity would itself appear to presuppose a standard of congruence. Given that AB is congruent to CD according to the choice of g_{ik}, a rod would be regarded as rigid if it were found to undergo no deformation when transported from the one to the other. Thus, it appears that one simply *stipulates* which line segments are congruent, with different stipulations generally giving rise to different geometries.

The possibility of a vast array of logically possible metric geometries posed a profound epistemological problem in the latter part of the nineteenth century. So long as there had been but one metric geometry, namely that of Euclid, it was naturally supposed that it faithfully represented the spatial properties of the objective world. Parenthetically, it should be recognized that it is a mark of Kant's genius to have realized before the discovery of non-Euclidean geometry that such an assumption was problematic. With the development of non-Euclidean geometry, philosophers were faced for the first time with the question: Of the various geometries, which one is to be regarded as the geometry of space? The properties of space were no longer to be regarded as given *a priori* but were to be determined by procedures of discovery.

One of the most celebrated protagonists of the philosophical crisis engendered by the new geometry was the illustrious mathematician, Jules-Henri Poincaré (1854–1912). Briefly, the problem confronting Poincaré is to the effect that although geometric assertions are prima facie factual, there are sufficient philosophic grounds to disqualify them from both the class of synthetic *a priori* truths and that of empirical generalizations. What, then, is their status? To this question, Poincaré offers his famous reply that they are conventions. However, Poincaré's doctrine of conventionalism has sometimes been misrepresented in the literature of philosophy. While it is true that he argued that the propositions of geometry have the logical character of conventions or definitions, which can never be overthrown on experimental grounds, he goes to considerable lengths to account for our geometrical beliefs in terms of their experiential origins. His position is to the effect that although experience can never impose a particular system of metric geometry, the content of experience is such that one geometric system will be adopted as the most natural one for the expression of physical laws, on the grounds of descriptive simplicity and convenience. Moreover, experience is such that Euclidean geometry, although conventional, is the natural choice.

To support this doctrine, Poincare describes possible worlds in which the order of phenomena, in particular the behaviour of transported rods and light-rays, is such that the natural choice, from the standpoint of descriptive simplicity, would be one of the non-Euclidean geometries. At the same time, he argues that even in such hypothetical worlds it would be possible to retain Euclidean geometry by making appropriate adjustments to the laws of physics. Accordingly, it may be seen that Poincaré's geometric conventionalism actually leads to the view that the laws of physics also contain a conventional ingredient, although, again, Poincaré would insist on their experimental origin.

In the final analysis, for Poincaré, the space of pure mathematics is completely amorphous. It makes no sense, in his view, to ascribe congruence relations to the mathematical continuum. Claims about the structure of space which are ostensibly factual are, in fact, assertions not about space in isolation but pertain to the combination of space and the patterns of phenomena, especially to the behaviour of our measuring instruments. Physics and geometry are inseparable. An experimental result which seems to confute the one may always be accounted for by a compensating adjustment in the other.

"Under these conditions does space possess geometric properties independent of the instruments used to measure it? It can, we have said, undergo any deformation whatever without our being made aware of it if our instruments undergo the same deformation. In reality, space is therefore amorphous, a flaccid form, without rigidity, which is adaptable to everything: it has no properties of its own. To geometrize is to study the properties of our instruments, that is, of solid bodies" (1963, p. 17).

In summary, Poincaré's position is to the effect that space is a theoretical construct whose metric properties are decreed by the intellect principally on the ground of descriptive simplicity. It is important that Poincaré's notion of simplicity be distinguished from the more controversial notion of *inductive simplicity* whose definition is a problem, some would argue a pseudo-problem, of inductive logic. Inductive simplicity may be illustrated in terms of the curve-fitting problem. Let us suppose that the totality of empirical data in a given area are represented by points on a graph. A "theory" consistent with those observation points is a continuous curve which is drawn through all of them and which would normally be expressed in the form of a mathematical function, e.g. $y = 3x^2 + 5$. Obviously, through a finite number of points it is possible to draw infinitely many smooth curves. This may be construed as the claim that there is a potential infinity of theories consistent with the data or the facts. Nevertheless, all such theories would be inconsistent with each other. One of the problems of inductive logic is to formulate a criterion of inductive simplicity, which would serve as a selection principle for the choice among these rival hypotheses. Now there are actually serious grounds for questioning the very relevance of this matter to the task of theory construction but the point we wish to emphasize is that this is not the sense of simplicity which Poincaré had in mind. According to Poincaré, the choice of geometry is not a matter of choosing from among several competing or factually incompatible hypotheses but rather that of choosing from among several factually equivalent theories. The difference between two such theories is not a factual one but a purely descriptive one. Accordingly, the choice will be based not on facts but on conventions. The issue as to whether space is Euclidean or non-Euclidean is, therefore, not a matter of fact but one of preference for one or another mode of description. In this sense, Poincaré believed that the greater intrinsic simplicity of Euclidean geometry would ensure its being first in order of preference, but that is not to be construed, in any way, as implying that space has a Euclidean structure.

The Geometric Empiricism of Hans Reichenbach

Since the advent of the theory of relativity, its most painstaking philosophic interpreter in the first half of the century was Hans Reichenbach. He wrote several books and papers on the content and epistemology of relativity theory. The most notable of these are: *The Axiomatization of the Theory of Relativity* (1924) and *The Philosophy of Space and Time* (1928). Although profoundly influenced by Poincaré's conventionalist philosophy of science, Reichenbach was much more careful than the latter in his attempts to distinguish, on the basis of detailed analysis, the conventional from the non-conventional ingredients in scientific theories. Such investigations into the structure of scientific theories led him to conclude that the conventional aspect is not as pervasive as Poincaré had supposed. It is for this reason that I have chosen to represent him, albeit somewhat nominally, as an exponent of geometric empiricism. However, as applied to Reichenbach as distinct, for example, from Russell, this label still represents an attenuated form of conventionalism.

Reichenbach once illustrated an aspect of his philosophy of geometry by means of the following oft-cited parable.

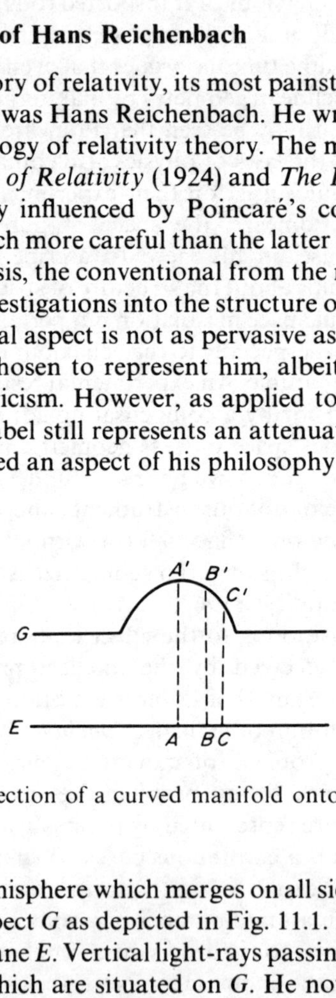

FIG. 11.1. Projection of a curved manifold onto a flat manifold

He considers a great glass hemisphere which merges on all sides into a glass plane. A cross-section would present the aspect *G* as depicted in Fig. 11.1. Parallel to the plane of *G* and underneath it is an opaque plane *E*. Vertical light-rays passing through *G* cause shadows to be cast on *E* of all objects which are situated on *G*. He now supposes that *G* is a world whose denizens determine by simple geodetic measurements that they live on a plane with a hemispherical hump. The measuring rods which are employed to determine the geometry of *G* cast shadows on the surface *E*. Such shadows will obviously be deformed in the central region of *E*. Now suppose that *E* is also inhabited, and that an imperceptible force causes a deformation of the measuring rods of the *E*-people, such that their length is always equal to that of the shadows of their counterparts on *G*. From our external standpoint, the distance $A'B'$ is equal to $B'C'$, whereas the projections AB and BC are unequal. However, in consequence of the aforementioned force, the *E*-people find AB to be equal to BC. In effect, the results of measurements on *E* would be identical to those on

G. Thus, the *E*-people should also conclude that they inhabit a plane with a hemispherical hump. According to Reichenbach, in such a situation, the question as to whether the *E*-people or the *G*-people are right is, strictly speaking, meaningless. "We may just as well say that *G* is the surface with the 'illusion' of the hump and *E* the surface with the 'real' hump. Or perhaps both surfaces have a hump" (1958, p. 13).

Thus far, the moral of Reichenbach's parable would seem to be in precise accord with the conventionalist doctrine of Poincaré. However, Reichenbach would contend that an important methodological ingredient has been neglected in the foregoing account. A scientific theory, he would argue, unlike a system of pure mathematics, requires that its basic concepts be related to the physical world. More precisely, such a theory stands in need of what he calls *coordinative definitions*. Unlike the usual dictionary definition, a coordinative definition does not relate a new concept (*definiendum*) to antecedently understood concepts but establishes a relationship between a concept and a thing. For example, the coordination of the concept of unit length with the standard platinum metre bar in Paris is an instance of a coordinative definition. What is more to the point in the present context is that if geometry is to acquire the status of a system of statements about the world, it too must be augmented by such a semantic interpretation. Such notions as "congruence" and "straight line" must be linked to the physical world by means of coordinative definitions which typically involve material measuring rods and light-rays.

As an example, we may consider the well-known theorem of Euclidean geometry that the angle-sum of a triangle is equal to two right-angles. From the standpoint of pure mathematics, this theorem is simply a logical consequence of the Euclidean axioms (and definitions), no more and no less. The question of its truth is otiose. However, we saw earlier that if the set of axioms is modified, the theorem no longer holds. Moreover, one is free to select any set of axioms subject only to the usual constraint of internal consistency. Up to this point, Poincare and Reichenbach would still be in agreement. However, according to Reichenbach the crux of their difference is that firstly he would argue that the status of geometry is no different from that of any semantically interpreted system and secondly that when supplemented by appropriate coordinative definitions, the aforementioned theorem of Euclid is transformed into a full-fledged empirical statement which may be confirmed or falsified like any other. The correct geometry is then not a matter of convention or subjective preference but one of empirical fact.

There is certainly an aspect of Reichenbach's philosophy of geometry which appears to be unarguable. This is the obvious point that there are, in principle, infinitely many pure geometries or "abstract spaces" and no apparent internal criterion for determining which of them is isomorphic to physical space. No matter how one casts around, there would seem to be available no objective basis for preferring one metric geometry to another. Indeed, if the reader cares to ponder over the question of whether the angle-sum of a triangle is "in reality" equal to two right-angles, he is likely to conclude that the question makes no sense.

Now Poincaré had argued that, as a matter of fact, we are free to select any system of metric geometry and to *stipulate* that it is to be *the* geometry of physical space. He did attenuate this claim by adding that although all of the available choices are logically equivalent, it will be found in practice that Euclidean geometry is preferable not only in terms of its internal simplicity but also in virtue of its leading to the simplest formulation of the laws of physics. To that extent, experience does play a role in shaping our geometric preferences. However, that is emphatically not to say that one geometry more accurately represents reality than another.

Now Reichenbach advances considerably beyond Poincaré in the attenuation of geometric conventionalism. His position is that once one's geometric vocabulary has been endowed with physical significance, it is indeed possible to express various geometric "facts" about the world. However, his position is somewhat subtle and deserves to be considered in detail.

Suppose that measurements of a sufficiently large triangle disclosed an apparent deviation from Euclidicity; let us say that the triangulation of the peaks of three distant mountains revealed an angle-sum of 180.05°. Would this be a sufficient justification for abandoning Euclidean geometry? Reichenbach argues that it would not on the ground that it would be possible to explain the discrepancy by invoking *universal forces* which distort our measuring rods and modify the paths of light-rays. Such a universal force, as distinct from differential forces, is one which affects all bodies in the same way, regardless of their chemical composition, and which cannot be shielded.

Suppose that the measurement of the aforementioned triangle had been carried out by optical means. In effect, one would have been investigating a triangle whose sides are composed of light-rays. Consequently, the outcome of the experiment could be interpreted as signifying not that Euclidean geometry is false but that the paths of light-rays are not Euclidean line-segments, being subject to the distorting influence of universal forces. What was measured, having curvilinear sides, was simply not a Euclidean triangle. In short, one would appear to be saving the geometry by modifying the physics in the manner of Poincaré. However, Reichenbach would argue that such a procedure fails to constitute a genuine alteration of the laws of physics. "The assumption of such forces means merely a change in the coordinative definition of congruence" (1956, p. 133). So Reichenbach now appears to be saying that it is possible to preserve the free choice of geometry by resorting to an appropriate semantic reinterpretation.

Reichenbach's departure from the position of Poincaré goes far beyond the relatively minor points to which we have so far alluded. He proposes that the various ways of describing an objective state of affairs which simply differ in virtue of a semantic interpretation be called *equivalent descriptions.* Let us symbolize the system of Euclidean geometry by G_0 and a given system of non-Euclidean geometry by G_1. Moreover, P_0 will symbolize the standard or normal system of physics while P_1 will stand for a system of physical laws which incorporates non-vanishing universal forces. We may now express Reichenbach's doctrine of equivalent descriptions by the formula: $G_0 + P_1 = G_1 + P_0$. This simply means that the empirical state of affairs may be equivalently described by asserting either that the geometry is Euclidean and there are universal forces which deform both measuring rods and light-rays or that the geometry is non-Euclidean and there are no such distorting forces. According to Reichenbach, the two descriptions of the numerically identical data belong to an equivalence class of empirically equivalent descriptions.

The origin of Reichenbach's view seems to be a strict verificationist theory of meaning in accordance with which two rods may be said to be of equal length only when they are juxtaposed. When one of them is transported to another region of space, one may no longer be said to *know* that they are of equal length and may only continue to speak of them as equal by introducing a semantic convention to that effect. In short, his position seems to rest on the presupposition that any aspect of our language that is not based on the immediately given must be treated as playing a stipulative role.

Reichenbach regards the theory of equivalent descriptions as having originated from the insights of Poincare. However, there is more to be said on the matter. Firstly, he points out that from a given equivalence class of descriptions one may be able to single out a privileged representative. This will be the member that does not invoke universal forces; it is called the *normal system.* Furthermore, the geometry which is incorporated in the normal system will be regarded as the *natural geometry*. Thus, Reichenbach may be seen to accept the view that it is an empirical fact that a particular geometry is the natural one for the description of an aspect of the physical world.

His sharpest departure from Poincaré lies in his contention that there are other equivalence classes of descriptions which are disjoint from the one that we have been considering. Thus, the system which asserts both that the geometry is Euclidean and that there are no universal forces is the normal representative of an equivalence class of descriptions whose referent is a factual state of affairs which is essentially different from the previous one. Reichenbach's complete doctrine may be schematically summarized as follows:

$$G_0 + P_1 = G_1 + P_0$$
$$G_0 + P_0 = G_1 + P_1$$

but

$$G_1 + P_0 \neq G_0 + P_0.$$

A thoroughgoing geometric conventionalist would simply be construed as arguing that regardless of the empirical state of affairs, it will always be possible to find a P_k such that $G_0 + P_k$ will adequately describe it. However in the light of Reichenbach's analysis, such a contention seems to be of little philosophic merit.

I do not propose to labour through an extensive critique of Reichenbach's theory. Its principal weaknesses are much the same as those to which we alluded in the course of discussing the conventionality of simultaneity. Like most verificationists, Reichenbach would appear to harbour a characteristic distrust of theories, which are, virtually by definition, transempirical. It is quite true that theories assert and entail propositions which cannot be known with certainty. No responsible theorist would, in fact, claim certainty for any theory. However, although theories typically go beyond immediate experience, they will always, considered as systematic wholes, be subject to the test of experience. In the final analysis, even transempirical concepts must indirectly withstand the test of experience. In the case of general relativity, the specific parts of the theory that may have to be revised in the light of experience include those elements which are commonly regarded as geometric. Accordingly, once one has advanced beyond the level of philosophic abstraction, one finds it difficult to treat the thesis of the conventionality of geometry with much seriousness.

In spite of his considerable knowledge of theoretical physics, Reichenbach appears to hold a somewhat simplistic view of the methods of experimental physics. He conveys the impression of being sceptical, *in principle*, of the validity of measuring instruments. It is clearly the case that some measuring rods, clocks and similar devices are more reliable than others. The experimental physicist, when employing a measuring instrument to test a theory, will place reliance on that instrument in virtue of theoretical considerations which are independent of the theory which is under investigation. We earlier introduced the concept of a standard clock as one which measures equal intervals of time along its own world-line. But that a given physical clock is a realization of a standard clock is not a matter that is simply stipulated. On the contrary, one may, for example, favour the use of a caesium clock for purposes of testing general relativity on the basis of theoretical considerations which derive not from general relativity but from atomic physics. Now Reichenbach has argued that there will always be a conventionalist move at one's disposal whereby one may discredit any set of measurements independently of considerations pertaining to the physical constitution of the measuring instrument on which those measurements are based. However, once it is admitted that such moves are merely linguistic manœuvres, one may be forgiven for treating them as trivial if not tendentious.

Despite our stated antipathy to Reichenbach's philosophical position, we recognize that he performed a valuable task in unpacking Poincaré's version of geometric conventionalism. He makes a valuable point in recognizing that there are objective and epistemically accessible differences between a world in which the natural geometry is Euclidean and one in which it is non-Euclidean, even though the former could be described with the aid of a non-Euclidean geometry. While Reichenbach's position is in agreement with Poincaré's in so far as it excludes the possibility of providing any answer to the question concerning the true geometry of space, it at least accords factual significance to the question of the natural geometry of the physical world. We shall subsequently make use of this important insight.

Geochronometric Conventionalism

Geochronometric conventionalism is the name which Grünbaum attaches to his own brand of conventionalism, which is distinct from both of the versions which we have so far examined. In particular, whereas Reichenbach bases his views on epistemic matters, Grünbaum claims his own version of geometric conventionalism to have an ontological basis. That is to say, that he is able to envisage a world in which geometric conventionalism would be false as a matter of ontological fact. However, it happens, again as a matter of ontological fact, to be true. In what follows, we shall attempt to elucidate the main features of Grünbaum's philosophy of geometry principally as they were worked out in the first edition of his *Philosophical Problems of Space and Time* (1963). However, it is only fair to add that he has since then offered numerous clarifications and modifications of his basic position. These are to be found chiefly in the second edition of that work, which was published in 1973, and also in *Geometry and Chronometry in Philosophical Perspective* (1968).

The crucial feature of Grünbaum's position is that the thesis of geometric conventionalism should be construed neither as an epistemological claim nor as a semantic one but as a claim about the very nature of space itself. He draws a distinction between those properties of an entity which are intrinsic to it and those which are extrinsic or relational. What he has in mind may be very simply illustrated by the example of the predicate "is a man", which is intrinsic to the individual John Smith and such a predicate as "is an uncle" which may be truly applied to John Smith but which is dependent on the existence of another individual—namely, a nephew or niece of John Smith. The former predicate refers to an intrinsic property of Smith, whereas the latter refers to a property which is extrinsic. Grünbaum would argue that the metric properties of space are all of the extrinsic variety. By this, he has in mind that it is conceivable that space might have had a "built-in" metric in a sense in which the natural numbers may be said to possess an intrinsic or built-in metric whereby the interval between 7 and 3 is objectively equal to that between 18 and 14. He argues that there is no counterpart to this natural standard of equality in the case of space due to the following considerations. In the first place, every spatial interval is continuous. The continuity of space entails the impossibility of comparing the magnitude of two such intervals by a process of counting, since all continuous intervals have the same cardinality. That is to say, that technically they all contain the same number of points—namely, a non-denumerable infinity of them. Should this appear counterintuitive to the reader, he may consider the case of the set of all positive integers and the set of even integers. Although the latter is obviously a proper subset of the former it is still the case that the two sets have the same cardinality, since their respective elements may be put into a one-one correspondence—namely, 1–2, 2–4, 3–6, ... It is a peculiarity of an infinite set of any cardinality that it will have proper subsets of the same cardinality as itself. Continuity is a necessary but not a sufficient condition for the extrinsicality of the spatial metric. For example, although the real number system is continuous it is, nevertheless, possible to make an objective comparison of the size of two intervals in the real number system. This is because the elements of the set of real numbers are intrinsically distinct from each other. The real number 6 is greater than the real number 2, which is why one may say that the interval from 0 to 6 is three times greater than the interval from 0 to 2. On the other hand, there is no intrinsic difference among the various points of space. All spatial points, or the unit sets which contain them, are qualitatively homogeneous. Thus, in virtue of the combination of continuity and homogeneity, there is no intrinsic basis for spatial congruence comparisons. That is to say, that space has no intrinsic metric. As Grünbaum puts it, space is metrically amorphous.

The foregoing considerations do not, of course, entail that metric comparisons or judgments of congruence are not or cannot be made. What is entailed is that all such judgments must be made on the basis of an appeal to a standard which is external to space. To ascertain the equality or inequality of two spatial intervals it is necessary to have recourse to the sort of external standard which is provided by a rigid measuring rod, which is transported from one region of space to another. In the event that a rod which coincides with the interval AB also coincides with the distinct interval CD, the two intervals are said to be of equal length or congruent.

The necessity for making an appeal to an external standard is not, by itself, sufficient to establish the conventionality of congruence. All that has been established is that one must appeal to something other than space itself to establish the equality of two spatial intervals. Such a situation is extremely common. For example, the fact that one must have recourse to some form of thermometric device to ascertain the temperature of a body does not imply that temperature comparisons are conventional. However, the two situations are not precisely analogous. Grünbaum couples the preceding considerations with the standard conventionalist claim that it is impossible without circularity to demonstrate that the transported standard has not contracted or dilated, i.e. that its length is not a variable function of its position. In short, that the rod coincides with both AB and CD indicates nothing about the putative equality of those intervals in the absence of a stipulation or decree to the effect that the measuring standard is "self-congruent". On the ground, therefore, of the dual claim that the basis for the ascription of equality of length to two line segments is *both* external and conventional, Grünbaum concludes that the choice of a spatial metric must be conventional. It must be emphasized that the significance of Grünbaum's claim is not merely epistemological. It is not to the effect that the true metric of space is empirically inaccessible but that space has no true metric: it is metrically amorphous! Furthermore, it should not be construed as asserting that space is a non-entity or methodological fiction but merely that it is not the sort of entity which could possess objective metric properties.

To emphasize the ontological or factual character of his thesis, Grünbaum employs an argument which is derived from Riemann. He contrasts the case of continuous physical space with that of a hypothetical space which is discrete or granular. In such a space, every interval would comprise a countable number of basic space atoms or quanta. Space, itself, would then be disposed to admit of an intrinsic standard of length which would be defined as the arithmetic sum of spatial quanta. Space intervals would then be compared in the same manner as arithmetic intervals. Grünbaum intended the foregoing to serve as an illustration of an intrinsic and factual metric as distinct from one which is extrinsic and conventional. However, as several critics have remarked, although such a standard of congruence would be in some sense intrinsic, it could still be held to be based on the convention that all space atoms have the same magnitude. Any attempt to establish that fundamental equality by some form of measurement carried out by means of an extrinsic standard would presumably deprive the metric of its strictly objective character. However, this criticism does not seem to be particularly damaging to Grünbaum's principal thesis.

In recent years an extension of general relativity has been propounded by the eminent physicist, J. A. Wheeler. His theory, which he calls *geometrodynamics*, is an attempt to reduce all of reality to a monistic ontology of curved spacetime. It is beyond the scope of the present book to deal with this topic but it has been expounded in detail by Graves in *The Conceptual Foundations of Contemporary Relativity Theory* (1971). It is merely mentioned in the present context since its attribution of intrinsic curvature to space is obviously anathematic to Grünbaum. Since the curvature of space depends on the metric, and since the metric is not an intrinsic feature of space, it follows that space could have no intrinsic curvature.

This line of reasoning was challenged by Glymour (1972), who points out that even if it be conceded that space is devoid of an intrinsic metric it is still possible to associate a species of curvature with it. The reader will recall that the curvature of space is revealed by the non-vanishing of a geometric object called an affinity, which was represented by the symbol Γ^i_{jk}. Now the usual measure of curvature of a given manifold is the Riemann–Christoffel tensor, which was symbolized by $R^j{}_{pqr}$. Now it happens to be the case that the Riemann–Christoffel tensor is definable in terms of the affinity and its first derivatives without reference to the metric g_{ik}. In short, a metrically connected manifold is merely a species of the more inclusive class of affinely connected manifolds. The principal difference between them is that there is no standard for the comparison of lengths along distinct curves in the case of an affine manifold. However, such a manifold is endowed with well-defined geodesics. Accordingly, Glymour's point is simply that the extrinsicality or conventionality of the *metric* is perfectly compatible with an intrinsic curvature of space.

Glymour's argument was recently challenged by Salmon (1977), who offers an elaborate reply in which his main point is to the effect that since the choice of an affinity is just as arbitrary as that of a metric, it follows that affine curvature is no more objective or absolute than metric curvature. That is to say, that one may add to the dictum of Grünbaum that space is affinely amorphous. The point which Salmon is making is the purely mathematical one to the effect that a neighbourhood of a differentiable manifold may be endowed with an affinity such that the mixed curvature tensor $R^j{}_{pqr}$ vanishes but by the very same token may be endowed with some other affinity such that $R^j{}_{pqr}$ is non-vanishing. In the one case, the manifold is said to be affinely flat and in the other is said to be affinely curved.

The question which is of interest to the philosopher of physics is whether there are physical considerations which override the foregoing mathematical ones. According to Salmon there are not. He considers the case of a physical theory which employs an affinely connected manifold whose geodesics are interpreted as the paths of gravitational test particles. The adoption of a different affine connection would simply result in a redescription of the selfsame physical situation such that the trajectories of the particles would no longer be geodesic curves. He then argues that there is no property which is intrinsic to the notion of physical space which compels the adoption of the one affinity in preference to the other, any more than there is an intrinsic property of space which compels one to maintain that a rod has or has not remained "self-congruent" under transportation. The salient point is that test particles, like measuring rods, are *extrinsic* to the manifold.

Salmon concludes his discussion with an attempt to elucidate the relationship between alternative metrizability and metric amorphousness. His fundamental claim is that a sufficient condition for the metric amorphousness of space is what he calls its *unconditional alternative metrizability*. Suppose that an n-dimensional differentiable manifold X_n is coordinatized by a coordinate system X^i and endowed with a metric tensor field $g_{hk}(X^i)$. Now X_n, or more strictly speaking an open neighbourhood of X_n, is recoordinatized by the system $\overline{X}^j$, where $\overline{X}^j = f^j(X^i)$ is a diffeomorphism. X_n is then endowed with another metric tensor $\tilde{g}_{hk}(\overline{X}^j)$ in accordance with the rule

$$g_{hk}(X^i) = \tilde{g}_{hk}(\overline{X}^j). \qquad (11.1)$$

The significance of (11.1) is that $\tilde{g}_{hk}$ has the same functional form in terms of the new coordinates as g_{hk} has in terms of the original ones. For example, let us suppose that X_i are rectangular Cartesian coordinates (X, Y) and X_2 has the customary Euclidean metric δ_{hk}. X_2 is then recoordinatized by polar coordinates (r, θ) and endowed with the new metric: $(ds)^2 = (dr)^2 + (d\theta)^2$. That is to say, that the new metric tensor has the same functional form as the original one. If I understand Salmon's train of argument correctly, he is here trading on the point, which was originally made by Grünbaum, that the intrinsic curvature of a manifold as measured by the curvature tensor depends only on the functional form of the metric tensor and not on the choice of coordinates. In short, if X_2 is a portion of the Euclidean plane in the first case, it will continue to be Euclidean when endowed with the new metric. Nevertheless, such a remetrization is far from being trivial. In particular, since $(dx)^2 + (dy)^2 \neq (dr)^2 + (d\theta)^2$, it follows that line segments which are congruent with respect to the first metrization will be incongruent with respect to the second one. In other words, the two metrizations will represent two disparate partitionings of point pairs into equivalence classes of congruent intervals. Since X_2 has remained the same throughout, the change could only represent a deformation of our measuring rods or standard of congruence. Now Salmon states that any manifold which is such that it has no property which precludes a remetrization according to the rule (11.1) is to be deemed unconditionally alternatively metrizable. Finally, he declares that unconditional alternative metrizability is a sufficient condition for the metric amorphousness of the manifold of interest.

A Critique of Geochronometric Conventionalism

In accordance with the general theme of this book, my present intention is to assess geochronometric conventionalism in the light of the theory of relativity. The conclusion which I shall attempt to establish is that this doctrine is not factually incorrect so much as conceptually misconceived. However, I shall begin by outlining the respects in which I take it to be correct. Despite the enormous critical controversy over Grünbaum's distinction between intrinsic and extrinsic features of the manifold and granting the difficulties attendant on a formal characterization of this distinction, I believe that its intent is reasonably clear. There are properties which a thing may possess in its own right and others which it may only possess in relation to something else which is external to it. Another way of putting it is simply that things possess both absolute and relational properties. Certain properties, such as chemical composition, are clearly absolute, whereas others, such as "being to the left of" are just as obviously relational. In other cases, arguments may be made for either possibility. However, it does seem sufficiently clear that Grünbaum and his disciples are correct in claiming that space and time and spacetime are devoid of metric and affine properties in any absolute or intrinsic way. To me at least, it makes no sense to claim that space and spacetime are intrinsically curved or, for that matter, intrinsically flat. There simply is no basis for attributing a "built-in" metric to spacetime any more than there is a basis for endowing it with built-in coordinates. Furthermore, it should be obvious that the lack of an intrinsic metric for spacetime is precisely the message of general relativity. To use the terminology of Anderson, the metric of general relativity is not an absolute object but a dynamical one. That is to say, that the space-time metric is determined by the matter-energy of the universe. The metric and affine properties of spacetime are, on this theory, clearly relational. That is not to say that I am advocating a far-reaching relational theory of spacetime in the tradition of Berkeley, Leibniz and Mach. I do not wish to advocate the doctrine that spacetime is nothing but a system of relations among bodies or events. My thesis is the much weaker one that only the *metric properties* of spacetime are relational.

One of my concerns about this aspect of geochronometric conventionalism is not the question of its truth but rather its triviality. We examined three versions of geochronometric conventionalism, and there are others, through which there runs a common thread. It is brought out quite clearly, I think, in Reichenbach's claim that one may alter one's geometry by making compensating changes in one's physics. Grünbaum conveys a similar message in arguing that the ascriptions of equality or congruence to spatial intervals depend on one's conventions concerning the behaviour of measuring rods. The implication seems to be that one is confronted on the one hand with space to which metric properties are to be attributed, albeit conventionally, and on the other with physical objects to which again metric properties are to be attributed, also by convention. It seems to me that this way of describing the relationship between geometry and physics is misconceived.

Suppose that someone who knew nothing of the geometry of the earth's surface had catalogued the distances between several pairs of cities which are distributed over the terrestrial globe. He then proceeds to construct a map of their location on a piece of paper, i.e. on a two-dimensional Euclidean manifold. He eventually discovers that the set of measurements is apparently inconsistent in virtue of its failure to conform to the structure of a Euclidean manifold. For example, it is found that three non-colinear cities cannot be mapped onto the apexes of a triangle. He finally concludes that the geometry of the earth must be non-Euclidean and actually approximates that of a two-dimensional spherical surface. Taking our cue from Reichenbach, let us now consider how a geometric conventionalist might analyze such a situation.

One may surmise that he would argue that the claim that the earth's surface is spherical is inconclusive. Reichenbach's doctrine of equivalent descriptions would suggest the following alternative. The geometry of the earth is actually Euclidean, whereas space is non-Euclidean. The non-Euclidicity of space obliges one to revise one's standards of congruence, since it has a systematic deforming influence on one's measuring instruments. The result of such deformations is that the distance assigned to the spatial separation of any two cities fails to represent the shortest distance between them. Hence, one is presented with the descriptively equivalent alternatives of declaring the earth's surface to be non-Euclidean while space is Euclidean and of declaring the earth's surface to be Euclidean while space is non-Euclidean. The reaction of a cartographer to such a suggestion would no doubt be to the effect that he has no interest in the geometry of space and even fails to understand what the mapping of space could possibly mean. His sole concern is to construct a map of the surface of the earth, and this he is able to do by methods which are *intrinsic* to that surface in the sense of their being independent of the structure of the space in which the earth resides.

The point of the parable, of course, is that the outlook of the cartographer reasonably represents that of the physicist. All of the talk in the philosophical literature concerning the nature of the spatial metric is a red herring for which Poincaré should assume the ultimate blame. Now there is a certain artificiality about the foregoing parable, which stems from the fact that it deals with the case of a two-dimensional manifold embedded in one of three dimensions. A less contrived illustration would be the case of a descriptive astronomer who attempts to construct a three-dimensional map of the local star system with the aid of various optical instruments. Suppose that he likewise concludes that the only obvious way in which the results of his measurement of interstellar distances may be rendered consistent is by mapping the stars on a non-Euclidean manifold. The geometric conventionalist will now respond to the effect that the astronomer's measurements have been affected by a systematic deviation from Euclidicity in the paths of light-rays. In short, it is possible to retain Euclidean geometry by a suitable adjustment of one's optics.

In this case, the appropriate counter-response may be somewhat more subtle than before. In the case of the cartographer, the situation was clear since the various cities were physically constrained to be on the surface of the earth. The cartographer was mapping a certain physical structure comprising a surface endowed with geodesics and some arbitrarily selected nodal points called cities, which were located at the intersections of certain of the geodesics. The conventionalist move amounted to arguing that the nodes could be treated as elements of another structure called "space". Similarly, the astronomer is engaged in mapping a structure consisting of light-rays and stars. He carries out this task by associating the former with geodesics and the latter with points. Then, given sufficient data, he is able to select a metric, i.e. an abstract space, which is determined up to a coordinate transformation. To be sure, the physical structure which is to be coordinated with the geometric one is not chosen at random. The scientist is preoccupied with systems that are of interest in virtue of their lawfulness. To confirm the legitimacy of his coordination he will perform additional experiments that are independent of the original ones. Thus, the astronomer may, in principle, verify his optical measurements by equipping a space vehicle with an inertial guidance system and a clock, and have it travel among the stars at some predetermined velocity. Then the various distances would be calculated on the basis of the elapsed time as measured by the clock. Should the two independently determined sets of measurements be in agreement, he would have excellent grounds for being satisfied with his choice of geometry. It is precisely such universal concordance of results that is required by, and predicted within the framework of, general relativity.

At this point, an unrepentant conventionalist would still be at liberty to argue that the situation could be redescribed with the aid of a different geometry by recourse to the standard stratagem of invoking a force which causes light-rays to deviate from the postulated geodesics while also causing a corresponding retardation in the rates of clocks. This is why the anti-conventionalist philosopher, Sir Karl Popper, has observed that conventionalism is logically unassailable. I will avoid the temptation to dismiss the conventionalist move as obviously *ad hoc*, since conventionalists have correctly argued that such a charge misses the point. It is to the point, however, to argue that the move is unphysical in so far as it treats the question of the geometry of space as though it were on all fours with the question of the geometry of the star system. The argument which I wish to press against geometric conventionalism is that the physicist is always concerned with the geometry of a physical system of some kind and never with the geometry of pure space. To argue, for example, that the surface of my desk is Euclidean if the geometry of space is flat, but curved if the spatial geometry is curved is, in my opinion, to commit a methodological fallacy. I shall now amplify on my allegation that the doctrine of geometric conventionalism is trivial.

Salmon's treatment of affine and metric amorphousness is, I think, a case in point. He argues that given a differentiable manifold which is endowed with a particular affinity, there is nothing to prevent one from replacing the latter with some other affinity. This, of course, is true but utterly trivial. All that has been shown is that a differentiable manifold, qua abstract mathematical space, is not disposed to accept one affinity more than another. In short, a differentiable manifold is affinely amorphous—the affinity is not built in. That, however, is not to say that an affinely connected manifold is affinely amorphous. By the same token, a differentiable manifold is obviously metrically amorphous but that is not to say that a manifold which is endowed with a metric is metrically amorphous. Such truisms fail to carry over to physical cases. Consider his argument that a theory which employs an affinely connected manifold in which the trajectories of test particles are associated with affine geodesics may be recast in a form in which the trajectories are no longer geodesic. That is not, I would argue, merely to redescribe the same physical situation but amounts to the more radical step of abandoning a putatively good theory with no apparent justification.

It was argued in the opening section of this chapter that the referent of a physical theory is always some sort of physical entity. The applied geometer will be concerned with such physical entities as table tops, spheroids such as the earth and, in particular, physical fields such as that of gravitation. He is not directly concerned with investigating the properties of mathematical manifolds. Such properties are simply decreed or freely chosen. Therein lies the essential truth of geometric conventionalism. It does not follow that the structure of a physical entity is determined by convention. It is not the case, for example, that the gravitational field is metrically amorphous and, hence, accessible to conventional metrization. To the extent that gravitational phenomena are successfully coordinated with a particular metric space, the field itself may be said, qua physical manifold, to possess that metric objectively. One may, of course, argue that a theory which associates the gravitational field with a particular metric space may be abandoned in favour of another theory which associates that field with a different metric. However, the familiar circumstance of one theory's being supplanted by another should not be construed as evidence favouring conventionalism. The latter is not an instance of freely replacing one convention by another but is rather a case of one factual conjecture or postulate being replaced by another which is deemed to be more accurate. The fact that the metric of a physical manifold may not be known with certainty should not be taken to indicate that such manifolds lack an objective or determinate metric. The appropriate place for conventions is in pure mathematics not in factual science. Indeed, one may suspect that geometric conventionalism has been fostered by the all too frequent confusion that exists between a physical theory and the mathematical theory in terms of which it is expressed. While allowing the correctness of the claim that there is nothing in the nature of space, qua differentiable manifold, to tip the balance in favour of any particular metric, we maintain that in the case of a physical manifold there will be abundant grounds for selecting one metric in preference to another. At the same time, in fairness to those we oppose, it must be added that conventionalists would reply to this that there are indeed grounds for preferring one metric convention over others, but that fails to alter the fact that they are conventions. However, if this were a correct assessment of general relativity, it would be difficult to say precisely what a practitioner of that theory is trying to discover about the world.

Once it is recognized that the referents of scientific theories are physical structures rather than conceptual ones, the cogency of Salmon's rule of unconditional alternative metrizability may be called into question. Firstly, the reader should be aware that the rule (11.1) is not a theorem of Riemannian geometry. The implication of Salmon's argument, I believe, is that one is free to choose any metric *cum* coordinate system. Now it is true that there are infinitely many ways in which to coordinatize a given manifold. It is also true that given a particular coordinate system, there are infinitely many ways of selecting the functions g_{ik}. However, as the reader knows, once the g_{ik} have been chosen, a definite geometry is thereby determined. Salmon seems to be suggesting that any pair from the Cartesian product of coordinate systems and metric tensor fields is as good as any other. In the case of applied geometry, however, the situation is altogether different. One starts with a given physical manifold which is represented by the curvature tensor B_{jklm}. One is then obliged to choose a metric tensor *cum* coordinate system, $\langle X^i, g_{jk} \rangle$, which is compatible with the curvature tensor. In short, one selects from the triple Cartesian product of curvature tensors, coordinate systems and metric tensors. Although there are infinitely many ways in which this choice may be made, there are also infinitely more ways in which it may not be made. In fact, any two choices must be related in accordance with the keystone rule of Riemannian geometry:

$$(ds)^2 = g_{ik} dX^i dX^k = \bar{g}_{ik} d\bar{X}^i d\bar{X}^k. \tag{11.2}$$

The effect of (11.2) is to restrict one's choice of $\langle R_{jklm}, X^h, g_{pq} \rangle$ so that when one recoordinatizes a particular manifold one must make a compensating choice of metric tensor to ensure the invariance of ds. In short, one's choices of metric *for a given manifold* must all be isometric. One may relax the restriction to allow a change of scale but that is all. This suggests that although (11.1) may possibly make sense in the context of mathematical paradise, it is irrelevant to the world of applied mathematics or physical theory. All that Grünbaum and Salmon have succeeded in showing is that the mathematical notion of a differentiable manifold is metrically uncommitted. But that is, to me, so obvious as hardly to warrant comment, not to mention elaborate philosophical exposition.

I have disregarded Grünbaum's ingenious argument to show that there are infinitely many non-isometric metrics which yield the same manifold geometry. Indeed, Salmon's proposal would seem to be derived from it. However, this possibility again exceeds the framework of Riemannian geometry by introducing so-called non-customary standards of congruence. These, in effect, are measuring rods whose length is a function of their position and orientation. I take the position that there are objective grounds for preferring customary standards of congruence, e.g. steel rulers in preference to strips of elastic, which have nothing to do with geometry, but pertain to such matters as the molecular structure of different materials. Indeed, one should drop the philosophical pretense that physicists ascertain the structure of the gravitational field with the aid of measuring rods. The latter are nothing but a metaphor for the actual standards of congruence employed by experimental physicists such as the wavelength of the orange line of krypton and the frequency of the caesium atom. These are selected as standards of length and time precisely because there are overwhelming grounds for supposing them to be genuinely constant or, metaphorically speaking, rigid. Moreover, to select a non-customary standard of congruence is to deny the constancy of our customary standards. But that is to call into question the status of many if not all of the fundamental constants of nature, at which point the entire edifice of physical theory falls apart.

General Relativity and Space-time Structure

The reader may have been slightly puzzled by the drift of my argument in the preceding section. I began by affirming my general agreement with Grünbaum's basic ontological tenet concerning the metric amorphousness of space but proceeded to engage in an anti-conventionalist polemic. My point, however, is that I do not believe that what I take to be an essentially correct account of the nature of space serves to illuminate the philosophical foundations of general relativity. In line with the general view of applied mathematics which I propounded in the first section, I interpret the theory of general relativity to be essentially an attempt to coordinate a conceptual or mathematical entity, namely a pseudo-Riemannian manifold of variable curvature and Lorentz signature with a physical entity, namely the gravitational field. Unfortunately, the implications of general relativity have been obscured by a confused semantics in which the concepts of "gravitational field" and "metric field" are used interchangeably. This has led to the supposition that gravity and geometry are one and the same. However, I see no compelling reason to regard general relativity as different from any other field theory in this regard. For example, the electromagnetic field may be represented by an antisymmetric tensor of the second rank. That should not be taken to signify that the electromagnetic field is literally a tensor field in the mathematical sense. Most reasonable people who are acquainted with the theory of electromagnetism take the field to be a physically real energy field. Similarly, the single most persuasive reason for treating the gravitational field as a physical entity is precisely that it plays the role of a carrier of energy. In particular, when a gravitational source acts on a distant particle, the effect is transmitted with a finite velocity. In a pure action-at-a-distance theory a conceptual problem would arise as to what happens to the energy, more strictly speaking the energy-momentum, which is expended by the source prior to its acting on the passive particle. To ensure that the fundamental principle of the conservation of energy-momentum is not violated, one postulates an intervening field as the carrier of energy-momentum. That is why a relativistic theory of gravitation is bound to be a field theory.

Having said that, however, one must certainly allow that general relativity is probably unique in the degree to which it successfully coordinates the structure of the gravitational field with a mathematical structure without recourse to additional physical principles. In virtue of the identity of gravitational and inertial mass, the motion of a particle under the influence of a gravitational field may be formally treated as a force-free or "natural" motion. The geodesic hypothesis is an immediate consequence. The latter states that the trajectory of a gravitational monopole is a geodesic in spacetime. Consequently, as we have seen, the gravitational potential may be represented by a space-time metric tensor. As Grünbaum puts it, the metric tensor does double duty. Now if it is reasonable to grant my methodological distinction between the metric structure of a physical object such as a table-top and the mathematical structure of a Euclidean manifold which is coordinated with it, I would contend that it is equally reasonable to draw a similar distinction between the gravitational field, on the one hand, and the mathematical structure of a space-time manifold, which we conveniently call "the metric field", on the other. I earlier contended that it makes no literal sense to attribute flatness or curvature to space, itself. They are properties which characterize various physical entities viewed as manifolds. *Mutatis mutandis*, it does not make literal sense to attribute curvature to spacetime. To do so is to commit what Whitehead called "the fallacy of misplaced concreteness". What may truly be said to exhibit curvature is not spacetime but simply the gravitational field. To speak of the curvature of spacetime is as misguided as to speak of the flow of time, as though time were the sort of entity or process to which one might attribute a velocity.

The basic confusion between physical processes and entities, on the one hand, and mathematical continua, on the other, has given rise to two grave philosophical misinterpretations of general relativity. Firstly, there is the "hyper-realism" of Wheeler, who regards, or once regarded, spacetime as the fundamental substance of which all aspects of physical reality are but modifications. Secondly, there is geochronometric conventionalism which illicitly extrapolates correct claims concerning the metric amorphousness of spacetime to the incorrect doctrine of the metric amorphousness of the gravitational field. Various workers in the field of general relativity have proposed that the appropriate way to measure the gravitational field is not by means of rods and clocks, as suggested by Einstein, but by means of freely falling massive particles and photons, which reveal respectively the timelike and null geodesics of the field. Grünbaum has argued that such methods should be interpreted not realistically but conventionally. His point is that the aforementioned geodesics are not to be regarded as constitutive of the gravitational field. Rather, it is *by human decree* that the trajectories of photons and gravitational monopoles are associated with geodesics. The reader will recall essentially the same point having been urged by Salmon. The rationale of Grünbaum's argument seems to be that the aforementioned probes of the field, like measuring rods, are extrinsic standards. Hence, just as it is conventionally decreed that two intervals of space which are successively occupied by a measuring rod are congruent, so in this instance it is decreed that the space-time trajectory of a free particle is a geodesic. Like the stipulation that $\varepsilon = \frac{1}{2}$, it may lead to the simplest possible description of the field but that does not alter its essentially conventional status. Grünbaum grants that general relativity imposes a definite metric on spacetime but argues that it is to be construed not *descriptively* but *normatively*. I take it that he here uses the epithet "descriptive" in the sense of objective rather than in his customary use of it when speaking of descriptive simplicity.

It would seem that Grünbaum may be guilty, perhaps unknowingly, of a misinterpretation of the methodology of general relativity. This misinterpretation probably runs along the following lines. One observes the trajectories of massive and massless particles and adopts the convention of treating them as timelike and null geodesics, respectively. One then asserts that the metric or gravitational field is whatever manifold happens to possess that particular geodesic structure. I cannot say with authority that Grünbaum treats general relativity in the manner I have indicated but find the evidence rather compelling. I think that it may be shown to be erroneous on at least two counts.

In the first place, on this view the geodesic hypothesis, which is an essential ingredient of the theory, would fail to have the status of a physical law but would be a mere convention such that the theory as a whole would be rendered practically immune to revision. For example, one of the classical predictions of general relativity was the bending of starlight in the sun's gravitational field. Now on the interpretation which I have putatively attributed to Grünbaum, this prediction would not have constituted a critical test of the theory, since whatever the trajectory of a light-ray might be, that trajectory would have been conventionally decreed to be a geodesic. The famous expedition of Eddington in 1919 which was conducted for the purpose of testing Einstein's prediction would therefore have been pointless. What is overlooked is that Einstein predicted the path of starlight before the fact and with a high degree of numerical precision. The geodesic trajectories of the theory are deductive consequences of the field law, which is the heart and soul of the theory. If the status of these trajectories as geodesics is merely conventional, then the entire theory must be little more than a web of conventions. But Grünbaum would surely wish to reject such an unbridled conventionalism as that. However, one of the inherent dangers of conventionalism appears to be that once one has put a foot on its path one can hardly stop until one is caught in the web of Quinean pragmatism.

In the second place, such an interpretation fails to take account of the rich experimental resources of general relativity. In particular, there are now instruments at one's disposal by means of which the Riemann–Christoffel tensor may be directly measured. Next to the metric tensor from which it is formed, the Riemann–Christoffel or curvature tensor provides the most information about the structure of the manifold. In general relativity it plays the role of representing the so-called tidal forces. Due to the inhomogeneous character of the gravitational field, two freely falling particles which traverse initially parallel paths will eventually cross or diverge from each other. The reader may gain an intuitive understanding of this by picturing two particles which are released from a great height above the surface of the earth and gradually draw closer as both are attracted to the earth's centre. In the context of general relativity, this phenomenon is called *geodesic deviation*. Let $x^{\mu}(\tau)$ and $x^{\mu}(\tau)+\xi^{\mu}(\tau)$ be the time tracks of two neighbouring particles in free fall. Then the geodesic deviation is given by the second intrinsic derivative of ξ^{μ} along x^{μ}.

$$\frac{D^2\xi^{\alpha}}{D\tau^2} = -R^{\alpha}_{\beta\gamma\delta}\xi^{\gamma}\frac{dx^{\beta}}{d\tau}\frac{dx^{\delta}}{d\tau} \tag{11.3}$$

The derivation of the equation of geodesic deviation will be found in any of the more recent treatises on general relativity. Now there are instruments, such as the gravity gradiometer, which actually measure geodesic deviation or tidal force. Hence, it is in principle possible to construct the curvature tensor field. From this field, within a latitude that need not concern us, one may then derive the geodesics. Accordingly, the geodesics can be no more conventional than the observable phenomenon of geodesic deviation. (11.3) also provides a basis for the possible detection of graviational waves, another phenomenon which only makes sense on the basis of a realistic as distinct from a fictionalistic interpretation of the gravitational field. I suppose that the conventionalist might still reply to my criticisms by insisting that it is, nevertheless, by an act of will that the paths of gravitational probes are treated as geodesics. However, to treat them as anything other than geodesics is not simply to opt for a semantically equivalent redescription of physical reality but is rather to take the more radical course of abandoning the theory of general relativity. This is the basic reason why geochronometric conventionalism must be judged a failure in its capability to provide a philosophic interpretation of general relativity.

Once one grants the validity of the distinction between the gravitational and the metric field, viewing the latter as the mathematical representative or model of the former, the conclusion emerges that spacetime has no metric structure of its own. I am now making a tripartite distinction of physical gravitational field, metric field and *spacetime* as the arena in which gravitational and other processes take place. The recognition of the last of these as a distinct mode of existence has long been highly unfashionable, although much less so in the more recent philosophical literature. I shall have more to say about it in the next section. By locating the metric in the gravitational field I am opting for a relational theory of the metric. That is to say, that I am viewing the metric not as an intrinsic property of spacetime but as a relational property of the field which is determined by the distribution of matter-energy. If matter-energy were non-existent, i.e. $T_{\mu v} \equiv 0$, then there would be no gravitational field and consequently no metric.

The more commonly held view among physicists is that matter is not the source of the metric, *per se*, but rather the cause of its curvature or deviation from flatness. Thus, it would be held that if $T_{\mu v}$ were everywhere zero, the universe would be represented in the form of flat Minkowski spacetime. The most obvious physical ground for this view is that electromagnetism in the absence of gravitational influences is generally represented against the background of absolute Minkowski spacetime. Special relativity is the limiting case of general relativity. However, while practicable, this account of the matter is slightly unrealistic since the electromagnetic field must itself be a gravitational source, however negligible. Accordingly, one cannot form a consistent picture of a universe with electromagnetic fields and no gravitation. Another argument which favours an autonomous Minkowski metric is that flat spacetime is imposed as a boundary condition for the celebrated Schwarzschild solution of the field equation for an insular spherically symmetric mass. The solution was obtained by assuming that at infinity the geometry of the field would be Minkowskian. Since that time, incidentally, it has been shown that such an assumption is superfluous, since the Minkowski condition at infinity is a necessary feature of the physical arrangement. In any event, the implication seems to be that a completely empty universe may be viewed as having a well-defined metric structure. Specifically, it would possess the geodesics of Minkowski spacetime. My view of the gravitational field is quite literally that of a guiding field that constrains particles to traverse space-time geodesics. Consequently, if an hypothetical particle of inappreciable mass were introduced into the empty universe it would be guided along a timelike geodesic. This is just an operationalist way of saying that an empty universe with a Minkowski metric would contain a gravitational potential.

Various problems emerge from such considerations. For example, there is the vexatious fact that there are actually non-flat solutions to the field equation in the absence of matter-energy. Moreover, if the test-particle moves in any way at all in the empty universe, what is the ultimate cause of its motion? However, I would come back to the fundamental problem that in the absence of matter-energy there is no gravitational field and, accordingly, no meaningful way to ascribe free motion to a particle. Put differently, if there are geodesics, they must be geodesics of something. But in the absence of gravity, there is nothing of which there could be any geodesics. A drastic but possible solution would be to treat them as properties of empty spacetime. But that creates as many problems as it solves, since it amounts to the reinstatement of an absolute metric and the consequent abandonment of the relational theory of gravitation.

Such conceptual difficulties may be avoided by reverting, once again, to the characterization of general relativity as the coordination of a mathematical structure with a physical one. I have intimated that the degree of structural similarity will vary from theory to theory. In the case of general relativity, the remarkable closeness of the fit has been constantly emphasized. Nevertheless, it would be mistaken to suppose that there is an isomorphism down to the last detail. This mistake is commonly committed precisely because the two have been treated as one. As we have seen, the standard interpretation of general relativity treats the metric field not as a mathematical representation of the gravitational field but as numerically identical to it. Once this Pythagorean error is exposed, many of the aforementioned problems may be viewed from a different perspective. In particular, if the metric field is treated as a mathematical model of a gravitational field whose source is the matter-energy in the universe, it follows that if all that matter-energy were spontaneously annihilated there would be nothing left to model. In short, general relativity, that is to say, the physical theory as distinct from its ingredient mathematical formalism, is a theory which is designed to describe and explain the nature of a universe in which $T_{\mu\upsilon}$ does not vanish everywhere. Apart from that there is nothing to which the theory should or could be applied. Unconventional as this interpretation may be, it seems both methodologically sound and in accordance with one's rational intuitions. To take a rather simple, although I hope not unduly simple-minded, analogy, let us suppose that one has an instrument for ascertaining the geometry of various physical objects. One is confronted with a solid cube and asked to determine its geometry, which one does. The cube is then replaced by a solid sphere and one is again required to determine the geometry, once more successfully complying with the request. Finally, an object is removed and replaced by nothing, but one is again asked to ascertain the geometry. The appropriate response is, of course, to point out that the question is no longer a well-posed one. To make a measurement, there must be something to measure. By the same token, I suggest, to ask for the metric of an empty universe makes no sense. It is rather like the question posed by the proverbial Buddhist sage: What is the sound of one clapping hand?

Let me conclude this section by underlining its principal philosophic message. It is that talk of the curving of space and the slowing down of time in the context of general relativity ought to be treated as a metaphorical or indirect way of referring to various properties of the gravitational field. Like Grünbaum, I am convinced that spacetime is, in itself, metrically amorphous. Where we would perhaps differ is that whereas he would seem to hold that in the absence of an objective metric for spacetime it is necessary to introduce a freely chosen metric convention, my own view is that since spacetime lacks metric properties, it is simply a mistake to ascribe them either conventionally or otherwise. Talk of the curvature of spacetime is not a convention but rather a *façon de parler*. On the other hand, in my view, general relativity provides abundant support for the view that the metric of the gravitational field is not conventional but objective. In this regard, we would seem to differ rather emphatically.

The reader may wonder about the implications of these doctrines when applied to the everyday world, which is just the weak-field approximation to the universe of Einstein. Surely, one may ask, if a clock does not measure time then what could it possibly measure? In the first place, to speak of the measurement of time seems to suggest that time has a rate of change. As the reader knows, to speak of the rate at which something is changing is a way of referring to the functional dependency of one variable on another. Usually, the independent variable is time itself. But to pose a question about the rate of time is to treat time as the dependent variable. If time had a rate, it would be with respect to itself and would always be given by $dt/dt = 1$. Another way of putting it is to say that a variable is always constant with respect to itself. What then are clocks for if not to measure time? A clock is very simply a standard periodic sequence of events, which are typically ticks or oscillations *in* time. Given such a standard sequence of events, one may meaningfully compare other such sequences with it as, for example, the number of rotations of the earth or the number of its revolutions about the sun, and so forth. The slowing down of time in general relativity is simply a metaphor, and unfortunately a misleading one, which refers to the rate at which a clock at a given gravitational potential is observed to tick when compared to a clock of an observer who is, in principle, infinitely removed from the gravitational influence. There is no harm in representing it analytically by $d\tau_1/d\tau_2$, so long as one does not interpret that expression as referring literally to time as distinct from g_{00}. Admittedly, this account fails to provide as much grist for journalistic mills as the customary metaphor but is still sufficiently remarkable to be a source of legitimate philosophic wonder.

The Ontology of Spacetime

For several decades following the inception of the two theories of relativity, the majority of scientists and well-informed philosophers supposed that the question of the absolute existence of spacetime as debated by Leibniz and Clarke had at last been laid to rest, with Leibniz the obvious victor. We have seen that Maxwell's aether, which was the embodiment of physical space, was shown in the context of special relativity to have been a gratuitous notion. The ultimate *coup de grâce* for the notion of absolute or substantial spacetime was taken to have been delivered in the theory of general relativity, which Einstein originally presented to the public as the scientific implementation of Mach's philosophical claim that the inertial properties of material objects, including those of reference frames, could be accounted for in terms of the total distribution and motion of matter in the universe, without reference to the action of space. Einstein's arguments for the *relativity* of space and time were unhesitatingly accepted as arguments for their *relationality*. Today, however, increasingly many philosophers are coming to recognize that the situation is not as clear as had once been supposed. In particular, it is now generally agreed that general relativity fails in various respects to implement Mach's principle. For example, the theory retains the notion of an invariant rest-mass, which is an inertial property of matter. Moreover, it is possible to define an absolute four-dimensional rotation vector on the space-time manifold of general relativity. In fact, Gödel has shown that the notion of an intrinsic rotation of the universe as a whole is compatible with the formal structure of general relativity. It is obvious that if spacetime were entirely dependent on matter, then it would simply make no sense to ascribe a rotation in spacetime to the universe as a whole.

A balanced interpretation of the available evidence would seem to suggest that although general relativity fails to eliminate absolute space, it rather convincingly supports the relational or non-absolute character of the metric. Accordingly, a clue to the inherent nature of spacetime is available to us. Spacetime must possess intrinsically all of those properties which it clearly must possess but which are not due to the influence of matter. In effect, we are suggesting that the way to discover the intrinsic properties of spacetime is by "factoring out" all those properties which are effects of matter-energy. Now the latter are precisely those features which are represented by the affine and metric properties of the geometric description of the universe. When these are removed, one is not left with a void or strict nonentity, as relationalists would contend, but with the topological structure of a *differentiable manifold M*. I take *M* to correspond to the amorphous space-time continuum which seems to be presupposed in Grünbaum's philosophy of spacetime.

What can be said about this empty continuum beyond the purely negative claim that it is affinely and metrically amorphous? In order to deal with this question, one is obliged to venture beyond hard science into the realm of speculative metaphysics. However, I now wish to show that the practice of metaphysics, while indeed speculative, may be undertaken with the same mathematical precision that is customary in scientific subjects. In brief, the procedure of the metaphysician should be to attempt to coordinate a mathematical model with physical reality. The justification for this is not a mystical commitment to the mathematical character of ultimate reality but merely the recognition that it is with the aid of mathematical conceptualization that one's ideas are sharpened and rendered sufficiently precise that they may be understood and rendered accessible to criticism. In the case of metaphysics, however, the aspects of reality which are of interest are generally experimentally inaccessible. The brake on wild speculation is provided for by the requirement that one's metaphysical construct be conformable with one's scientific knowledge of the world. This strategy is the basis of Mario Bunge's programme for exact philosophy in general and scientific metaphysics in particular. I should add that scientific metaphysics is not a body of doctrines but a methodology. Bunge, himself, would certainly disagree with the substance of the position which I intend to define. Since I am now venturing beyond the primary goal of this book, I shall limit my treatment to a brief sketch.

Let us begin with an account of the mathematical resources that are at our disposal. A differentiable manifold may be viewed as a homogeneous and continuous collection of points on which it is possible to define various differentiable functions. It has a dimensionality and may be coordinatized. Thus, every point may be labelled by an *n*-tuple of numbers x^i. Lines or curves may be defined on it, although they would lack an intrinsic length. A scalar field could be constructed by associating each point with an invariant quantity, $\Phi(X^i)$. One may also construct a covariant vector field, the gradient of Φ, by differentiating the scalar field: $\partial\Phi/\partial X^i = \Phi_{,i} = A_i$. It would also be possible to construct contravariant vectors (tangent vectors) although this requires mathematical techniques which are beyond the scope of this book. Clearly, such a vector could not be treated as a directed line between two points, since such a line lacks a definite length. Furthermore, the raising and lowering of indices would be out of the question in the absence of a fundamental or metric tensor.

The reader will recall that the differentiation of a vector field requires an affinity, which is not now at our disposal. Although, therefore, the derivatives of vectors cannot, in general, be defined on a bare differentiable manifold, there is one differential operation which does make sense in the present context. This operation, which we have not hitherto encountered, is somewhat analogous to the divergence $\nabla \cdot \mathbf{V}$. It is called the *curl* of a vector field and is classically symbolized by $\nabla \times \mathbf{V}$. In component notation, it is defined by

$$\text{curl } \mathbf{V} = \varepsilon_{ijk} \nabla_j V_k. \tag{11.4}$$

The reason that the curl of a vector is available to us is that (11.4) is equivalent to $V_{j;k} - V_{k;j}$. The reader may verify that when this is written in its expanded elementary form, the terms which depend on the affinity cancel out. If $\mathbf{V}_k$ is the gradient Φ, then the curl of $\mathbf{V}_k$ will be $\varepsilon_{ijk}\nabla_j\nabla_k\Phi$. Since, as you will recall from the discussion in Chapter 1, $\varepsilon_{ijk} = -\varepsilon_{ikj}$ and since j and k are dummy indices, which are therefore interchangeable, it follows that:

$$\varepsilon_{ijk}\nabla_j\nabla_k\Phi = -\varepsilon_{ikj}\nabla_k\nabla_j\Phi. \tag{11.5}$$

But this is only possible if (11.5) is identically zero. It follows that the curl of a gradient always vanishes.

The curl may be physically interpreted in terms of the *circulation* of a vector field. Consider, for example, the velocity field of a fluid. If an object is allowed to float on the surface of the fluid and is carried along by it *without rotating*, it follows that the net effect of the velocities or momenta of the various fluid particles around a closed path is zero. Now if this were the case everywhere, the velocity field would be said to have zero circulation. In general, however, the circulation of a vector field will have different values around different closed paths. If by the familiar limiting process we consider the circulation around an infinitesimal area, we arrive at the notion of the curl of a field. Intuitively, it is its circulation at a point and may be thought of as a minute vortex. A field whose curl is identically zero is said to be *irrotational* or to have zero vorticity. You may recall that a zero divergence is an indication of the absence of a source. The vanishing of the curl is to be similarly interpreted, although the source in the latter case would be of the vectorial rather than scalar variety.

Now that we have the rudiments of the mathematical theory of differentiable manifolds, we proceed to consider the nature of the entity, empty spacetime, with which it is to be coordinated. Now we must begin the speculative part of our venture. The points of M are associated with events or, perhaps, the *loci* of events. Similarly, the curves of M are the *loci* of the world-lines of particles. The import of this is that matter and events are not required to "create" spacetime but are located *in* an ontologically pre-existing spacetime. Material events do not constitute the intervals of spacetime but the measure of such intervals. In the absence of events in the physical sense, spacetime may be assumed to be homogeneous and isotropic. Hence, although one might choose to ascribe a scalar field to it, such a field would be constant. Hence, spacetime has a zero gradient and a zero curl everywhere. In the absence of the derivatives of vectors, it follows that linear velocities and accelerations cannot exist in relation to spacetime but only in relation to matter. To that extent, the doctrine of the relativity of motion is vindicated. On the other hand, the curl is closely associated with the notion of *angular* velocity or rotation. While a linear velocity is simply undefined for empty spacetime, an angular velocity may be defined in terms of the curl and even given the determinate value of zero. The reader should appreciate the profound difference between the claim that a certain geometric object is undefined and the claim that such an object has an identically vanishing value. Analogously, in a flat affine space, the affinity vanishes, but that is not to say that flat space lacks an affinity. Accordingly, one may make an objective claim to the effect that spacetime does not rotate. Consequently, it is possible to assign an absolute significance to the rotation of matter—namely, that a material object may have a non-zero angular velocity with respect to irrotational spacetime. We see, therefore, that Newton's bucket experiment was not as ill-conceived as Mach claimed. More significant, however, is that a reason for the failure of general relativity to account fully for the effects of rotation along Machian, i.e. strictly relationalist, lines can be provided in terms of our spacetime ontology. By the same token, the invariant rest-mass of matter, which cannot be accounted for along relationalist lines, can be made sense of in the proposed ontological framework.

Thus, the ultimate question to be faced by a philosophy of spacetime—namely, whether spacetime is nothing but a system of relations among material events or whether such events presuppose spacetime for their very possibility may finally, although not dogmatically, be resolved. Our picture of the world includes both relational and absolute aspects. In brief, all those properties which depend on the existence of a metric affinity may be deemed relational. The manifold properties of spacetime such as dimensionality, continuity and differentiability are absolute. Spacetime exists.

Bibliography for Chapter 11

ANDERSON, J. L. (1967). See reference in Chapter 9.

EARMAN, J., "Who's Afraid of Absolute Space", *Aust. J. Phil.* **48**, 287 (1970). A vigorous statement against the claim that general relativity overthrows the concept of absolute space.

GLYMOUR, C., "The Epistemology of Geometry", *Noûs*, **11**, 227 (1977).

GOENNER, H., "Mach's Principle and Einstein's Theory of Gravitation", in *Ernst Mach–Physicist and Philosopher* (eds. R. S. COHEN and R. J. SEEGER). Boston Studies in the Philosophy of Science 6, Reidel, Dordrecht, 1970.

GRAVES, J. C., *The Conceptual Foundations of Contemporary Relativity Theory*, MIT Press, Cambridge, Mass., 1971. An excessively realistic interpretation of general relativity and geometrodynamics, but beautifully presented and argued.

GRÜNBAUM, A., "Absolute and Relational Theories of Space and Time", in *Foundations of Space-Time Theories* (eds. J. EARMAN, C. N. GLYMOUR, and J. J. STACHEL), Minnesota Studies in the Philosophy of Science, University of Minnesota Press, Minneapolis, 1977. This volume contains several papers which indicate many of the trends of contemporary research.

GRÜNBAUM, A., "Geometrodynamics and Ontology", *Journal of Philosophy*, **70**, 775 (1973).

NERLICH, G., *The Shape of Space*, Cambridge University Press, Cambridge, 1976. A sustained argument for the reality of space and time including an intrinsic metric.

POINCARÉ, H., *Science and Hypothesis*, Dover, New York, 1952 (orig. 1902).

POINCARÉ, H., *The Value of Science*. Dover, New York, 1956 (orig. 1905).

POINCARÉ, H., *Last Thoughts*, Dover, New York, 1963 (orig. 1913).

REICHENBACH, H., *The Philosophy of Space and Time*, Dover, New York, 1958.

REICHENBACH, H., *The Rise of Scientific Philosophy*, University of California Press, Berkeley, 1956.

SKLAR, L., *Space, Time and Spacetime*, University of California Press, Berkeley, 1974. A good general introduction to the philosophy of space and time, which includes a balanced discussion of conventionalism.

TORRETTI, R., *Philosophy of Geometry from Riemann to Poincaré*, Reidel, Dordrecht, 1978. A detailed and sometimes technical but highly readable account of the history of non-Euclidean geometry and the changing philosophy of geometry which accompanied it.

Index